Particle Physics
for Non-Physicists:
A Tour of the Microcosmos

Steven Pollock, Ph.D.

THE
GREAT
COURSES

PUBLISHED BY:

THE GREAT COURSES
Corporate Headquarters
4840 Westfields Boulevard, Suite 500
Chantilly, Virginia 20151-2299
Phone: 1-800-832-2412
Fax: 703-378-3819
www.thegreatcourses.com

Steven Pollock, Ph.D.

Associate Professor of Physics, University of Colorado, Boulder

Steven Pollock is Associate Professor of Physics at the University of Colorado, Boulder. He did his undergraduate work at MIT, receiving a B.S. in Physics in 1982. He holds a Master's and a Ph.D. in Physics from Stanford University, where he completed a thesis on "Electroweak Interactions in the Nuclear Domain" in 1987. Professor Pollock did postdoctoral research at NIKHEF-K (the National Institute for Nuclear and High Energy Physics) in Amsterdam, Netherlands, from 1988–1990 and at the Institute for Nuclear Theory in Seattle from 1990–1992. He spent a year as senior researcher at NIKHEF in 1993 before moving to Boulder.

Professor Pollock's research work is on the intersections of nuclear and particle physics, with special focus on parity violation, neutrino physics, and virtual strangeness content of ordinary matter. Professor Pollock was a teaching assistant and tutor for undergraduates throughout his years as both an undergraduate and graduate student. As a college professor, he has taught a wide variety of university courses at all levels, from introductory physics to advanced nuclear and particle physics, including quantum physics (both introductory and senior level) and mathematical physics, with an intriguing recent foray into the physics of energy and the environment. Professor Pollock is author of *Thinkwell's Physics I*, a CD-based "next-generation" multimedia textbook in introductory physics.

Professor Pollock became a Pew/Carnegie National Teaching Scholar in 2001 and is currently pursuing classroom research into student attitudes toward physics in large-lecture introductory courses. He received an Alfred P. Sloan Research Fellowship in 1994 and the Boulder Faculty Assembly (CU campus-wide) Teaching Excellence Award in 1998. He has presented both nuclear physics research and his scholarship on teaching at numerous conferences, seminars, and colloquia. Professor Pollock is a member of the American Physical Society, Nuclear Physics Division, and the American Association of Physics Teachers.

Table of Contents
Particle Physics for Non-Physicists:
A Tour of the Microcosmos

Table of Contents

Particle Physics for Non-Physicists:
A Tour of the Microcosmos

Particle Physics for Non-Physicists:
A Tour of the Microcosmos

Scope:

The buzzwords appear with regularity in newspapers and magazines—
quarks, *neutrinos*, the *Higgs boson*, *superstrings* … It's the lingo of particle
physics, the study of the deepest, most fundamental constituents and
interactions of the physical world. This course will offer a tour of the
particle zoo and the ideas and phenomena involved in qualitatively
understanding current concepts of modern physics. No math involved!
What's so strange about strange quarks? Why didn't we build a
superconducting supercollider? Should we believe in particles that no one
has ever seen—and never will? We'll learn about the most fundamental
constituents of nature and the forces they feel—the history and discoveries,
the apparatus and ideas, some of the curious characters involved, and the
research and mysteries that are still being pursued today.

We begin on a fairly historical track. From the ancient Greek philosophers,
we jump to Renaissance scientists whose work formed the starting point of
physics. The scientific method developed by Isaac Newton more than 300
years ago continues to serve us well. Many of the physical insights Newton
had, although now deepened and improved by modern physics ideas, are
still relevant for understanding how the world works, which is the goal of
this course and, indeed, of all physics. We jump again, to the start of the
20th century, when new discoveries forced a radical shift in thinking about
the behavior of matter—the dawn of quantum mechanics. We will not cover
(or assume) any detailed knowledge of these laws of physics, discussing
only the key ideas we need for the rest of this course. A "classical"
approach to particle physics, although not technically correct, will serve us
quite well. Students will find that understanding the basic elements of
particle physics using common sense and classical intuition is possible,
provided that we keep our minds open for the occasional quantum
weirdness!

We will follow the early developments, both theoretical and experimental,
and see how they lead us to an organizing scheme for matter at the smallest
possible scale. The idea of seeking the fundamental constituents and the
forces they feel will guide us through the rest of the course. We will learn
about the growing "particle zoo," with just enough vocabulary to talk

sensibly about the fundamental objects discovered from the early 1900s up through the most recent findings. This will lead us to quarks and neutrinos, force carriers and Higgs bosons, squarks and Zinos. Along the way, we will learn qualitatively about the theories required to describe such creatures—quantum fields, virtual particles, and all!

All this leads to what is now known as the *standard model of particle physics*—really the standard *theory*—a complete and self-consistent description of the apparently fundamental, point-like building blocks of everything, their interactions with one another, and the "rules of the game." We then move to more modern questions: Do we really have a fundamental theory at hand? Does that question even make sense? What is going on today in the world of particle physics? What are the central issues? What's hot these days? As we move through the course we will continually ask ourselves two key questions: Why is this idea interesting, and how do I know that it is true? Both are at the heart of appreciating science.

Lecture One
The Nature of Physics

Scope: What is the world made of? How do the constituents "fit together"? What are the fundamental rules that these constituents obey? These are the broad and deep questions addressed by the branch of science called particle physics. We will begin with a discussion of the history and development of human understanding of atoms and "subatoms." We will articulate some of the primary current ideas in particle physics. A central theme will be the idea of *reductionism* in science, which leads to the concept of simplicity as a guiding principle in seeking truths about nature. We will discuss the broad relevance of particle physics to science, and to our lives, asking the key question (as we will throughout the course) "Why should we care about this?"

What background do you need to know to follow this course? Very little—mostly just some curiosity and common sense. The content will be entirely nonmathematical, but we needn't be afraid of introducing the occasional number when it helps our understanding.

> People are always asking for the latest developments in the unification of this theory with that theory, and they don't give us a chance to tell them anything about one of the theories that we know pretty well. They always want to know things that we don't know. So, rather than confound you with a lot of half-cooked, partially analyzed theories, I would like to tell you about a subject that has been very thoroughly analyzed. I love this area of physics and I think it's wonderful ...
> —Richard Feynman (*QED: The Strange Theory of Light and Matter*, chapter 1)

Outline

I. Physics as a whole is the study of what the world is made of, how it works, and why things in the world behave the way they do.

A. Particle physics is the branch of physics that tries to understand the world at its smallest level—the simplest building blocks of the world and the most fundamental rules that govern how things work.

B. The concerns of particle physics can be summed up in five words that address questions specific to this field:

    1. *Force* and *energy*: How and why do things interact?

    2. *Matter*: What is the world made of? What are these things that are interacting?

    3. *Space* and *time*: What is the framework in which these things exist and interact?

C. The physical world seems very complicated; it contains many types of objects and exhibits many types of phenomena, but these can be tied together. Particle physics shows that everything physical arises from a simple set of building blocks and rules.

II. What is the world made of? Greek natural philosophers from about 400 B.C. debated this question.

A. Democritus put forth the idea that everything in the world was made of little "chunks," which he called *atomos*, meaning "unbreakable" or "uncuttable."

    1. To illustrate this idea, think of a stick of butter. It is a physical object that has certain characteristics: It is yellow, soft, and so on. If you cut the stick of butter in half, you still have butter. You haven't changed its essence; you just have less butter.

    2. Democritus believed that if you kept cutting the butter in half, again and again, you would ultimately find a physical and real but uncuttable chunk. In other words, matter is not infinitely divisible.

B. Aristotle disagreed with this idea.

    1. For Aristotle, the world was made up of the four elements: earth, air, fire, and water.

    2. These elements are not physical chunks. They are characteristics or qualities, which means that they are infinitely divisible.

C. Aristotle's view prevailed for millennia but has been discredited now.

1. Neither of these philosophers was a scientist in the modern sense. They believed that the world could be understood by just thinking about it, not by collecting data.
2. The puzzle of how the world works was unsolved for a long time. Now, although it may seem arrogant to say so, this puzzle is largely solved. For example, we have physical evidence that the world is made of atoms, and we have physical evidence of subatomic particles, even smaller constituents of atoms.

D. The ancient Greek philosophers also asked themselves about the earth's place in the universe: Is the earth flat or is it a ball? Is the earth at the center of the universe or is something else at the center? It was not until 2000 years later, during the Renaissance, that thinkers began to collect data from nature to answer these questions.

1. Now that we have evidence about the solar system, we have developed a framework, or mental model, that allows us to understand certain phenomena.
2. Our goal in this course is to develop a similar mental model of particle physics.

III. One of the guiding principles of physics is *reductionism.*

A. Reductionism means "understanding" complex physical systems in terms of their constituents: what they are made of.

1. A doctor might "explain" the human body by saying that it is a collection of organs that are fairly universal and that function in certain ways. In other words, the doctor "reduces" a complex system to its fundamental building blocks.
2. A particle physicist tends to think even more reductively. When thinking about the human body, a physicist might ask, "What is a heart?" The answer is that it is a structure made of cells of a certain type that interact in well-defined ways. Again, a complicated system is understood by reducing it to its components and how they interact.
3. We can reduce the complexity even further by asking, "What is a cell? How does it work? What is a cell made of?" As we answer these questions, we ultimately arrive at universal components, atoms, which are constituents of everything in the world.

B. Particle physicists ask how far the complex system of the world can be reduced.

C. Reductionism is one approach to studying phenomena, but it is not the only approach, nor is it applicable to every field of human endeavor.

IV. Why should we study particle physics?

A. Often, when people ask this question, they are looking for technological applications.

1. Physics offers those applications in the form of televisions, microwave ovens, cell phones, and many other innovations.

2. Particle physics is responsible for the development of particle beam cancer therapy, superconducting magnets, PET scans, relativity corrections in global positioning satellites, and so on.

3. These applications are fascinating, but they are not the primary reasons for studying particle physics.

B. Particle physics also offers connections to other branches of science. We understand chemistry and, by extension, biology, astronomy, and cosmology, in part through the ideas of physics. No idea in science is independent of others.

C. Another reason to study particle physics is the innate human desire to know.

1. Particle physics is fundamental science; it looks at what's at the bottom. Almost any chain of "why" questions in science leads to the particles and interactions that build the world.

2. Our appreciation of the world is also enhanced by understanding its ultimate physical nature, its simplicity, regularity, and interconnectedness.

D. What is the role of particle physics in the broader world of science?

1. Particle physics is, in many respects, the most fundamental science there is. Although abstract, the ideas are elegant, simple, and compelling.

2. By no means has particle physics developed a complete description of the world. As mentioned earlier, however, physicists have built up a fairly deep understanding of the world, which is called the *standard model of particle physics*.

V. Where are we headed in this course?

    A. Without getting heavily into mathematics, we will create a conceptual model of particle physics that we can then apply to understanding other aspects of our lives and the world around us.

    B. We will use a quasi-historical approach that will evolve into a more conceptual one.

    C. No background in physics is required for this course. We will learn what the world is made of, how the fundamental objects fit together, and what the big concepts are to arrive at the standard model of particle physics.

Essential Reading:

Kane, *The Particle Garden*, chapter I.

't Hooft, *In Search of the Ultimate Building Blocks*, chapter 1 (short).

Weinberg, *Dreams of a Final Theory*, chapter III. (Long but "breezy." You may want to start at the beginning of this book because chapters I and II are listed for the next two lectures.)

Recommended Reading:

Lederman, *The God Particle*, chapter 2. (See comments on Weinberg's book above.)

Wilczek and Devine, *Longing for the Harmonies*, Prelude Two.

Questions to Consider:

1. How do you *know* the earth is round? Why do you believe that? What evidence can you think of that might convince a nonbeliever?

2. What is more important to you, the practical applications of science or the aesthetic and philosophical pleasure of learning "truths" about the world?

3. Does science in fact find truths, or are we merely making successively improved approximations? Or, to go to an extreme, could scientific truth be merely social construct?

Lecture One—Transcript
The Nature of Physics

Greetings, and welcome to particle physics. My name is Stephen Pollock. I'm a professor of theoretical nuclear and particle physics at the University of Colorado in Boulder.

Those words may or may not mean a whole lot to you right now: theoretical nuclear and particle physics. It's a branch of physics. And, by the end of this course, you're going to have a very good sense of what those words mean, what it is that I study, and what we're going to be learning about in this course.

Particle physics is a branch of physics. Physics, as a whole, is just the study of what the world is made of, how things work, how they behave and why they behave that way. It's trying to understand things and phenomenon, both natural and technological.

In the ordinary world, you toss a ball up in the air and it follows an arc. We can understand that path. It makes sense. It follows certain laws of physics. Particle physics is a branch of this study where we really try to get down to the bare bones. What is the world made of at the smallest level? What are the simplest building blocks and the most fundamental rules and ideas which govern how things work? Why do they behave the way they do?

Particle physics is about the particles that make up the world that we live in. I could describe particle physics in five words. Those five words are somewhat technical. They're defined by physicists to have very definite meanings, but they also connect with our everyday intuitions and uses of those words. Force, energy, matter, space and time: those are the five building blocks of the field of particle physics and, in fact, of much of physics itself.

Force and *energy* refer to how and why things interact. *Matter*—what are the things? What is the world made of? *Space* and *time*—what is the framework in which all of that stuff is going on? These are the concepts of particle physics. As we go on in this course, we will be dealing especially with the middle— matter—and also with all of the other elements: force that matter feels, energy that is involved in these interactions and the space and time in which they live.

Look around you for a moment. Think about the world that you are in right now, the physical world—the objects and the phenomenon. There are hard things, wherever you might be, that are so hard you might try to break or bend them, and you couldn't. There are also soft things, like your hands or your skin. There's ethereal stuff like the air that you live in, that you can just barely detect or sense. Then there are the things going on in that world—the sound that's hitting your ears and the light that's striking your eyes, the colors, the textures. All of that stuff is the realm of physics.

You could imagine that this enormous diversity of phenomenon is overwhelmingly complicated and you can have this sense of the world as being filled with a whole bunch of different things. We could study any one of them that might be a branch of science, and it's hard to imagine that you can tie it all together, but we can.

This is one of the wonderful things that we have learned about the world. Everything physical—from objects in this room, objects in outer space, objects in the deepest, deepest galaxies and everything in between—all arise from a very simple set of rules, and a very simple set of ideas and fundamental building blocks. That's where we're going in this course.

What are those ideas? What are those building blocks? How can we make sense out of them? Let's go back 2,500 years. This is an old question people have been puzzling about—what the world is made of—from time immemorial. I guess they didn't write it down in the days of cave people, but we do have records of Greek philosophers thinking about this—what is the world made of?

We're going to be learning in future lectures about some of these folks and their ideas. But first, let me summarize two viewpoints that arose 2,500 years ago. There was Democritus, a Greek philosopher from about 400 BC, who had this idea. He wasn't the originator of the idea; he was following in a long tradition of other philosophers who were thinking about how the world works. His idea was that everything is made of little chunks. He called them 'atomos,' which means unbreakable or "uncutable."

Let me give you a kind of a concrete metaphor to think about Democritus's idea in real terms. Let me hand you a stick of butter, okay? This butter is an object; it's a physical object and it has certain characteristics. It is yellow and it's soft. It melts if you heat it, and it has a certain heft to it. You can describe it. You can study it and understand it. It's a material object.

Now what if I chop it in half and throw away half, and you're left behind with a half a stick of butter. I think we would all agree that there's still butter there. It's less massive, but it's still definitely butter. We haven't changed the essence or character of this material.

So if you want to understand butter, it doesn't matter how much I give you. So what if I chop it in half again, and again and again? So now you've got a pat of butter, still butter. There is no difference between the pat of butter and the original stick, except for how much I gave you.

And now Democritus is imagining, what if I keep cutting it in half, and in half again and again? Ultimately, believed Democritus 2,500 years ago, we will come down to a chunk, an individual, indivisible, "uncutable" chunk of butter. It's physical, it's real, it's a little thing that you could imagine seeing or looking at; although it turns out that it's much smaller than anything you could actually see with your eyes. And this is Democritus's idea of atoms. They live in the void, they move around, they interact with one another; they're real, solid, physical objects.

Okay, that's an idea, it's a philosopher's idea, and there were other philosophers at that time who disagreed completely. In fact, Aristotle, probably one of the most famous philosophers of that era, disagreed a lot. And, in fact, most of what we know about Democritus's ideas come from Aristotle's critiques.

Aristotle had an idea, which by modern standards is a little more nebulous. He believed the world is made of earth, air, water and fire, but they're not physical chunks. They're more like characters or qualities, and so, matter is infinitely divisible. You can keep cutting that butter into smaller and smaller and smaller bits forever but it's butter all the way.

Aristotle's ideas survived for thousands of years. He was a very brilliant man in many respects. He had a lot of influence in the fields of religion, religious philosophy and philosophy in general. But many of his scientific ideas, this one in particular, are really quite discredited now.

There is no evidence of what Aristotle had to say, and there is a *lot* of evidence for what Democritus believed. But neither of those fellows was a scientist in any modern sense of that word. They believed that you could understand the world just by thinking about it—just be a philosopher and ask yourself, how should the world be for it to make sense?

That turns out to be a very unproductive way of understanding how things work. It's much better to go into the laboratory and ask nature how things are, not what do we believe is the way it should be. Just take data, collect information and ask yourself, how does butter behave? You can discover (it took over 2,000 years, it was in the 1800s that people began to come up with concrete evidence) that, indeed, butter is composed of what we would now call molecules.

Then, at a later time, we discovered that those molecules themselves were understood to be built out of even smaller objects, which we now call atoms, after Democritus's original idea. So this puzzle about how the world works has been around for a long time, and it remained unsolved for a long time.

I would argue that it has been solved today. I know that sounds a little bit like an arrogant physicist statement—that we understand the world—and I'm going to put lots of asterisks on it. As the course goes on, we're really going to be talking about what we *do* know and what we still *don't* know— what are the remaining puzzles and what are the deep questions that we have not yet resolved?

But to a very large extent, we *do* know that the world is made of atoms; we have direct physical evidence. We have electron microscopes that can image individual atoms now, and even sub-atomic particles. The atoms are themselves made of smaller constituents. The true building blocks of nature are smaller than atoms. That's a fairly recent idea, and one that we're going to be visiting over and over again in this course.

Even with those sub-atomic particles, we have direct evidence: they leave little trails behind them in particle detectors, kind of like the trail you see when you look at a jet airplane. Even if you can't see the jet, you *can* see direct evidence of its passing.

In this course, we're going to learn about that data and the ideas that we have built up around the data to try to make sense of it, to try to understand it.

Let me give you an analogy so that you can get the sense, as I do, that this idea of atoms and sub-atomic particles is really well understood. There's another scientific idea that these Greek philosophers debated, which was our place in the universe, a very different question. For instance, do we live on a flat earth, or is it a ball? Are we the center of the universe, or is something else the center, or *is* there no center?

These were all legitimate questions for people to be asking and to be curious about. And again, there were different philosophical camps 2,500 years ago. There were, in fact, Greek philosophers who believed a contemporary version of the way the world works, but they weren't really collecting data and making measurements, watching the planets move through the night sky. It really took 2,000 years, from the Renaissance and afterwards, for people to begin to make definite statements.

Today we all have this worldview. It's taught to us as little children in elementary school, that the sun is at the center of the solar system and the planets go around in orbits around it. We can make sense of a phenomenon—as, for instance, that the sun rises in the east, travels across the sky and sets in the west—without invoking this kind of primitive idea that we're at rest and the sun is going around us once a day. I think we can say with conviction that that's not a scientifically valid idea, and I think pretty much everybody will agree with me.

If you disagree with me, it's worth thinking about why we know that. And even if you *do* agree with me, it's worth asking yourself *how* you know. Why is it that you believe this picture that was given to you, handed to you by your elementary school teachers, that it's not the case that the sun is rotating around the earth once a day? There are many good reasons to believe that, there's a lot of data, there are many experiments and measurements that you can make—in fact, even ones that you could *really* make—and it's fun to think about. It's something that we're going to think about throughout this course.

When I tell you about scientific fact and these facts are all based on evidence, you should always be asking yourself, *How do I know? Why do I believe that?*

The goal of this course, then, is to give you the framework. It's much like this framework where I could describe the solar system and you could have a mental model. Once you have a mental model of the solar system, you could understand news reports about satellites in orbit around the earth, or rocket ships heading out towards Saturn to explore it. You have a mental model, a framework, into which you can fit contemporary ideas, explorations and discoveries and make sense of them.

What I would like to do is to give you such a mental framework to understand the contemporary ideas that we have about particle physics, about the fundamental building blocks. It's a model which is not so common, although I think lots of people do have some idea, some crude intuitive sense given to us in our elementary through high school training about what an atom is; but we're going to go deeper. We're going to look beyond atoms and ask what they're made of, and what they're made of is made of, and we're going to follow this chain as far as we can.

Particle physics' goal is to try to get to the bottom, and this leads me to an idea; it's a kind of a philosophical idea, an approach to understanding the world, which I call reductionism (actually, everybody calls it reductionism). It's the idea that when you look at something complicated, there are many ways to try to understand it, and one of the ways is to try to ask, what is it made of? What are the kinds of more universal building blocks? That's a reductionistic approach, because it reduces a complicated thing to a simple thing.

Now this is not the only way to approach science. If you are looking at human beings interacting with one another and you say, what's going on, how do I understand the conflicts and the discussions of human psychology or sociology? You could, in principle, think about that by investigating human beings and looking at various aspects of their interactions. Or you could be a reductionist and say, let me study the human brain and try to understand human behavior in terms of the chemical and physical makeup of the brain.

Now that's certainly not the only way to understand human behavior, but I would argue it's a *useful* way. We could come up with drugs, perhaps, if somebody is dysfunctional. In some cases it really does boil down to a chemical imbalance. In other cases it requires psychotherapy. OK, there're lots of possible approaches that you could imagine, the reductionist one being only one of many, but a very powerful one.

If you ask doctors, how does the human body work? Especially western doctors, I think their approach is to a large extent reductionistic; they say, well the human body is a collection of organs, and those organs are fairly universal; people have hearts, dogs have hearts and worms have hearts. So we can understand complicated systems in terms of the fundamental building blocks, and how they fit together. You've got a heart and lungs and bones and skin, and we need to understand not just what those parts are, but how they hook together. Now it's not the only approach to medicine, and it may ultimately just be one of many useful approaches, but I think that we can argue that it's definitely a useful, valuable scientific approach to do that, to think reductively.

As physicists, we tend to think reductively to deeper and deeper levels. I mean, what's a heart? How does it work? Well, to me, the way a heart works—primarily it's made of a bunch of cells of a certain type, they have a certain structure and they interact with one another in well defined ways. So again I understand this complicated thing by reducing it to its components and how they fit together.

We can keep asking that question, OK, what's a cell? How does it work? Why does a cell behave the way it does? The answer is, at least one answer is, a cell is made of various protein molecules which fit together in a certain configuration. Great, so what's a protein molecule? Well, it's a bunch of atoms—hydrogen and carbon, nitrogen and oxygen and whatever it is that makes up a protein molecule—fitting together in a certain way, obeying certain rules. So we keep understanding complicated things in terms of simpler ones which are more universal.

By the time we get down to these atoms that make up the molecules, these atoms are constituents of everything in the world. So if you understand how they work in a protein, you also understand how they work in the frame of a car. So there is an appeal and a value to following this reductionist approach, even if it's not the only way.

Particle physicists just ask, how far can we go? We've got these atoms; what are they? Well, now we have a mental model; they're made of a nucleus, and electrons which orbit around it. It's a kind of classical picture which we will be talking about in future lectures in this course.

And then we can ask, OK, what's the nucleus made of? We understand it in terms of smaller particles still, protons and neutrons, which themselves are made of something called "quarks", a bizarre word which we're going to be learning a lot about.

So anything that's unfamiliar at the beginning of this course, hang on. We're going to be defining what's a quark, what's an electron, what's a lepton. All the buzzwords of this field are going to make sense, because it's really just a simple story.

In order to make sense of them, it's nice to have some mental framework, some model, and that model can be quite simplistic. It's perfectly legitimate to have a kind of a naïve view to start with, to give you the essence of the story. I have to make an apology for reductionism because there's lots of discussion in the scientific communities about how useful and how important reductionism is.

For example, suppose I want to teach you about music. I say I want to understand Mozart. There are various approaches. One approach might be to sit down at a piano, listen to the Mozart music and figure out what the individual notes are that I'm hearing. I might write them down on a piece of paper with some lines, using a scientific notation, to encode which notes the music is made of, how long they last and how loud or soft they are.

Of course that's the way musicians get the music; a pianist sits down and reads that notation and can play Mozart. So basically, as a reductionist, I've said Mozart is just 88 notes on a piano. I have reduced music to the simple, fundamental building blocks, those 88 keys, and then the various properties they can have—how loud and soft, how long they last.

That's a reductionist approach and, of course, it's not completely satisfying. It's not just the names of the notes and the order; there's something deeper about Mozart's music that's compelling and beautiful. So there's more to understanding the music than just knowing the notes; but, clearly, knowing the notes is useful, important and valuable. So this is where we're headed. We are going to be learning about the building blocks and how they fit together.

But let me also in this first lecture, before going into any of those details that come later in this course, ask the kind of big question, why is it that we care? I can't just give you a hard and fast answer to that question, although it's perfectly legitimate. People will ask why you are you taking this course, why would you spend time and energy and money to try to understand particle physics? There are many reasons that I can think of. Many of them are personal reasons, and as the course goes on, I'm going to be asking this question at various levels over and over again.

Why do we care about this stuff? It's interesting at many levels and in many ways. Let me throw out a couple of ideas to you today, and it's something that you should be thinking about for yourself as we go along.

People say to me, why do you study particle physics? If I'm sitting in an airplane and the person next to me learns that I'm a physicist, after their initial gut reaction they want to know what we get out of this; what's the reason for studying particle physics? And I think that they're thinking about technology. That's what scientists are famous for.

What has come out of this particle physics research that affects our everyday lives? I could list some things. Particle physics is a bit of an esoteric field. We're really trying to look at microscopic objects, and many times these objects are so disconnected from the ordinary, everyday world that it's a little bit hard to point your finger at some specific application.

There have been a lot of medical technologies that make direct use of what we've learned about particle physics. From radiation therapies, proton beam therapies, to imaging of human brains with antimatter, P.E.T. scans—all of these are really, in a direct sense, offshoots of the study of particle physics.

If you allow me to generalize particle physics to modern physics, which includes the building ideas of quantum theory and relativity, then all of a sudden I can point to just about all of technology. From television sets to microwave ovens to cell phones, all of that stuff relies upon, and was developed from, ideas of quantum theory.

If you really focused on the particles and the particle physics, well, we've developed magnet technologies from the accelerators that we use to study the particles, but the particles themselves don't always enter into our lives. So it may be that I have to kind of back off and say, you know, maybe this is not the reason why we study particle physics.

Let me ask you this, why did we go to the moon in the late '60s and the early '70s? When people ask NASA what was the point of going to the moon, there are many different answers that they can give, and every now and again you hear them say, well, gosh, the space program helped us developed Tang and Teflon. All right, that's a nice idea, and, you know, I'm grateful when I'm camping that we have Tang, but it's really not the reason to me at all for why we went to the moon. As you can imagine there are much deeper more profound reasons.

Now you can think about spin offs in another sense. The space program teaches us about astronomy. It teaches us about materials. It teaches us about geology. In a similar way, particle physics can spin off.

What we learn when we're studying particle physics is a mathematical description of how particles work together, interact and form more complex structures. If you generalize that idea, you realize that you can apply it to something other than the fundamental building blocks. A biologist might take the same mathematical structures and apply them where your building blocks are now cells. We can understand how an organism built out of cells functions in terms of a theory, which is really the same theory, the same mathematical framework as the particle physicists have developed. This has happened over and over again.

In history, all of science is tied together. There is no branch of science that's completely disconnected from particle physics, nor is particle physics completely disconnected from particle physics, nor is particle physics completely disconnected from geology, astronomy or chemistry; it all ties together.

So one can talk about applications in this scientific or intellectual sense. And sometimes it's even fairly applied. We have discovered particles of nature in particle physics, exotic particles with crazy names like muons and neutrinos, and we have begun using those, for example, as astronomical tools. We can use them to look at distant objects in a way that we were never able to before.

We're going to learn about that in this course. Those mysterious particles all *do* make sense; there is a simple picture that you can have in your mind of what a neutrino is, what a muon is. And then you will be able to understand, at some fairly deep level, *what is that technology and why did it work?*

So these are still spin offs, in some abstract sense. But I think we can motivate studying particle physics at a level that's really not even about applications. It's not about applying the technology or about applying it to other fields; it's our innate human curiosity about the world.

We are puzzle solvers. I think it's genetic. If you go back a million years and there's some cave people, you know, I can imagine one cave person watches a rock tumbling down the cliff and saying: hmmm, rocks fall in parabolic arcs. The next time a rock is bouncing toward me, I know which way to step so it doesn't whack me in the head. and the one who wasn't curious gets whacked in the head and dies. I mean, one could imagine that curiosity is here for a reason; it's innate.

We want to know what the world is made of, at least I do, and since you're watching this course, probably you do, too. What are those building blocks and can we understand them? It's the same kind of appeal of a crossword or jigsaw puzzle, except this one is real, it's not artificial. The world has provided us with this puzzle, and we want to know how we put it together, how we solve these mysteries of nature.

I think there's another level, which for me is, I don't know, some kind of personal satisfaction. So when I look at a rainbow, it's a beautiful phenomenon. Some would even say it's magical, and I understand the use of that word. I feel that way when I look at a rainbow; I feel that it's magical and I love it. I will point it out to my friends, "Hey look, there's a rainbow!" Right, and we all enjoy it!

However, I think I get a deeper satisfaction from my understanding of a rainbow as a physical phenomenon. I know what's going on. There's little water droplets in the rain, the sunshine is reflecting internally, and I understand enough about optics to know why the colors appear, in the directions that they appear, at the angle thbt it appears, and that doesn't take away any of the magic. For me it adds to the pleasure and experience of this physical phenomenon because, even though it is magical and mysterious, it also makes sense.

The world is really fundamentally simple. At some level everything from a rainbow to music to hard and soft objects, anything you can think of, all boils down to fundamental particles interacting with these fundamental laws. It's not about the *facts* themselves. It's about putting together those facts into a framework.

Particle physics is in many ways regarded as the most fundamental. It's also for that reason the most abstract of any sciences. If you're a reductionist and you're trying to understand complex phenomenon, they're interesting in their own right. A biologist trying to understand a human body is doing something very useful and very fascinating; the human body in all its complexity is great science. To me, it's a little overwhelming. I'm a simpleton. Physicists go for the simple underlying truths, and particle physicists really try to get to the very bottom as best we can.

Now I don't want to argue that we're done. I don't want to argue that today we know what the bottom level is. I began this lecture saying we do have a deep understanding of the world, and you will learn in this course about that deep understanding; it goes by the title, "The Standard Model of Particle Physics." So this course is really about learning the standard model of particle physics.

I'm going to try to convince you that it is a deep, true statement (in the scientific sense of the word "truth") not that it's the end of the line, nobody knows; maybe we will understand these deep truths in terms of yet deeper ones. That's certainly possible, and one of the open questions which we'll be heading toward at the end of this class is, how can we go beyond our present state of understanding?

Our goals here, without getting heavily into mathematics (in fact, without getting into mathematics at all), are so you can still create for yourself an understanding, a conceptual model—the standard model of particle physics—so that you can make sense of it when people are telling you, in the news, about some new discovery, when you are relearning something that you learned in your youth, it all fits together.

The picture is really quite straightforward. In this course, we're going to begin with a sort of a historical format. I believe there are many ways to approach a broad field of human understanding. I think that it's nice to look back at the historical evolution of these ideas.

Some of these ideas are pretty wild. Some of them are pretty abstract. Some of this stuff will feel a little bit like heavy going, and when that happens, we're going to just pause and look at the people who discovered it and the experiments that they did which convinced them of these facts of nature.

We're going to be looking at a progression of human thought which has reached a truly spectacular level. This is very accurate. This is quantitative physics, which allows you to make calculations and predictions to design new apparatus. So we're going to be following this historical path, but not entirely. From time to time, there's a mystery that appears in the 1930s, that we've solved today, and I'm certainly not going to leave you hanging. I want the story to begin to form a coherent whole right from the start. So we will be progressing along fitting all the ideas into this standard model of particle physics.

As I said, you do not have to be a physicist. In fact, I don't expect that anybody knows any particle physics, nuclear physics; you don't even really have to know what an atom is, except for some vague memories from your youth. Particle physics is a big field of knowledge.

Okay, this first lecture has just tried to paint, in very broad strokes, what is physics? What is particle physics? And very important, I've been talking about motivation, why are we interested in this stuff? Now, given the complexity of this topic, I think it's really worth our while to continue at this introductory level for one more lecture.

Okay, we'll start to get into the meat of the material in lecture three, but next time, I want to continue; first of all, I want to be very specific about what is particle physics. Let's define it; and I want to also talk a little bit more specifically about how big are these things we're talking about. How small is an atom? How small is a sub-atomic particle? It's nice to set real scales so we can think about this in an intuitive sense. I want to talk about the structure of an atom in a little bit more detail. I'd like to begin to introduce some of the buzzwords that we'll be using throughout the course, next time, so that we have a framework, some place to hang all these ideas that we're going to be discussing.

At that point, at the end of the next lecture, I think we'll be equipped to create an outline for the course. I really haven't done that yet. I want to talk about where we're headed, what's the structure of the course, how it is organized. At that point we will really be able to begin our detailed study of the standard model of particle physics.

Lecture Two
The Standard Model of Particle Physics

Scope: We begin by "zooming in" to the micro-world with a discussion of the distance scales involved. Where do we stand today in our understanding of the smallest building blocks of the world? The standard model of particle physics is one of the greatest quantitative success stories in science. What are the players; what are the forces; what are some of the concepts and buzzwords? This lecture begins our exploration of these questions and offers some teasers for what's to come in the course. In what sense can or should we think of particle physics as "fundamental"?

> The fact that, at least indirectly, one can actually see a single elementary particle—in a cloud chamber, say, or a bubble chamber—supports the view that the smallest units of matter are real physical objects, existing in the same sense that stones or flowers do.
> —Werner Heisenberg, in an uncharacteristic quote (*The Physicist's Conception of Nature*)

> Artists, like physicists, may not always be able to make themselves understood by the general public, but esotericism for its own sake is just silly.
> —Steven Weinberg (*Dreams of a Final Theory*)

Outline

I. The first step in our construction of the standard model of particle physics is to develop a sense of the *distance scales* of physics.

   A. The central idea of particle physics is that all objects in the physical world can be understood to be constructed from a small underlying set of particles, interacting via a small, well-understood set of forces.

      1. We need some frame of reference for the small scale of particle physics in relation to the ordinary world.

2. Picture a meter stick, which is similar to a yardstick, in your mind. A meter stick is usually divided into a thousand little lines that represent millimeters. That distance, 1/1000 of a meter, is about the smallest distance that people can easily visualize.

3. To describe such measurements, scientists usually use a system of scientific notation, in which 100, for example, is described as 10^2, or 10×10. This notation is a shorthand for describing very large and very small numbers.

4. Under this system, 1000 is 10^3, or $10 \times 10 \times 10$, or in metric terms, a kilo. A millimeter is 10^{-3}. The minus sign means $1 \div 1000$, or .001.

5. If we wanted to talk about 1/1000 of a millimeter, which itself is 1/1000 of a meter, that measurement would be one-millionth of a meter, or a micrometer, abbreviated as the term *micron*, or 10^{-6} meters.

6. A micron is hard to visualize, but if you looked through a good microscope, you might see a bacterium or a cell that is about a micron in size.

7. What if we shrink down by another factor of 1000? This measurement would be a billionth of a meter, or 10^{-9} meters. The metric prefix for this measurement is *nano*, and a nanometer is currently about the smallest distance scale that human beings can manipulate in technological devices.

8. If we shrink down by a factor of 10, not 1000, but just 1/10 of a nanometer, or 10^{-10} meters, that measurement is about the size of an atom, and all atoms are about the same size. This measurement is sometimes called an *angstrom*.

9. One way to visualize the size of an atom is to go through all these steps, as we have done. Another way to visualize this size is to think of an apple and the following analogy: An atom is to an apple as the apple is to the planet earth, also a factor of about 10^{10}.

10. We will look inside the atom and compare distance scales in the subatomic world to this distance scale of 10^{-10}m.

B. How else do we think of an atom? We might think of its "logo," or the image of an atom that began to appear in textbooks in the 1950s.

1. This image shows a dot in the middle with about a half dozen other dots whizzing around it in elliptical orbits.

2. This is a primitive representation of what an atom is made of: a nucleus in the middle and the electrons orbiting around it, in orbits of about 10^{-10} meters in diameter.

3. This is not necessarily the correct way to think of atoms, but it is an adequate representation. It doesn't show all the subtleties of quantum mechanics, but it is a good concrete mental image that will work for our purposes.

II. How do we understand this mini–solar system?

   A. First, we might ask, "Why don't the electrons simply fly off?" The answer is: simple electrical attraction, the same kind of attraction that is created when you rub a balloon on your shirt and stick it to the wall. Electrostatic forces hold things together when they are electrically charged.

   1. The electron is negatively charged; the nucleus is positively charged.

   2. Positive and negative electric charges attract each other, and that's what holds the atom together.

   B. Next we might ask, "What is the nucleus?"

   1. In the "logo," it is usually shown as a blob made up of smaller blobs, which are protons and neutrons.

   2. Protons are positively charged objects. The neutrons are similar to the protons, but they are electrically neutral.

   3. The protons and neutrons bundle together to form the nucleus.

   4. What is the size of the nucleus? Typically, 10^{-14} meters, or 10,000 times smaller than the atom itself.

   5. To visualize this system another way, picture a football field. The electrons are out in the end zones. In the middle of the field is a grape—that's the size of the nucleus.

      a. If you can visualize this, then you begin to get a sense that an atom is mostly empty space. Even hard, solid objects are mostly empty space.

      b. Electrons are far away from the middle and sense the presence of other electrons. Most of chemistry—most of the everyday world that we live in—arises from the interactions of electrons in one atom with electrons in another atom.

C. What holds the protons and neutrons together?

    1. Protons and neutrons stick together by a very strong force that is not electrical. Electricity would actually make them fly apart, because like charges repel each other.

    2. This force is called the *strong force*. It sticks the protons and neutrons together like pieces of Velcro. The strong force, or nuclear force, is responsible for nuclear energy and the powering of the sun.

D. The *standard model of particle physics* is the name for the framework in which we understand these particles and the forces by which they interact.

III. In the next lecture, we will begin to look at the evolution of ideas in classical physics.

    A. Classical ideas arise from everyday phenomena, such as tossing a ball, riding a bike, and so on.

        1. These experiences lead us to an understanding of our world, the construction of the laws of physics, that is very accurate. This understanding is also accurate in describing the micron-sized world and the nano-sized world.

        2. However, as we begin to approach the distance scale of the atom, we see that these laws of physics are no longer completely accurate in describing the world.

        3. We need to construct a new set of laws that describes the behavior of elements in this distance scale. These laws go by the names of *quantum mechanics* and *relativity*.

    B. We will work our way quickly through the scientific ideas of the 1600s–1800s, then begin to focus on the evolution of ideas in the 20th century.

    C. Along the way, we will need some vocabulary, particularly the terms *quark*, *lepton*, and *force carrier*.

        1. Quarks are very heavy objects with, as far as we know, no internal structure. Quarks also interact very strongly; they bind together incredibly tightly to form the nucleus of atoms. With the quark, we may be looking at a bottom-most layer of the world.

        2. Leptons are a class of particles; the electron is the most familiar of these. Other leptons are the neutrinos and muons.

3. The idea of a force carrier is a modern way of thinking about interactions between particles through yet another particle. Thus, a photon is the force carrier of electricity and magnetism. A gluon is the carrier of the very strong force that binds the quarks together.

D. Our name for what we will study in this course is *relativistic quantum field theory*.

1. *Relativistic* comes from Einstein's theory of relativity.
2. *Quantum* is the word for the evolution in our way of thinking from the classical ideas about how objects push and pull against one another. Those classical ideas are modified when the objects are really tiny.
3. *Field* refers to force fields, a way of thinking about electric, magnetic, or nuclear forces in the world.

E. In our next lecture, we begin to look at the development of the ideas of classical physics and the causes of the transition to quantum physics.

Essential Reading:

Barnett, Muhry, and Quinn, *The Charm of Strange Quarks*, chapters 2.1–2.2 (up to p. 18) and 2.9.

't Hooft, *In Search of the Ultimate Building Blocks*, chapter 2.

Weinberg, *Dreams of a Final Theory*, chapter I.

Recommended Reading:

Calle, *Superstrings and Other Things*, chapter 1.

Lederman, *The God Particle*, chapter 1.

Questions to Consider:

1. What do you think is the smallest object you could see directly with unaided eyes? What limits your ability to literally see something smaller than that?

2. To what extent can physicists ever prove a theory? Is it proven that the sun will rise tomorrow? Is your confidence that the sun will rise tomorrow improved or unaffected by your awareness of a simple explanatory model of the solar system (round, rotating earth orbiting the sun)?

3. Consider these two statements: The sun will still rise every day next year. The sun will still rise every day 20 billion years from now. What affects your confidence in the truth of the two statements?

Lecture Two—Transcript
The Standard Model of Particle Physics

The central idea of particle physics, the one idea at the core of this branch of physics, is the following: all objects, all physical, material objects, are understood to be made up, built out, of a small set of fundamental constituents—little, elementary particles. Those little, elementary particles interact with one another through a small, well understood set of forces; and that's it—every object, every material thing can be understood, in this microscopic sense, as arising from some small set of particles and some small set of forces. In this course we're going to learn the stories and the words that explain in more detail what those forces are, what those particles are, and how everything ties together in a kind of a neat mathematical framework—without talking about the mathematical details—from which we can then understand, in principle, more complicated phenomenon in the world.

In order to talk about particle physics, and these things are small; they are *really* small, I would like to talk not about the math but about the *numbers*. It's nice to have a sense of scale so that when we're talking about these little microscopic objects, you have some reference frame to connect with the ordinary world. Now I could give you the size of one of these objects in some unit system, and the first problem is that scientists always work in the metric system, not in the American system of units. So that's already a level of unfamiliarity. But, you know, a meter, the standard unit of length, is about the distance of a human scale. Hold out your hands; that's a meter, and for the purposes of understanding the size of atoms, that's good enough. It's a nice, intuitive scale from which to start.

So picture a meter stick, which looks very much like a yardstick, and now look carefully at it, and you'll see that they're usually divided into a thousand little lines. The distance between those little lines is a millimeter, and the distance between those little millimeter lines is about the same. If you look at a regular yardstick, there are 32nds of an inch; on a meter stick there are millimeters. To me, that's sort of the smallest kind of distance that I can easily visualize. I can focus down on the size; it's about the width of a nail, a millimeter. "Milli" is a metric prefix; it means one-thousandth.

Throughout this course, from time to time when I talk about big or small numbers, I'll use the system that scientists use: scientific notation. So, for example, if I want to describe a hundred, I'll describe it as 10^2; 10^2 means

10×10. You multiply two tens and that's 10^2. So 10x10 equals 100; it's a one followed by two zeros. Ten squared is 100. It's just a convenient, shorthand way of describing large numbers. A thousand, 10^3 or $10×10×10$, is a one followed by three zeros. That would be, in the metric system, a kilo; so a kilometer is 1,000 meters.

Going the other way, the millimeter is 10^{-3} meters. The minus sign means it's one divided by a thousand, or .001. There's a one in the third decimal place in 10^{-3}. So now, when we talk about a thousandth of a thousandth, we're trying to get smaller and smaller. I'm going from the meter stick to a millimeter, and now I want to go down another factor of a 1,000, the same relative amount. That would be a millionth of a meter or a micrometer (abbreviated as a micron). So in scientific notation we're talking 10^{-6} meters. And now you understand why it would be really hard to write down .0000001; I could easily slip a zero in there, but if I write it as 10^{-6}, I know exactly what I'm talking about. 10^{-6}, a micron, is really small and hard to visualize. You could just barely; you can't see that with your eyes very easily, but if you look through a microscope, a good microscope, you could see an object like a bacteria or maybe a little cell, which is about a micron in size. I guess that's why it's called a *micro*scope, because it can get down to sizes of a micrometer.

It's not the smallest that we can go. Let's go another factor of a thousand—shrinking down, zooming in—so that we can begin to have a sense of the how the world of atoms and sub-atoms compares with the macroscopic world we live in. Another factor of a thousand, a thousandth of a millionth, is a billionth: 10^{-9}. The metric prefix is the lovely word "nano", as in "Mork and Mindy." A nanometer, 10^{-9} meters, is about the smallest distance scale that human beings can currently manipulate in technological devices. So, for example, if you're building a very, very sophisticated computer chip, and you're laying down little, teeny-weeny wires, the smallest wire thickness that you can presently imagine making would be nanometers. In fact it's tens or even hundreds of nanometers today and getting smaller all the time—hence the name 'nano-technology,' which refers to the growing scientific enterprise of trying to build things on that tiny distance scale.

A nanometer is inconceivably small in our human lives, and if you go one more power of ten—not a factor of a thousand, but just one more power of ten smaller, a tenth of a nanometer—that's 10^{-10} power meters. There's not really a good name for that; some scientists call that an "angstrom," but I won't use that measure very often; 10^{-10} meters is now the size of one atom, a hydrogen atom, a carbon atom, a nitrogen atom, a fundamental building block of chemistry; they're all about the same size. This little factoid may or may not seem obvious, and in upcoming lectures—in fact already starting in the next lecture—we're going to talk about the size of atoms, the scale of atoms and why we believe in atoms, but for now, just take this as a fact of nature; 10^{-10} meters is the typical scale of an atom.

So to a chemist a long time ago, who was just beginning to formulate ideas about how the world works, the world was built out of little building blocks. In fact, the word "atom" comes from the indivisible or "uncutable" elements of the world, and they're all about the same. You can imagine it's a lovely picture for visualizing complicated objects. You've got these little building blocks that are all about the same size (of course carbon is much heavier than hydrogen), so you can build very complicated and interesting things out of something the size of 10^{-10} meters.

How do you visualize that thing? At first I sort of imagined in my mind's eye just a little ball, a little round billiard ball, but it's really small, just 10^{-10} meters; 10^{-10} meters is—how do you visualize this? One way is going down these steps. We could make the leap in one. So an atom is to an apple as an apple is to the planet Earth. That's also a factor of about 10^{10}—10^{9}, give or take; it's a crude analogy. Some people like that sort of analogy to help them visualize how tiny an atom is. To me, I don't really quite know how big the Earth is; so for me that's actually a better way of visualizing how big the Earth is, rather than how small the atom is. But whatever it takes, you don't really have to visualize this because we're going to be looking inside the atom and comparing distance scales in the sub-atomic world to this 10^{-10} distance scale.

So, for example, let's talk about how we nowadays think about an atom that goes a little bit deeper than just saying it's some object, some round ball, that's 10^{-10} meters on a side. In the 1950s, in various high school text books and in logos for the Atomic Energy Commission and so on, an image sort of appeared in American culture, and I think we still have that image today. It's the image of the atom, that logo. It's got a little dot in the middle that's not a point; it's spread out, and then there are these little dots that whiz around it in these kind of elliptical orbits, and there's usually about a half a dozen of them in the typical logo.

I think you can visualize this thing. It's a representation, a primitive representation, of what an atom is, what it's made of. There's a nucleus in the middle, and the particles that are whizzing around. The electrons, those are in orbit—of about 10^{-10} meters in diameter. So I'm not telling you that this is the *right* way to think of atoms, but it's an *okay* way. It doesn't involve all the subtleties of quantum mechanics. These aren't really particles. Nowadays we kind of think of them as a kind of combination of particle and wave, which is very hard to visualize, but it's a good, concrete mental image which works for me, and it will, I think, work for anybody who is trying to understand what the world is made of.

Now, how do we understand this little system, this little mini solar system, with a sun in the middle and a little electron planet running around on the outside? Well, first of all, why doesn't the electron run away? I mean, it's a particle, it's cruising along at high speeds. Why doesn't it just travel in a straight line and run away from the nucleus? The answer is simple electrical attraction. It's the same kind of attraction if you take a rubber balloon and rub it on your shirt and it gets charged up with static electricity. It works better in the winter because the air is dryer. I remember this from my birthday parties when I was a little kid; you rub the balloons on your shirt and you stick them to the wall. There is a force of nature, electro-static forces, that hold things together when they are electrically charged.

The electron, we say, is negatively charged, and the nucleus is positively charged. Positive and negative electric charges attract one another, and that's what holds the atom together. So there's this fundamental force of nature—static electricity, and these particles, the electrons and the nucleus—that we're visualizing in our minds' eye as explaining what an atom really is.

How about that nucleus, what is that? In the picture, it's always a little blob. Sometimes it's a little blob made up of even smaller blobs, and those little, smaller objects are protons and neutrons. These are the buzzwords, and we're going to be talking a lot more about what protons and neutrons and electrons are as this course goes on.

The protons are little, positively charged objects. The neutrons are very, very similar to the protons, but they are electrically neutral, hence the name "neutron." Those bundle together, and they form a tiny object, but it's not a point.

Okay, what's the size of the nucleus? Well it depends on the nucleus; all nuclei are not the same size, even though the atoms themselves are about the same size. A typical nucleus might be 10^{-14} meters, which means it's ten-thousand times smaller than the atom itself. So I'm visualizing a little solar system where the sun is an object that is about ten-thousand times smaller than the distance out to the planets. That's actually not so far off from our solar system—maybe a factor of ten off—but it gives you a kind of a sense of the scale of things.

Another way to visualize that is to picture a football field, and that's where the electrons are, out in the end zones, then in the middle of that football field, right in the middle, is a little grape; that's the size of the nucleus. That's 10^{-4} power or 10^{-4} of the size of an atom, 10^{-14} of the human scale, the meter.

If you can visualize this, you're beginning to sense that an atom is mostly empty space, and that's really an accurate description of the universe. Most of the world, even really hard, solid objects, is mostly empty space, and this helps to explain an awful lot of physical phenomenon that we'll be talking about. These electrons are far away from the middle, and it's the electrons that sort of sense the presence of other electrons. Most of chemistry, most of the ordinary, everyday world that we live in, arises from the interactions of these electrons in one atom with the electrons in another atom. We're going to be talking soon about the ideas of atoms, but for now I just want you to sort of try to sort of put together a picture of the—now we're talking *sub*-atomic world.

What about those protons and neutrons? They're sticking together by some very strong force. It's not electrical; electricity would actually make them want to fly apart, because positive charges, like charges, repel one another. It's called the "strong force" of nature. It's a nice name because it *is* a very strong force, and it sticks the protons and neutrons together like little pieces of Velcro. They stick very, very tightly, and it is that strong force (also called the nuclear force) which we will see is responsible for nuclear energy and the powering of the sun.

"The standard model of particle physics" is the term for the framework in which we understand these particles and the forces by which they interact. The standard model has built into it a set of what are the fundamental, elementary objects. The electrons belong to that class of fundamental particles. The protons and neutrons do not. If you could imagine zooming in (we've been zooming in by 14, 15 orders of magnitude), if you could get down all the way to a proton itself, it's 10^{-15} meters, okay? We can get smaller still.

Modern accelerator technology, sometimes called "atom smashers," (although I think physicists don't really like that terminology) is really a giant microscope that we have constructed, and we have constructed many of these. We'll be learning about how we built these giant microscopes, how they work and what they teach us, but fundamentally, at some level all they're doing is zooming in and looking at what is a proton made of. The buzzword is "quarks. A proton is made out of three, tiny little objects which we call quarks, and they're in very, very tight orbits around one another. They're flying around like a little star system, but it's not gravity that's holding them together; it's some new, fundamental force of nature.

As this course progresses, we're going to be learning about these particles, the quarks and how they tie together. In fact, let's talk a little bit right now about how this course will be structured. We're going to begin, starting in the next lecture, with a kind of an historical approach. I want to talk about the evolution of ideas, starting from what you might call classical ideas— ideas that arise from ordinary, everyday phenomenon: tossing balls around the room, riding a bicycle, driving a car. Those kind of experiences lead us to an understanding of the world which is very, very accurate in the world that we live in, and it turns out that it works very well to describe smaller objects, micron-sized objects, even nano-sized objects. But right around when you get to that distance scale, as you approach the size of individual atoms themselves, these laws of physics begin not quite to describe the world right.

It's a fascinating dilemma. We thought we understood the world, and then everything that we study, everything that we study is well described, until we finally get to this extreme, extreme limit of very tiny objects. Now we've found little particles like electrons and quarks, which no longer obey the same laws of physics, so we need to construct a whole new set of laws, and those go by the names of "quantum mechanics" and "relativity." Okay, those are words which we will be talking about, and I want to explain, at least in some fundamental or basic ways, what is quantum mechanics? what does it teach us about the world?

We're not going to go into a whole lot of detail. Quantum mechanics is in fact a whole separate study of physics, a whole different branch of physical research. We're just going to talk about quantum mechanics enough so that we can make a little bit of sense of what's going on, the funny business that's going on inside the atoms; and, nonetheless, for the most part we can get away without it.

We can think of the universe as being made up of little teeny objects, almost like little billiard balls. It is (I will sort of keep reminding you when it's important)—sometimes it's okay to think of electrons as little billiard balls and sometimes it's just inappropriate. It's not an accurate description of how they really work.

We will be progressing historically then, starting very briefly with the sort of ancient ideas—the pre-scientific ideas—and working our way up through the 16, 17, and 1800s very quickly at the beginning of the course. Then we'll really start to pay attention and focus on the evolution of ideas that began in the 20th century, starting around 1900 and working up to present day. We learned a lot in those early days of the 1900s, from about 1900 to about 1930, so we'll spend a few lectures talking about that progression of ideas, and then we will continue.

There were lots of new discoveries, new technologies, which helped us to deepen our understanding of the world. The discovery of the quark that I just mentioned didn't really happen until the 1960s. We've discovered new quarks in the 1970s and we're still looking for some of the fundamental building blocks of nature. So as we go through this course, we're going to begin with this historical approach and we'll take interludes from time to time.

When we need to talk about the concept of symmetry (that's a very beautiful idea, and one which is very powerful when you're trying to understand how the world works, to talk about the symmetries of the world), we'll leave this historical progression briefly and talk about this concept. Same thing when we want to talk about forces and force fields between particles—a very useful concept which developed over many years and which helped us to put together all these ideas.

There is going to be a lot of vocabulary. (Well, *is* it a lot?) Sort of a dozen words, words like quarks and leptons, which I haven't explained at all yet, and one of the goals in this course is for you to kind of create a mental picture. What is a quark? Well, you have a little carton picture in your head that's based partly on the experiments that helped discover it, partly on the experiments that helped to verify it's existence, and in the end, these buzzwords should begin to form a nice, coherent story.

It's kind of like when you open up a new Shakespeare play (well, new to you). There's still a few left that maybe we haven't read yet, and what do you do on page one? At least what *I* do is I look at the cast of characters in that play.It doesn't mean a whole lot to me at first, but I still like to read it. I want to know the names, and usually on that front page, there'll be a kind of a list of some little, brief summary of who the characters are: the prince, or the king, or the king's daughter—right? There's always some explanation. As the play goes on and you hear the story, you begin to make connections, and those words become meaningful to you. "Harry" becomes the prince, the king's son, who's going to become the king and it all ties together; you know who his friends are, you know who his enemies are. In the world of particle physics, as we progress, all of these buzzwords, the "quarks" and "leptons" and the "force carriers" are all going to (I hope) become part of a simple, coherent story. Then, from that story, you can then build onto new ideas as you hear about novel discoveries. After this course is over, when we make some breakthrough in the world of physics, you will have some sense about how to tie it in with this worldview that I'm trying to describe.

Let me say a few introductory words about these particles. The quarks are very, very heavy objects. They are little dots. They have, as far as we know today, no internal structure. We can't say what's a quark made of at this moment in time. That's an interesting question to ask. Is there another level? Are we going, are we peeling layers off the onion? As we look inside an atom, and then inside the proton, could we look inside of a quark? Not yet, in fact, the picture that we have is so self-contained, so mathematically elegant and so descriptive, that there's not really a strong sense that we need to look any deeper.

As we will see, there were very compelling reasons, right from the beginning, to believe that an atom had something inside of it, electrons and a nucleus. There is no such compelling reason right now that says we need to look any deeper than the quark. So I believe that at least we can have this sense that we might be looking at a bottom-most layer.

The quarks are very heavy and they interact very, very strongly. They bind together incredibly tightly and form the nucleus of atoms. That's where most of mass lives in the universe. Your mass, when you step on the scale—it's all the fault of the quarks; they are responsible for most of what we weigh. The electrons, which buzz around on the outside, are very light. In fact, the word "lepton" comes from a Greek word, "leptos," which means light. Leptos, leptons—that's a class of particles. The electron is one of them. It's the most familiar one; it's the carrier of electricity. It's the thing that defines chemistry and chemical properties. When you plug your plug into the wall, what's happening is that there's some metal, and the electrons, which used to belong to a nucleus, are free, and they run through that metal, carrying their little electric charge with them and lighting up a light bulb. So the flow of electricity is really boiling down, at some microscopic level, to little electrons moving around for the most part.

There are other kinds of particles in the world besides the quarks and the electrons. The lepton family contains some very exotic members, and we'll be learning about them. The neutrinos are wispy little particles that are very real, and they're present in our world but we don't feel them (at least we don't feel them very strongly). We'll talk about what they are, why we believe they exist and what impact they have on our lives. There are other exotic particles, like muons. We'll start to learn these names as we go on, and I think I want to stop with the specific details in this lecture.

As we continue, when you first hear the name of a new particle, you have to try to fit it in. Does it belong to the leptons? Is it like an electron? Does it belong to the quarks, is it heavy and strongly interacting? These are ways to think about the world that will help you to organize this image that we're developing. We're going to be talking a lot about the forces, what binds these things together?

There's a novel way of thinking about forces that I really like. It's a part of quantum mechanics, which we will be talking about (in fact we will devote a whole lecture to this very idea). Let me introduce it today. Suppose you have two particles like two electrons, and of course they have the same charge, so they electrically repel one another. How do you *think* about that force of nature? How can you imagine two particles, which are separated from one another, nonetheless pushing on one another?

It's kind of a wild idea and a fun one to think about. How do you get action at a distance in the real world? There are many ways to think about it, and one of the ways to think about it, a quantum way of thinking about it, would be, say, to imagine that electron number one is allowed by nature to spontaneously produce a little medicine ball out of nothing and toss it to its neighbor. If I picked up a ball and tossed it to you, as I tossed that medicine ball I would recoil backwards, because it's heavy. So as I pushed it, it pusheed back on me; I felt a force.

Now you might argue that, microscopically, it has nothing to do with that other electron, but there is this other particle on the other side that catches the medicine ball, and as it catches the medicine ball, it, too, recoils. Now this is a kind of simple way of visualizing how it is that particles could feel forces. They create force-carrying particles.

Now, it's not medicine balls in the microscopic world. And these force-carrying particles have themselves little, clever names made up by physicists trying to describe what's going on. The force carrier of electricity and magnetism, if you like, the little particle of light, is called a "photon."

You have photons to carry electric forces, you have "gluons" (a lovely name) to describe the very, very strong force that glues the quarks together. The gluon is the force carrier, and there will be even more exotic ones, called "w-bosons," for weak particles that are also present in the world.

These things, some of these more exotic particles, don't play a role in our ordinary, everyday lives. At least, we're not aware of them. So we're going to have to talk about the level of abstraction that's required. How do you think about an object that nobody has really ever seen directly? You will never see a quark in your life. You might see an image from some particle detector, which indicates the *presence* of a quark. But you won't be really seeing the quark with your eyes, because what you see with your eyes is electromagnetic radiation, which can only see objects down to a certain small distance and not any smaller than that. A little bit smaller than a micron would be sort of the limit of what you can actually see.

So when we're talking about these particles and their stories, we need to make enough sense of them that we can have a mental image of these abstract concepts, and that's part of why I like these little cartoon images. Although they're not exactly realistic, it's not really little billiard balls flying around one another, it's a helpful way, to me, of having an understanding of what's going on. The mathematics have fancy words associated with it, as well. All of this stuff is what we're going to be spending the course learning about.

Let me give you the whole, fancy title, "Relativistic Quantum Field Theory." "Relativistic" meaning that Einstein's theory of relativity will be important. We will not be learning about the theory of relativity in any detail in this course., and we really won't need it. When you need it is when you're really doing numerical calculations. We'll talk about it briefly, in its abstract sense of what it's about. This theory—relativity—tells you what's going on when particles move really fast, and in the little sub-atomic world, the electrons and the quarks are almost always moving really fast. So you need relativity—relativistic quantum field theory. "Quantum" is the word for this evolution in our thinking, from the classical ideas about how objects bounce and move when they push and pull on one another. Those ideas are modified when the objects are really tiny; and, of course, we're talking about the tiniest of all when we're talking particle physics.

Again, we're not going to be learning about quantum mechanics in detail, but just the ideas that we need to make sense of particles inside of an atom. "Field" refers to force fields. It's a way of thinking about electric or magnetic or strong nuclear forces in the world. A field is a sort of a visual, abstract impression. If I have a magnet, I can visualize the magnetic field. You can sort of picture it, you can sprinkle little iron filings on top of the magnet and you can really see those field lines in space. All forces and all particles nowadays are understood as being expressed in the world as fields.

So that's what we're going to be studying. That is the standard model of particle physics, a relativistic quantum field theory, which describes the little quarks and the leptons, and how they interact with one another. At the moment, all of that stuff should make very little sense to you; it's just words, and so, where we're headed next is beginning with a historical path, and then beginning to focus on the concepts. What do we know about the world, the sub-atomic world? Why do we believe it? Why is it relevant, important and interesting? That's where we're going.

Lecture Three
The Prehistory of Particle Physics

Scope: Our story began with the Greek philosophers, but we move quickly to the origins of contemporary science in the 1600s. We summarize some aspects of the scientific evolution of atomism, starting with pre-scientific ideas, then the "classical" worldview formed by Isaac Newton, to the modern ideas of "fundamental constituents." We examine some specific details leading to our understanding of atoms, such as the periodic table, the origins of chemistry, and early questions about the existence and utility of atoms. We conclude with a survey of the state of physics knowledge at the start of the 1900s. How could a famous physicist say that physics was "done" in 1899?

> If, in some cataclysm, all of scientific knowledge were to be destroyed, and only one sentence passed on to the next generation of creatures, what statement would contain the most information in the fewest words? I believe it is the atomic hypothesis (or the atomic fact, or whatever you wish to call it) that all things are made of atoms. Little particles that move around in perpetual motion, attracting each other when they are a little distance apart, but repelling upon being squeezed into one another. In that one sentence you will see, there is an enormous amount of information about the world, if just a little imagination and thinking are applied.
>
> —Richard Feynman (*The Feynman Lectures in Physics*, chapter 1, volume 1)

> So, naturalists observe, a flea
> Has smaller fleas that on him prey;
> And these have smaller still to bite 'em;
> And so proceed ad infinitum.
>
> —Jonathan Swift ("Poetry, a Rhapsody")

Outline

I. Particle physics is 20th-century science, but to understand it, we must look into the "prehistory" of science.

II. How do we know that atoms exist?

    A. As Feynman notes in the quote that introduces this lecture, once we appreciate the idea of atoms, we begin to understand a wealth of data and concepts about how the world works.

    B. As mentioned in Lecture One, Greek philosophers debated the existence of the atom 2500 years ago. The Greeks were not scientists, however; they were philosophers. In the 1700s and 1800s, science began to develop an approach in which questions were asked of the world, rather than posed among philosophers.

    C. This approach is the scientific method. We propose a hypothesis, then we try to determine the quantitative consequences of that hypothesis by testing it in the lab.

    D. With regards to atoms, that testing began in the 1800s with chemistry.

        1. John Dalton, the father of chemistry, began to realize that there is a great deal of order and regularity in what happens when you start combining chemicals.

        2. In particular, Dalton made careful measurements of weights of elements. For example, he found that if carbon monoxide is heated, it will always break down into the same ratio of carbon and oxygen, that is, 12 parts carbon to 16 parts oxygen.

        3. Dalton began experimenting with many different materials and realized that he could isolate and understand these elements based on their masses. Even though he didn't know what the unit of mass was, he understood chemistry as a combination and reorganization of these fundamental, unbreakable units.

        4. Once the processes of chemistry began to be understood, biology, geology, and other physical sciences also began to make sense.

    E. This "atomistic" idea originated with chemists but was developed further by physicists. Physicists were less interested in combining materials, however, than in adding energy, in the form of heat, to see how the materials would behave.

1. Data from these experiments fit beautifully with the idea that the material was made of atoms.
2. Scientists began to understand, for example, the ideal gas laws, a phrase that refers to the fact that if you compress a volume of gas, the pressure or temperature may increase. This phenomenon can be explained by the idea of atoms.
3. As time went by, the idea of atoms could be used to explain temperature behaviors, stresses, crystal properties, and other phenomena. The evidence for atoms was indirect, but we could see the *consequences* of their existence.

III. In the late 1800s, Ludwig Boltzmann, an Austrian physicist, developed the mathematics of thermodynamics and statistical physics.

   A. Boltzmann's idea was that we should be able to make quantitative predictions about materials based on probability and statistics.

   B. Boltzmann was a proponent of the idea of atoms, but in the late 1800s, the reality of atoms was still debated. Some believed that atoms were a fiction, a mathematical tool to help us explain the world, but not real.

   C. Boltzmann committed suicide in 1905. His reasons were unclear, but some believe that he was frustrated by scientific antagonism toward his brilliant ideas.

   D. Also in 1905, Einstein published a paper on *Brownian motion.*
   1. In the early 1800s, a scientist named Brown saw that if a grain of pollen, for example, was placed in a fluid, it jiggled around. At first, Brown believed that the grain of pollen might be alive, but the same motion also occurred with other materials, such as dust.
   2. In 1905, Einstein realized that Brownian motion provided direct evidence for the physical existence of atoms. He hypothesized that the movement of the tiny piece of material is caused by its being bumped by atoms surrounding it.

IV. Further evidence of the existence of atoms came from Dmitriy Mendeleev, a Russian chemistry professor.

   A. Mendeleev was trying to understand all the different elements that people knew about at the time, such as hydrogen, sodium, and lithium. He recorded the name of each element, its relative weight, and certain chemical properties of each element on cards.

B. Mendeleev laid his cards out in order of increasing mass and in columns according to the chemical similarity of the elements. For example, hydrogen, the lightest element, was in the first row, and lithium was placed underneath it. Lithium weighs about seven times as much as hydrogen but is chemically similar.

C. In this way, Mendeleev formed the periodic table, which organized the atoms in a very elegant way. In some places, the table had gaps, from which Mendeleev predicted the existence of new elements, that is, fundamental materials that would be indivisible by any chemical means. Mendeleev could also predict the weights and certain chemical properties of these new materials.

D. Within a decade, gallium and germanium were discovered by scientists attempting to fill in the gaps in the periodic table. The prediction and discovery of these new elements served as profound evidence of the idea and organization of atoms.

E. In the late 1800s, people were making crude estimates about the properties of atoms. The atomic hypothesis became firm science, but it was still not deeply understood in this era.

V. Let's take a step back from this early development to look at the scientific method and the development of another model, the planetary model.

A. In the 1500s, people debated whether the earth, the sun, or something else was at the center of the universe. Some weak data had been collected, but people still did not subject the data to rigorous mathematical testing.

B. This testing began with the work of a Danish astronomer named Tycho Brahe. Brahe took excellent data on the motion of planets through the night sky.

C. Brahe had a kind of confused idea of the solar system, but he was more interested in collecting the data than in proving his ideas. This is the first step in the scientific method.

D. Brahe's assistant, Johannes Kepler, had an idea of the solar system that was similar to the ideas of Copernicus; that is, that the sun was at the center and the planets revolved around it.

1. Copernicus believed that the planets moved around the sun in circles.
2. Kepler carefully analyzed Brahe's data and determined that the planets do revolve around the sun, but in elliptical orbits.
3. Kepler's achievement was in describing the data accurately; he did not attempt to explain the data.

E. If we truly want to understand something, however, we must go beyond just describing the data. The person who took this step in relation to the solar system was Isaac Newton.
   1. Born in the 1600s, Newton was perhaps the first real scientist. In the *Principia*, he laid out the scientific method.
   2. Newton applied the scientific method to the data that imply that planets move in ellipses.
   3. Newton had the idea of a universal force of gravity that kept the planets in orbit.
   4. He worked out a formula that tells quantitatively how gravity depends on the mass of objects and the distance between objects. He then applied this formula to the question of the motion of the planets.
   5. In other words, Newton started from scratch with Kepler's ideas on a deeper level; he *explained* Kepler's *description* of the data.

VI. In the late 1800s, people were satisfied with the picture of atoms as a description of nature, but they still lacked understanding of the underlying theory of atoms.

A. They were at the stage of Kepler in describing the atom, but the Newton needed to explain the atom hadn't come along yet.

B. To develop deeper understanding of the atom, we needed something beyond the physics of the 1600s–1800s.

C. What was missing was quantum physics and relativity. The transition from classical physics to modern physics had not been made, although there were some hints that this transition was soon to come.
   1. The first hint was the discovery of radioactivity, which didn't fit in with the classical scheme of atoms. Radioactivity— energy spontaneously generated and emitted from certain universal materials—was very mysterious.

2. Another hint was the inability to calculate, using the laws of classical physics, the *glow* of heated objects. Calculations in these experiments did not agree with the observed data.

3. In the next lecture, we will discuss these conflicts.

Essential Reading:

Kane, *The Particle Garden*, chapter 2.

Weinberg, *Dreams of a Final Theory*, chapter II.

Recommended Reading:

Calle, *Superstrings and Other Things*, chapter 7 (up to "First models of the atom").

Lederman, *The God Particle*, chapter 3.

Questions to Consider:

1. Do you believe in the objective reality of atoms? Why or why not? If so, what evidence do you consider the most compelling? If not, can you articulate why not?

2. Imagine a giant jar filled with tiny "super-duper" superballs that never lose their energy. They bounce around continuously; if you drop one on the floor it bounces back up to full height, over and over again. The jar contains many of these balls, bouncing around in every direction. Picture a piston at the top of the jar that can slide in or out without friction, but sealing the top of the jar completely. As the superballs bounce around in the jar, some will hit the piston from below, giving it a little push upward as they bounce off it. You have to push downward to prevent the piston from flying away. Now ask yourself the following:

   (a) What would happen to the force you need to apply on that piston if you suddenly doubled the number of superballs in the jar, everything else remaining the same?

   (b) What precisely would happen to your force if you suddenly doubled the speed of each superball? (This is tricky! They would rebound with twice the speed, but it would also take only half as long for them to reach the piston again.)

Lecture Three—Transcript
The Prehistory of Particle Physics

Particle physics is 20th century science, but in order to understand it, it's useful to go back in time a little bit and ask about the prehistory—the development of ideas that led up to our understanding of what the world is made of. So I want to talk today about atoms and the idea of atoms. It's going to be useful to ask, What do we know? What do we believe today and why, why do we believe these things?

I'd like to begin with a quote from a very famous physicist named Richard Feynman. Feynman was a physicist at CalTech, and he was a very influential player in the development of many of the ideas of modern physics. He was, in a certain sense, responsible for the ideas that we have today of light, matter and their interactions. Richard Feynman was also famous for his involvement in the investigation of the space shuttle explosion. He was the one who discovered that it was the O-ring that was faulty in the Challenger explosion.

Here's a quote from Mr. Feynman, "If in some cataclysm, all of scientific knowledge were to be destroyed and only one sentence passed on to the next generation of creatures, what statement would contain the most information in the fewest words?" This is typical Feynman; it's a very creative idea. What one sentence can you come up with that carries the most useful information for someone that's trying to redevelop ideas of science?

Here's his answer, "I believe it is the atomic hypothesis, or the atomic fact, or whatever you want to call it. But all things are made of atoms: little particles that move around in perpetual motion, attracting each other when they're far apart, repelling if you squeeze them too close together." That's it. In that one sentence you will see there is an enormous amount of information about the world, if just a little imagination and thinking are applied. That's Mr. Feynman's idea of the single most useful sentence about modern science—the world is made of atoms.

Once you appreciate this idea, you begin to understand a wealth of data, of ideas, of concepts about how the world works. You begin to understand why gasses behave the way they do, and liquids and solids, vapors. Just about anything that you can think of that describes how the world works, boils down to this idea that the world is made of atoms. This idea begins already

2,500 years ago. There were Greek philosophers who were debating about how was the world constructed. Can you divide the world up into an infinite number of little pieces? Or do you get to a bottom, solid object—the atom?

It was Democritus who had *that* idea first, and at that time it wasn't really science; it was philosophy. People were just debating and thinking about how they believed the world should be. It really wasn't until the 1700s and the 1800s that science began to develop an approach where you could ask the question of the world, rather than asking the question of your ideas about the world.

So this is the scientific method, right? You have a hypothesis, "I propose that the world is made of atoms." Let us try to figure out what are the consequences of that idea. Let's think about quantitative consequences and then test them. Go into the laboratory. That testing begins in the 1800s, with chemistry. Chemists are tinkering with combining materials and looking to see what happens.

I remember when I was a little kid I was given a chemistry set. You know, I was interested in science and my parents must have thought this would be a useful thing for me. I didn't really like it so much because I didn't know what was going on. I had these chemicals, little vials, and they had instructions. They would say to mix two parts of this one with one part of that one, and hey, look! The color changes! It was kind of fun, but I didn't get it; I didn't have any idea of what was going on.

Of the chemists of the early 1800s, John Dalton was perhaps the name most famous as the father of chemistry. He was doing the same kinds of things as I was, but I guess he was a little smarter than I was when I was a kid. He began to realize that there is a great deal of order and regularity in what happens when you start playing with chemicals. One of the things that he investigated in particular was making careful measurements of weights.

So, for instance, suppose you start with a vial of gas, which is identified to chemists in the early 1800s, nowadays we might call it a flask of carbon monoxide. If you heat it up, maybe add some electricity, that gas will separate, and out of it will come a little dust, some carbon dust, and another gas with different chemical properties, oxygen. You can weigh the oxygen that comes out and weigh the carbon dust that falls down, and what you'll discover is that the amount of carbon to the amount of oxygen always comes out in the same ratio: 12 parts carbon to 16 parts oxygen. So he started doing this with a whole array of materials. Of course, they didn't call it "carbon monoxide" back then; that's the modern terminology, which reflects our understanding that carbon monoxide is a gas containing a carbon atom and an oxygen atom.

What Dalton's brilliance in the story really boils down to is he realized that you could explain it always comes out to 12 parts carbon to 16 parts oxygen. You could understand that if you have the idea in your head that carbon is a little physical entity that weighs 12 somethings, and oxygen is a little physical entity that weighs 16 somethings. Dalton didn't know what that unit was, that quantity, the mass of an individual atom, but he knew the *relative* masses.

So as you go through all the different materials that were known, you begin to isolate various elements. Hydrogen is the lightest; hydrogen on this scale would have mass one, and then you have other elements: lithium, carbon, oxygen and nitrogen. All of the fundamental elements are synthesized, isolated, and all the processes that you can then do with these elements (like mixing them together and making the color change) now makes sense to anybody who's thinking about it, who talks with Dalton, reads his works. Because now you understand that the world is made of atoms.

Chemical reactions are nothing more than this: take an atom from here, an atom from there and combine them. The result will have a certain well defined mass because the constituents did. Once you understand this idea, chemistry begins to fall into place. Instead of being mystical—and a little bit magical, mixing beakers together—it just makes sense. Once chemistry begins to make sense, so does biology and geology—and pretty much any process where you're dealing with physical materials and the amounts and the volumes, anything that you can measure about them.

This idea was originated by chemists, but it was developed further by physicists. The physicists were more interested not in what happens when you mix stuff together, but for instance, you take a chunk of material and heat it up; if you add a certain amount of energy, if you do a certain amount of work. How much does the temperature go up? It's a perfectly well defined physical question, and people went into the laboratory so there was some data on this. The data, it turned out, fit beautifully with this idea that this material was made up of little atoms—little, individual atoms.

People began to understand, for example, the "ideal gas laws." That's a phrase that refers to the fact that if you take a beaker or a vial of gas with a piston, and you compress it, the temperature might go up as the volume goes down, or the pressure might go up. At first, that seems mysterious and a little bit magical. Then you say to yourself, "Look, let's imagine a balloon filled with a gas. Why is there a pressure, why do I have to squeeze to make that balloon compress?"

You could just say that's a property of balloons, but you could also say it's because there's gas in there made of little, teeny atoms—little balls flying around like crazy, and they're whacking into the walls of the balloon. There are lots of them, right? They're tiny. So what you feel when you're pressing on the balloon, you're feeling a whole bunch of little bumps which average out to a nice, smooth pressure, and if you squeeze it, then these little atoms have less room to jiggle around in. They're going to hit the walls more frequently, and you're going to feel more pressure. You can even get quantitative about this and put in some numbers. It all works, it makes sense.

So the idea of atoms is developing. The primary philosophy here is we have this hypothesis, we try to make quantitative predictions and then we verify them in the laboratory. The picture is really quite simple underneath. It's explaining, now, pressures of gasses, and as time went by, people argued, "Ah, look! It explains temperature behavior; as you heat something up it expands, stresses and strains, crystal properties...." It's a long list, and it continues to develop even today. The evidence for atoms is indirect; we're not actually seeing these little things; we're seeing the *consequences,* the scientific consequences of these things.

There was a physicist, an Austrian physicist in the late 1800s, Ludwig Boltzman. He was a very interesting character. He developed the mathematics of what's called Thermodynamics and Statistical Physics. Ludwig Boltzman said, look—if you have, for example, a gas, it consists of a whole bunch of little atoms in there. This is the mental model that we have. We should be able to make quantitative predictions, based just on probability and statistics. You've got a whole bunch of particles, for example, it's going to be just as likely that they're whacking on the right-hand side as that they're going to be whacking on the left-hand side. So that the pressure should be equal on both sides of your container, no matter what its shape is.

Ludwig Boltzman was a big proponent of the idea of atoms, and in the late 1800s, even though chemistry was well developed and physics was well developed, there were still big debates about whether atoms were real. It's a lovely question. What does it mean to argue whether atoms are real or not? Maybe they're just a fiction. Maybe it's just kind of a mathematical tool to help us to order and understand the world, but it's possible that it doesn't really make sense to talk about little microscopic things that are so small you can't even see them. Right? Nobody has ever seen an atom, even today. I can show you pictures, electron tunneling microscope pictures, but it's really a computer rendition of a scientific instrument's measurements. You can't see an atom with your eyes and you never will, because they're too small. They're actually smaller than the wavelength of visible light, and you can't see something that's smaller than the wavelength of the light you use to look at it. So this was a big debate in the late 1800s. Are atoms real, or are they some kind of mathematical fiction, which is just being used to order and explain all of this data?

Ludwig Boltzman was really trying to make convincing mathematical arguments, but he didn't convince everybody, and he was quite a tragic figure. Ludwig Boltzman ended up committing suicide in 1905. The reasons are a little bit mysterious, but some people think it was because he was so frustrated with the antagonism and arguments that he was getting in scientific conferences and with meetings. Because there was a bunch of big, powerful physicists, influential physicists, who just didn't believe that it made sense to talk about an object that was so small you could never directly see it.

It's ironic that Ludwig Boltzman died in 1905, because that was the year that Albert Einstein published a paper on something called "Brownian motion." Brownian motion is the following phenomenon: Take a little, teeny object—but one that you can still just barely see, like a little dust grain or a little piece of pollen—and put it in a fluid and watch carefully. Mr. Brown did this experiment in the early 1800s, and what he saw was that little grain would jiggle around, kind of like a drunken sailor, just this way and that way—kind of like the way I lecture to you, jittering around the lecture hall.

This motion, at first Mr. Brown said maybe it's because that little pollen grain is alive; maybe it's got some little flagella and it's moving around. It was a nice idea. Then he tried it with dust and dirt, and no matter what he put in there, they all behaved the same way. So his idea was—well, he didn't know. Mr. Brown just thought it was a mystery, a random motion of particles.

But Mr. Einstein in 1905 came up with a real, quantitative explanation. He said, "Look, if you have a little, teeny object, it's just barely visible, and it's surrounded by atoms, and they're real; they're whacking into it from all sides. Although it's random, on average there's going to be equal numbers of bumps on the right and equal numbers of bumps on the left. From time to time there might be, oh, a few more bumps on the right side—just briefly—

and so the thing will jitter to the left. Then, maybe later there would be a few more bumps on the top and it will jitter down." So that's why objects do a random walk. It's because they're so small that it begins to become noticeable that atoms are around it, and they're smaller than it, but not infinitely smaller than it. So now this is direct evidence of the physical existence of these little, teeny balls, whatever they are, these little atoms, whacking into objects.

There was further evidence and understanding of the physical reality of atoms around this period of time, the late 1800s, and this other evidence came from a chemist in Russia by the name of Dimitri Mendeleev. Mendeleev was a chemistry professor. He was trying to understand all the different elements that people knew about: hydrogen, lithium, sodium, and all of these different elements, the elements that we now see on what's known as the Periodic Table, the table that shows up in the corner of the lecture hall when you take a high school science class. It was Mr. Mendeleev who invented that table.

He had little cards, and on each card he would write down the name of the element, the known weight mass (or the relative weight, I should say) of that element, and certain chemical properties—how reactive it was, for example. He noticed that, if you just lay the cards out in increasing mass, and you've got, I don't know, 60 or 70 cards, these are all the fundamental atoms that people know about in the late 1800s. Instead of just laying them out in a horizontal row, he started organizing them in a table, starting from the lightest, working his way up, heavier and heavier. In the columns, he would make sure that the element going down a column would have the same chemical properties. So hydrogen is on the first row, and then—he didn't know about helium yet—the next element was lithium, which weighs about seven times as much as hydrogen, and it's very chemically similar; it's highly reactive. It's metallic, and hydrogen is gaseous, but the chemical reactive properties are about the same.

So in this way he formed a table, and this table organized the atoms in a very beautiful way. In fact, it was stunning that you could do this. Some people were arguing about whether it was just an accident. In particular, there were certain places on that Periodic Table where there was a hole. It just didn't make sense to put an element there, so you just had to leave a gap and start continuing the elements further along in the table.

Mendeleev thought that this table was significant. He didn't think it was accident or numerology. He thought that it was telling us something about the nature of atoms. And indeed it was. So, for example, for one of those gaps, he could say, "Look, we haven't found any material, but if my table is correct, I predict the existence of a new material—a new, fundamental material—which is indivisible by any chemical means, and it should have the following mass and the following chemical properties." He could really tell you what to look for.

So people went out to look for these things, and within a decade, gallium was discovered in the late 1800s (1875). Germanium was discovered in 1896. These were the gaps in his table. So the discovery of these new elements, predicted by a chemist, was really a radical and profound proof, evidence, that this idea that there are fundamental atoms—and that they are orderly and there is some sense to this organization—it was all beginning to come together.

Now, today, we still don't have direct, visual evidence of the existence of atoms, but we have incredibly powerful circumstantial evidence. Brownian motion, I think, was the one that convinced the world. Albert Einstein's paper quantitatively predicted how these objects *should* move if atoms were real. This is a game you play in science; it's a "what if?" If I believe in the existence of atoms, then I conclude that a pollen grain will jiggle in the following ways, with the following frequency and so on.

People were making estimates about the properties of atoms. They were still crude in the late 1800s, but they seemed to be objects with a definite size. That size was tiny; in the metric system it's 10^{-10} meters. It's a little ball, about 10^{-10} meters, and it didn't seem to matter which element you were looking at; they were all roughly the same size. All atoms are objects; they have different masses, but they are the same *size*. This atomic hypothesis becomes firm science, but it's still not understood in a deep way in this era. People are just describing what's going on, without really being able to say why.

I want to sort of take a step back from this early development and talk a little bit about the scientific method and the question of how do you know what you know? Why do you believe these things? So as a metaphor for this developing picture of atoms, I want to step back in history and look at the development of another model, one that's a little bit easier for us all to visualize, which is the planetary model.

Go back to the 1500s and people are debating about whether the Earth is at the center of the universe, or whether the sun is at the center, or whether there's something else going on. There are various, competing ideas. These ideas are backed up by some data. People have been observing the motion of the planets in the night sky, for example, but it's still kind of weak science; there isn't this idea of let's work out rigorous mathematical consequences of our ideas and then check and confirm whether the data agrees.

This process really began with the data taken by a Danish astronomer by the name of Tycho Brahe (I love that name). Tycho was an amazing character, a very dynamic man. He was in a duel in his youth and lost part of his nose, so he had a metal nose. He was favored by the king of Denmark, so this was the first example of big science. Here was this guy who developed a research program, he got funding from the government. He got a research laboratory, which was an island off the coast of Denmark, and he took great data; he took excellent data of the motion of planets through the night sky.

Now, Mr. Brahe had an idea about how the solar system worked, which was by modern standards kind of confused. He thought that the sun was at the center and that the Earth rotates around it, but then the other planets go around the Earth. It was some complicated system, but he wasn't really interested so much in proving his idea as he was in just collecting the data. This is the first step in a scientific approach.

It was his disciple, Johannes Kepler who really took the next big, important step in the development of science. He had to wait until Tycho died, because Tycho coveted his data. In those days you didn't "publish or perish", you just held on to your data; it was valuable. So, once Tycho was gone, Johannes Kepler was able to take the data and analyze it. Now, Kepler had this idea, which had been espoused earlier by Copernicus, that the sun is at the middle and all the planets are orbiting around it. Copernicus had thought that the planets were going around in circles. And Kepler said, "Let's ask the data. Let's not be philosophers about this; let's be scientists." So he looked at the data. He made incredibly careful measurements and calculations and determined that the data implied that planets are not going around us; they're going around the sun, and they're traveling in ellipses, not in true circles but in slightly squashed circles.

This is Kepler's brilliant achievement. It's just describing the data accurately. He's not explaining it. He doesn't tell us *why* does a planet move in an elliptical orbit. It's just saying it *does*; this is what the data shows. That's very important. When you are first looking at confusing data, just describe it.

However, if you really want to understand something, if you really want to do science, you have to go deeper than that. You need to look at the data and explain it, based on some fundamental principles, and the person who did *that* was Isaac Newton.

Isaac Newton is sort of the hero of any course on elementary physics. Isaac Newton was born in the 1600s, and he knew about the data of Kepler. He knew about the works of Galileo. He was, in a certain sense, the first real scientist. He, in fact, wrote a book, the *Principia*, in which he laid out what we now call "the scientific method." He said we need hypotheses. They need to have mathematical consequences. We need to be able to make predictions about what we will see in the laboratory. We need to do experiments to verify those predictions, and then we need to check that the whole story is consistent. If it's not, then we have to go back and fix up our hypotheses or check our calculations.

This method that Isaac Newton developed, he applied, in this case, to planetary motion. So he's got this data that planets are moving in ellipses and he's trying to understand why. Now, Isaac Newton is famous for sitting in an orchard and seeing an apple fall down, and realizing that the force of gravity, which pulls the apple down is a universal force, and if gravity can pull an apple, how high can it reach? Can it reach up to the moon?

This was Isaac Newton's really brilliant insight into how the world works. Why does the moon not fly away from planet Earth? Why does it stay in orbit? According to Newton, it's because there's a force of gravity holding it in. Why does the planet Earth not fly away from the sun? Because there's a force of gravity attracting us so that we go around in an orbit.

So now he says, "Okay, I've got this idea of gravity," and he worked out a formula that tells you, quantitatively, how gravity depends on the mass of the objects and how far apart they are. Then he applies this; he says, "Suppose that we have a solar system with a heavy sun and a massive planet Earth—what would our motion be?" So, in order to do this, he had to develop some mathematics. In fact, he invented calculus.

Very smart guy, Isaac Newton; in my mind, *the* smartest human being, at least the smartest *scientist*, who ever lived. He created science. He invented calculus. He discovered the laws of motion.

The physics that Isaac Newton discovered and wrote in the *Principia* has survived. Even today, it is the basis for pretty much all ordinary world science, and physics in particular, that we do. Amazing character; he had a brain the size of our planet.

He was also not a very nice fellow. He never married. He was kind of the archetypal nerd. He worked really hard. He had this incredible ability to focus on what he was thinking about. Not a nice guy; he had a lot of enemies. I think he was always worried that he might find somebody that was half as smart as he was.

So Isaac Newton has taken Kepler's ideas and started from scratch, at some deeper level. He has said, "I believe that there is a universal law of gravity which applies to all objects." By applying this law, he had deduced that planets move in elliptical orbits, and that they have the various propertie, the period and the radius relationship, that Kepler had also observed.

Isaac Newton is explaining Kepler's summary of the data, so now we have a much deeper understanding of planetary motion. It's no longer just a theory; it is now a deep, rigorous mathematical framework, and instead of reproducing the data that was already taken, you can now begin to make predictions. What would happen if we launched a rocket? What would happen if we tried to send people to the moon or to Mars? We can use Newton's laws and they will tell us exactly how to do that; and, of course, this has succeeded spectacularly.

I think that scientists were satisfied with the picture of atoms in the late 1800s as a description of nature. So, going back from Newton and his development of the scientific method to our development of the idea of atoms, people were satisfied with the idea that atoms are real. They're physical objects—little, teeny things that bounce around and interact with one another. They can explain an enormous wealth of data about physical properties of materials, chemistry and, from that, biology.

But there was still a kind of a lack of understanding of the underlying theory. It was like we were at the stage of Kepler, *describing* what was going on, but we didn't yet have our Isaac Newton to explain What is an atom? What is it made of? Why does it interact the way it does? Why does the Periodic Table appear (as we have seen it in our high school chemistry class) in these lovely rows and columns? Why is it that there is this ordering to the atoms?

In order to develop *that* idea, we needed something more, something beyond the physics of the 16, 17, 1800s. What was missing was quantum physics and relativity. The transition from classical physics of the Newtonian era, all the way up to the 1800s, to what we now call modern physics—which is these ideas of how things work when they're really tiny—that transition had not yet been made. There were some hints out there, and we're going to be talking in the next lecture about what those hints were—that physics was not quite done in the late 1800s.

Let me tell you about two of those primary hints. Number one, around the same time that the Periodic Table was being developed, published, understood, we also were discovering a new phenomenon of nature, radioactivity. Radioactivity didn't fit in with this scheme of atoms. Very mysterious. Energy just spontaneously being generated and spewing out of some certain, unusual materials didn't fit in with the classical picture.

That was one clue that we really didn't understand everything about the world. In fact, when you had radioactivity, you observed this bizarre phenomenon: that an atom, like a nitrogen atom, could change, if you bombarded it with radioactivity, into a different kind of atom, could convert nitrogen into oxygen.

This goes against all of the ideas of chemistry, which say the world is made of atoms. Nitrogen is one kind of atom. Oxygen is another kind of atom. They can bounce, they can stick together, but they can't change; they are indivisible and immutable. Remember, "atom" comes from the word "atomo," which means "uncutable". So radioactivity was a puzzle.

There were other puzzles at the time. For example, when you heat materials up and you look at their glow (this is an easy phenomenon to observe; look at your oven, it glows red-hot), if you try to calculate using the laws of classical physics, what's going on? You discover that they don't agree with the data.

So next time, we're going to talk about these conflicts between classical physics and the experiment that lead us to a deeper understanding of what are atoms? What are they made of? What are they, really, at some deep level? What's the theory?

Lecture Four
The Birth of Modern Physics

Scope: This lecture contains some stories of Planck, Einstein, Rutherford, and the early quantum physicists. What does quantum mechanics tells us about the world? How did Einstein help make the transition from 19th-century "classical" physics to 20th-century "modern" physics? This lecture offers answers to those questions and a brief guide to the radical shifts in the philosophy of science around the turn of the last century, including the important role of the discovery of radioactivity, the electron, and the fact that atoms are not indivisible. This period saw the beginnings of our primitive understanding of the realistic structure of atoms.

> A philosopher once said, "It is necessary for the very existence of science that the same conditions always produce the same results." Well, they do not! ... Do not keep saying to yourself, "But how can it be like that?" because you will get "down the drain," into a blind alley from which nobody has yet escaped. Nobody knows how it can be like that.
> —Richard Feynman, on quantum physics
> (*The Character of Physical Law*)

Outline

I. The ideas of classical physics developed steadily, starting with Isaac Newton in the 1600s and progressing through the 1700s, 1800s, and 1900s. This development included the concepts of e.g. forces, gravity, electricity, magnetism, energy, temperature, thermal physics, light, and sound.

II. The transition to modern physics was forced on the scientific community in about 1900 by inconsistencies between theoretical calculations and observed data. The first such inconsistency came from Max Planck's study of the *glow* of heated objects.

    A. This topic brings together many aspects of physics, including atoms, which move faster when heated; electricity and magnetism, which cause the glow when an object is heated; thermodynamics; and others.

B. In studying this problem, classical physicists found wild discrepancies between what their calculations predicted about the color and energy emitted by heated objects and what the data actually showed.

C. Planck started by playing the role of Kepler and describing, but not explaining, the data. He produced a formula that described the temperature and behavior of heated objects. The formula matched the data, but it was a description, not an explanation.

D. Later, Planck himself came up with an explanation. He challenged the prevailing theory that light was a wave of electromagnetic energy, postulating that instead of electromagnetic waves, atoms emit discrete pulses of light called *quanta*.

E. Planck found that his description of the data on hot objects in his formula matched his new idea, which contradicted the scientific thought of the day. He believed that he had found an interesting mathematical solution to a problem but did not really see the implications in his explanation of nature.

F. In 1905, Einstein took Planck's idea to the next step in a paper on the *photoelectric effect*. This was also the year in which Einstein published papers on Brownian motion and the theory of special relativity.

    1. In the late 1800s, scientists knew that if light is shone on metal, electricity is produced. At the time, the details of this photoelectric effect were a mystery, because people pictured electromagnetic radiation, or light, as a wave.

    2. In this view, we would expect that intense light would produce a great deal of energy in its electrons, but this is not the case.

    3. Einstein found that Planck's idea of light as quanta explained the data of what happens when intense light is shone on metal.

III. Another push toward the transition from classical to modern physics was the discovery of radiation and radioactivity.

A. The idea that energy could be spontaneously generated was very exciting in the late 1800s and spurred a great deal of study and rapid development of our understanding of this phenomenon.

B. One of the most important early experiments in radiation was conducted by a British physicist, J. J. Thomson, in 1897.

    1. Thomson constructed a device consisting of an evacuated glass sphere connected to a high-voltage battery on either side. Using this device, he produced radiation in the sphere, called *beta radiation*, or *cathode rays*.

    2. Thomson did further experiments with his device. He surrounded it with an electric field and watched the beam of radiation bend, demonstrating that the beam was electrically charged.

    3. Through these experiments, Thomson deduced that the rays were negative particles and that they were very light, 2000 times lighter than the lightest element known, which was the hydrogen atom.

C. Other experiments done with radiation at the time included those of Wilhelm Roentgen, a German physicist. Roentgen examined what happened when the beta rays hit the end of a device like Thomson's. He found that the beta rays stop, and another ray, a *gamma ray*, or *x-ray*, travels through the room. The x-ray is invisible but could be detected with photographic plates.

D. Ernest Rutherford, a great experimentalist, did research into alpha radiation.

    1. Rutherford found that alpha radiation was much heavier than beta radiation and positively charged.

    2. He also found that if alpha radiation hit a target, such as nitrogen, the nitrogen could change into oxygen. The fact that one element could change into another was a radical idea at the time.

    3. All these experiments combined were beginning to teach scientists that atoms were not indivisible and not immutable.

    4. Rutherford had a further idea to begin to understand atoms: Cover a source of alpha radiation, such as uranium, with lead so that the lead will absorb the radiation. Next, drill a hole in one side of the lead to allow a beam of alpha particles to escape and use this beam of alpha radiation to study atoms.

    5. Rutherford directed the alpha particles at a thin gold foil. According to Thomson's earlier "plum pudding model" of the atom, all the alpha particles were expected to pass through the foil and be detected on the other side. Rutherford found, however, that some of the particles bounced back, the equivalent of firing an artillery shell at a piece of tissue paper and having it bounce back.

6. To explain this occurrence, Rutherford came up with a new model of the atom: Suppose that the positive charge that we know must be contained in the atom is concentrated in the middle, and the negatively charged electrons are orbiting around it, like planets.

E. This new model contradicted classical physics and introduced questions about why the orbits of the electrons didn't deteriorate and result in the disappearance of atoms.

1. A young Danish physicist, Niels Bohr, came on the scene at this time and developed the first quantum model of the atom based on the work of Planck and Einstein.

2. Bohr theorized that the orbits of the electrons in the atoms are also quantized, which would explain why the atom was stable.

IV. People began to see that nature is different on the scale of individual atoms; nature does not behave classically on this level.

A. Louis de Broglie, a French prince, observed that scientists seemed to be seeing a duality—a world of both particles and waves.

B. He put forth the idea of a *wave-particle duality*: At the quantum level, an object can be both a wave and a particle at the same time.

C. If this is true for the quanta of light, which we now call *photons*— that is, if light is both a wave and a particle—maybe the same is true of electrons.

D. de Broglie applied this theory to electrons in the atom and discovered that Bohr's model of the quantum nature of electron orbits was explained. This was the first step from a crude theory to a more rigorous mathematical framework called *quantum mechanics*.

Essential Reading:

Barnett, Muhry, and Quinn, *The Charm of Strange Quarks*, chapters 2.2 (from p. 18 on), 2.3, and 2.4.

Schwarz, *A Tour of the Subatomic Zoo*, chapter 1, up to p. 9.

Weinberg, *Dreams of a Final Theory*, chapter IV.

Recommended Reading:

Calle, *Superstrings and Other Things*, chapters 21 and 23.

Greene, *The Elegant Universe*, pp 86–97.

Lederman, *The God Particle*, chapter 4 and the first half of chapter 5, up to "A peek under the veil."

Riordan, *The Hunting of the Quark*, chapter 1.

Questions to Consider:

1. If electromagnetic radiation (light) strikes a metal surface, electrons can be ejected. Two properties of these ejected electrons are relatively easy to measure: their energy, and their intensity (number/sec emitted). The incoming light also has two basic properties that can be easily adjusted: brightness (intensity) and color (frequency). These are completely independent properties.

    (a) Suppose light is a classical wave, like waves rolling up to a beach. (The ejected electrons are like pebbles being scattered.) How would both of the properties of the ejected electrons (listed above) depend on each of the properties of the incoming light?

    (b) Suppose light is, instead, a stream of "photon particles." How would your answers to part (a) change?

2. Suppose you were given three small radioactive samples: one is an alpha emitter, one is a beta emitter, and one is a gamma emitter. Imagine you were given the rather horrible order to eat one, put a second one in your pocket, and hold the third one in your hand. Which sample would you choose for which place and why? (Yes, I know, it's a rather disturbing choice, but there is a logical "best answer," assuming that you cannot choose "none of the above.")

Lecture Four—Transcript
The Birth of Modern Physics

Classical physics developed steadily, starting with Isaac Newton's ideas in the 1600s and progressing through the 1700s, 1800s, 1900s. The ideas of forces, gravity, electricity, magnetism, ideas about temperature, thermal physics, light and sound; all of this was being developed and understood. I think there was sort of a sense of contentment among the physics community members in the late 1800s, that we really had established a scientific method and a scientific framework that could describe the world, very deeply and very thoroughly. But there were some inklings in the late 1800s that there was something deep and important missing, and today I want to talk about this transition that was made pretty much right around 1900, from what we now call "classical physics" to what we now call "modern physics."

This transition was forced upon us by inconsistencies between theoretical calculations and the data that was observed in laboratories. The first one, or at least the one I think is perhaps the most influential in the transition from classical to modern thinking, was understood for the first time (or at least tentatively understood) by a fellow named Max Planck. Max Planck was a German physicist who went into college very young. He was a brilliant young fellow; he went to college at the age of sixteen, and an older professor told him, "You shouldn't go into physics because we pretty much know everything, it's a closed and almost dead field."

Fortunately, Max Planck wasn't paying much attention to this old guy. He started studying all of the classical physics that was known, and got very interested in the problem of what happens when objects heat up and start to glow. It just seemed like a ripe topic to put together all of physics. You've got atoms in the material that are jiggling because they are how, and then you've got to understand electricity and magnetism. Because, after all, when something glows what is really going on is there's little electric charges jiggling around, and they emit electromagnetic waves, electromagnetic radiation, which is light; and you need to understand thermodynamics. You really have to put together everything that you know about classical physics.

People had been trying to do this for quite some time, and they were discovering a big disaster. You would do the calculations, following the laws of classical physics and conclude that the color that you would predict was completely wrong. In fact, you get crazy answers; you get infinite answers for the amount of energy being emitted by some glowing object at ordinary, fairly high temperatures. So this is a big puzzle, it was some sort of a dilemma for classical physics.

Max Planck studied the problem and the first thing that he observed was there was no mistake being made. Classical physics was capable of solving this problem, and you worked out the details, and you disagreed radically, wildly with the data. So the next thing that Max Planck decided to do was kind of play the role of Kepler.

Remember, Kepler was looking at data describing planetary motion, and his idea was not to explain the motion, but just to describe it. That's always a good first step in science; just describe the data quantitatively.

So Max Planck sat down and looked at the data that had been taken. The data was sort of, you know, when something starts to glow it gets red, and when it gets hotter it gets bluish and then whitish. This was the phenomenon that he was really trying to describe quantitatively, and he was able to come up with a formula, which described the temperature and color behavior of hot objects. The formula matched the data, so it seemed to be in agreement with what people where observing in the laboratory, but it had no fundamental explanation. It was simply a description, but not an explanation.

Thus began what Max Planck called the most strenuous work of his life. Let me quote from him, "The whole procedure was an act of despair, because a theoretical interpretation had to be found at any price, no matter how high that must be." You've got to realize that this is a very insightful idea in the 1800s, when people are quite satisfied with the nature of physics and science. And this guy is saying, "Look, we've got an irreconcilable difference, we've got to understand it." So he's banging his head against this problem. (I love this; it's strenuous work, right? Theoretical physics, when you're sitting quietly at your desk thinking, it's really strenuous work.)

He did come up with an explanation, and he wasn't completely satisfied with it, but here was his explanation. Let me tell you something first about electromagnetic waves. When you jiggle a charge, little electric charges make electromagnetic waves which travel away, and that's what we call "light." Light is a wave. It's a wave like any other wave. Physicists understand waves very well. In the late 1800s, wave mechanics was a very well developed branch of physics. Imagine poking your finger into a pond; you'll create a wave. So your jiggling finger in the pond is like a jiggling electric charge, and water waves will emanate from that, and we understand how waves behave.

Waves are very special things. For instance, two waves can pass through one another. While they're passing through one another, they interfere with one another, but then they continue on their merry way. Waves can bend around corners. They have various properties that are very distinctive, and all of those properties have been experimentally verified, have been observed by light rays. Light rays can bend around corners. Light rays can pass through one another. Light rays can interfere with one another. So people really were convinced, from the early 1800s on, that light is a wave. And there were theoretical arguments that were very, very compelling that this was a fact of nature.

But Max Planck was willing to challenge the scientific framework of the time, and said, what if—this is his brilliance—what if atoms, little jiggling charges, do not emit electromagnetic waves, but instead emit light in little pulses, little flashes? So we now call those little chunks of light, "quanta" of light, for individual, little bundles of light. Now, this is a crazy idea; it doesn't fit in with the theoretical framework of the day. But Max Planck just asked, "What if?" He worked out the calculations, and surely enough, his description of this data of hot objects matches with this crazy idea.

So he has taken our idea about light and turned it on its head. He says, "No, it's not a wave after all; it's a little bundle of particles." When you look at a light bulb, think of a stream of little BBs of light coming at you. That's the new image of light that Max Planck is proposing. It's in contradiction with data from the early 1800s, but in agreement with this new data.

Max Planck got a Nobel Prize for this work in 1918, but he never really believed what he was describing was fundamentally a statement about nature. He thought that he had discovered a mathematical solution to the problem, a kind of a formal way of thinking about light that wasn't really describing light.

So that was 1900 when Max Planck published that paper. Five years later, in 1905, there was a young physicist who at the time was barely known. He had his Ph.D., but he couldn't get a job as a physics professor, so he took a job at the Patent Office, and he was a smart guy. This was Albert Einstein, and he spent most of his day thinking while he was at the Patent Office, because the Patent Office work was pretty easy.

In one year, 1905, Mr. Einstein published three papers. These were refereed physics papers in the journals, even though he didn't have a position at an academic institution. Any one of those three papers was so brilliant and so remarkable that he would be famous today if he had just written one of them; he would be easily as famous as Max Planck. But he wrote these three papers in what is sometimes called his "miracle year." These three papers were very influential in transforming our ideas about physics from the classical to the quantum.

So let me just briefly summarize. One of them was his paper on Brownian motion, which I talked about in the last lecture. It was the quantitative understanding of how atoms, bumping into small objects such as a little piece of yeast or pollen, would make it jiggle around. It was a quantitative demonstration that atoms are real. That was an important and influential paper.

Another that he wrote that year was his special theory of relativity, for which he is most famous: $E = mc^2$, and all that good stuff. I would love to tell you about relativity, but it's a whole course unto itself, so I'm just going to leave this theory of relativity. You'll have to go learn about it elsewhere, on your own. There are a lot of resources available. Relativity is tough going. It's kind of counterintuitive, but you can understand relativity without working through very much (or even any) mathematics, so I would encourage you to learn more.

What I want to say about it is simply that special relativity changed our worldview about space and time. Albert Einstein has explained to us in this paper that time and space are intimately connected. He called it "space-time," and this connected fabric—it's what we live in, it's what physics lives in, this space-time. It's not the way we had thought for hundreds of years. It's not universal and independent of observer. So this is Einstein's theory of relativity in a super nutshell. Although, in principle, you need relativity to understand particle physics, which is *this* course, in practice we can really understand most of what we need to know about particle physics without invoking any of the details of relativity.

What I want to focus on today is Albert Einstein's third paper from that year, which was on an effect called the "photoelectric effect." The photoelectric effect—"photo," "electric"—it comes from light and electricity. It was known in the early 1900s that if you shine light on metal that electricity would come out. This is extremely well known today, like solar panels; light comes in, electricity goes out. Also, you take your little remote control and you click the television set—infrared light hits some detector on the TV set, electricity comes out, gives the signal. Photoelectric effect is useful technology today. At that time it was a big mystery, because people thought that electromagnetic radiation, light, was a wave.

People were picturing a wave, like a wave at the beach, coming up and striking the metal, which is like the beach. So the pebbles on the beach are like the little electrons, and the wave jiggles the pebbles a little bit—that's your mental model in classical terms of light hitting metal, and what you would expect classically; you would make a number of predictions based on this worldview. For example, if you have an intense wave, like a tsunami, those pebbles are going to go flying, because there's going to be lots of energy. You would expect that electrons coming out from intense light should have lots of energy, but what you see is completely contradictory data.

It's kind of like Max Planck, who was looking at data from glowing, hot objects that were in contradiction with classical ideas. Einstein is looking at data which is contradictory to classical ideas, and he said, "Let's take Mr. Planck's idea seriously. Let us believe that electromagnetic radiation comes in the form of little bundles, little quanta of light." So now we're imagining a kind of a pellet storm hitting those pebbles on the beach, and any one, individual pellet of light can hit one pebble and knock it free. This makes different quantitative predictions and Einstein worked them all out; that was his paper.

This was the paper for which Einstein won his Nobel Prize, by the way, in 1921. It's a little bit interesting that Einstein did not get his Nobel prize for relativity, not for $E = mc^2$, not for his understanding of gravity, nor for all the stuff for which he's now most famous. It's an interesting social and political physics story, which I will leave to you to learn more about because it's sort of fun stuff.

So Albert Einstein has taken Max Planck seriously. Max Planck didn't take himself seriously, in the sense that he didn't really continue working on this idea of light quanta. But once Einstein weighed in on the topic, people began to take the idea a little bit more seriously.

Let me leave light quanta for a moment and talk about another phenomenon, which was in gross contradiction to classical ideas of the same era, the late 1800s. People were discovering radiation and radioactivity. In the late 1800s this was really exciting stuff. It was just so cool to find a little rock that glows, or that could expose film. It was almost like magic, but of course it's physics and people were trying to understand it. It just didn't fit in with the classical worldview. Energy was just coming out of nowhere. It was a crazy phenomenon, so it got a lot of attention.

People were studying radiation and radioactivity like crazy. The development of our understanding was very rapid. From the 1890s, when people first began to observe, measure and understand radiation, to the, say, early 1900s, people began to categorize radiation. Now, you have to name this stuff; everybody always wants to give a name when you discover something new, and the Greek alphabet was used. Alpha, beta, and gamma are the first three letters of the Greek alphabet, so radiation was called alpha radiation, beta radiation and gamma radiation. Nowadays we've used up the whole Greek alphabet. We'll be talking about that in future lectures.

People were trying to understand what this radiation was. One of the most important early experiments was done in 1897, by a British physicist named J.J. Thompson. He had this device, which we still use today; it's basically an evacuated glass sphere or chamber. He sucked all the air out of it and attached a big, high-voltage battery, one terminal on one side and the other terminal on the other side. Then some little rays appeared – radiation; it's called "beta radiation." (Sometimes it was called "cathode rays," because the chemists used to call one of the poles, the negative pole of the battery, the cathode.) The rays seemed to go from the negative side to the positive side, and in 1897, J.J. Thompson did a series of experiments that really pinned down what this beta radiation was.

"Cathode rays," by the way, is still the name we use if you're looking at your TV set, or a computer monitor; that's called a "CRT," a cathode ray tube, so these things are still in use. It's really cathode rays or beta rays, coming from one side to the other and hitting a screen, making it glow.

Mr. Thompson did a series of experiments. For example, he put an electric field around his device and watched the beam bend. So *that* demonstrated that it was electrically charged. Then he used a magnetic field and watched the beam bend so he could figure out properties of the beam. He could figure out which way it was going. He deduced that the rays were negative particles; they were negatively charged objects (that's negative referring to the electrical charge of the objects), and that they were very, very light in that they were 2,000 times lighter than the lightest thing known at the time, which was a hydrogen atom.

Remember, atoms are supposed to be indivisible, and the lightest, smallest atom of all, hydrogen, has one unit of mass, some system of measuring weights. Now we found something lighter still by a big factor, so this is a huge puzzle. J.J. Thompson did lots of measurements and discovered that, no matter what you made the tubes out of, no matter what the residual gas was, they were always the same particle, the same electric charge: these little, negative electrons that were very light. So these seem to be in everything; no matter what you make the device out of, there's electrons in there.

There were other experiments being done at that time. For example, Mr. Roentgen, a German physicist, was looking at what happened when these beta rays hit the other end of the device they stopped. Then there's something, some other mysterious ray, which has been called variously gamma rays or X rays. They were truly invisible as they would travel through the room, but you could detect them with photographic plates. For example, if you put your hand in the path of these X rays, you'll see a shadow image of your hand. Your bones will absorb the X radiation and the skin will not, so you see what is now used in medicine, at your dentist's, just an X ray. At that time we were calling them gamma rays. All of this stuff was beginning to come into some sort of order, but people really didn't have a deep understanding of where this stuff was coming from, what its origins were.

One of the big players in this story was Ernest Rutherford. He was born in New Zealand, spent his life in England, formed what is now known as the Rutherford Laboratories. He did a whole series of measurements over the long course of his career. He was a truly amazing experimentalist. He really was, in some sense, the one who figured out what radiation was. Though J.J. Thompson figured out what beta radiation was, Ernest Rutherford figured out, for example, what alpha radiation was—little particles, but heavier; in fact, four times heavier than hydrogen. They were very, very heavy compared to beta rays, and they were positively charged rather than negatively charged.

Mr. Rutherford figured out lots of things. He figured out that if alpha radiation was to hit a target, like nitrogen, it could change from nitrogen into oxygen. This is a radical idea in the early 1900s, right? That one element, nitrogen, could be converted into another element, oxygen. There is something going on here that is teaching us that atoms are not indivisible after all. They're made up of smaller pieces and they're not immutable; you could change one to the other. But still, there was no deep understanding of what the atom really was.

So Rutherford, a brilliant experimentalist, said, "Look, we have to figure out what an atom is." So here's his idea, to take a source of alpha radiation, just a natural chunk of material like some uranium, and cover it up with lead so the lead will absorb that alpha radiation, so nothing's coming out; it's just a little bit warm. So we're going to drill a little hole in one side, and we have a little beam of alpha particles that can come out of this thing.

So this is the essential ingredient for all physics experiments, particle physics experiments. Even today you need a source and a beam of particles. Now, what are we going to do with that beam? Instead of studying the beam, we're going to use the beam as a probe to study atoms. So now we need some atoms. He took a foil, like silver foil, except he used gold foil, and the reason he used gold is that it's a lovely material to work with. You can make it very, very thin. You've got a little, thin layer of gold atoms, and we're going to shine these alpha particles at the gold and ask, "What do they do?"

Now, in general, when you do an experiment like this, you have some idea in advance of what you think is going to happen. So what do you think is going to happen? J.J. Thompson (remember the guy who discovered the electron in 1897?) he has an idea that he's published. He calls it the "Plum Pudding Model." (I love that—the Plum Pudding Model of the atom!) So what do we know in the late 1800s, early 1900s about atoms? We know roughly how big they are; they are about 10^{-10} meters in size. All of them are about the same size. We know that they contain electrons inside of them, tiny light, objects which are negatively charged; but we know that atoms themselves are electrically neutral, so there must be some positive charge in there.

Mr. Thompson's Plum Pudding Model held that the positive charge is a smear; it's spread out over the entire side. That's why the atom is as big as it is, because the positive charge is a goo that spreads out over 10^{-10} meters, and that's the Plum Pudding Model. The "plums" are the little electrons in there—the little, hard nuggets that are very light—and you can knock them out in various experiments.

So Mr. Rutherford takes his beam of alpha particles, which he is envisioning as little, sub-microscopic particles that are very massive, and he runs it into the foil. Now, what would you expect if you took a little BB and you ran it into some plum pudding? Well, what you would expect is that you would swoosh through the plum pudding and go pretty much in a straight line, maybe deviate a little bit off to the side. Then he sent his graduate students down into the basement where he had set up the source and the target; they needed a detector. He used some phosphorescent material that, when an alpha particle hit it, it flashed a little bit. So these poor students were down in the dark, watching individual flashes, counting them, trying to see how many particles come through at various angles, to verify this theory of J.J. Thompson's.

One of those students was Geiger, and Geiger became famous for developing the Geiger counter, because he got so sick of sitting there, counting with his eyeballs. In later years he developed this electronic device to do the counting.

Mr. Rutherford was a smart guy. He was a good physicist, probably one of the greatest experimental physicists in history. One of the things that occurred to him was maybe this model was wrong. So, he said, "You guys go down there and look at the alpha particles at all angles; go out 30 degrees, 40 degrees; in fact, go back *behind* your gold foil target and just see if anything bounces backward."

Now, that's totally nuts, right? If this stuff is plum pudding, everything is pretty much going to go straight through. But they looked and they found a few events bouncing backwards. This is kind of crazy, I have a quote from Ernest Rutherford, "It was quite the most incredible event that ever happened to me in my life. It was as if you fired a 15-inch artillery shell at a piece of tissue paper, and it came back and hit you." Hey, it's radical. This is the man who has essentially discovered radioactive transmutation. He's done a lot of pretty wild stuff in his career, and to him this was the wildest event that he'd ever seen. So how can we understand it? Rutherford says, "I think, even though J.J. Thompson is, you know, the old senior physicist— brilliant, he discovered the electron, but—I think he's wrong. I don't think we're looking at plum pudding here."

So what *could* we be looking at? What's the model of the atom? Rutherford says, "I have a different idea. Suppose that the positive charge, which we know has to be in there, is very concentrated. It's a little dot right in the middle—massive, that's where the mass of the atom lives—and the electrons are these little, super light things that are sort are orbiting around like planets around the central sun, which is the nucleus." So you've got negative charges in orbit and a heavy, massive, point-like nucleus in the middle, and that's the new, planetary model of the atom. This was Mr. Rutherford's great idea.

Almost immediately, people said, "Sorry Mr. Rutherford, but this just can't be right, because we know classical physics." Classical physics says that if an electron that's a charged particle is going around in a circle, orbiting around; anytime you have a charged particle going in a circle, it's accelerating around the curve. It's radiating electromagnetic waves. And if you radiate away energy, your orbit's going to decay, like an old satellite hitting the Earth's atmosphere and spiraling in, ultimately crashing into the Earth. So if atoms were the way Mr. Rutherford had suggested, then do the calculation and you discover they would disappear. They would disappear in a puff of greasy, black smoke, within about a microsecond, and there wouldn't be any atoms left. So this model was no good. People didn't know what to do; it was a good model in some respects, but it had this contradiction with classical physics.

Along comes a young theoretical student. His name was Niels Bohr. He came from Copenhagen; he was a Danish physicist. The tradition then was the same as it is now; after you get your Ph.D., you do what is called a "post-doc", post-doctoral research. You go for a couple of years to a big-name laboratory with some great physicists. You work with them. You learn the ropes. Then if you do well, you can go off and become a professor on your own.

So that's what Niels Bohr was doing at the Rutherford Labs. And Niels Bohr said (he's a theorist, so he's trying to interpret this result), "Look, we know that classical physics seems to be breaking down in the work of Max Planck and the work of Albert Einstein, when we had the interaction of radiation with atoms. Well, here's another case where we've got atoms, and maybe *classical physics* is breaking down. Maybe the interaction of electrons with electromagnetic radiation isn't the classical idea that we've been thinking for so long."

So Niels Bohr has a new model. It's the first quantum model. He's taken the idea of chunks of radiation from Max Planck and from Albert Einstein, and now he's applying it to the world of atoms, trying to understand their structure. So he says maybe the electron's orbits are also "quantized"; it's not just radiation. Maybe the planetary orbits that are allowed also come in chunks, and that idea helps to explain why the atom is stable, because you can't just go from one orbit to one slightly smaller.

In fact, it explains lots more data that was known at the time about the light that is emitted from, say, hydrogen atoms when you heat them up. Hydrogen atoms glow, and the color of the glow can now be understood from Mr. Bohr's quantum model. This is a new idea, and once again it's flying in the face of classical theories; and once again, we have invoked the idea of chunks—quanta of energy, chunks of matter. It's a strange idea, and it's not really a very deep or rigorous theory yet; Niels Bohr's model is just a kind of *description* of what we're seeing. Like Kepler, we are describing atoms, but we don't really understand them yet.

People started thinking about these ideas, though. Now we've got lots of evidence, ranging from Planck, Einstein, Bohr and all this radiation phenomenon, but something different is going on. Nature is different, when you get down to the scale of individual atoms. It's not behaving classically like it does when you're throwing tennis balls, cars, bicycles and just ordinary world stuff. Something new is going on.

A French prince named Louis de Broglie was working on his physics Ph.D. He took this idea so seriously, he said, "Look, what we seem to be observing is a strange duality: things are particles and they are also waves." We can't really picture this. When I look at a water wave, it's a wave. It's not a bunch of particles; it's really a wave phenomenon. Sound, any wave phenomenon that you can think of, is not a bunch of little BBs flying about. So it seems that waves and particles are two contradictory possibilities in the world. You've got to be either one or the other.

Yet people, especially Louis de Broglie, are proposing now a "wave-particle duality." That's the buzzword. It says you can't quite picture it. It's not a classical idea that you can have a mental model of directly in your brain. But maybe at the quantum level, at the level of tiny things, you can somehow be both and neither at the same time.

I wish I could make this more concrete for you. I cannot draw you a picture of something that's a wave and a particle at the same time. You just kind of have to accept this as a crazy but true (experimentally true) statement about how little, teeny objects are. De Broglie said, "If it's true for light and the quantum of light, which we now call photons—if light comes in chunks, which are called photons, maybe something else which comes in chunks, like electrons—the little particles that are running around in the atom— maybe those things are also waves." After all, if a wave can be a particle, maybe a particle can be a wave.

Once you accept this idea, you kind of have to run with it, it's your obligation to do a "what if?" So, what if electrons are waves? Well, if electrons have a wave nature, think of the wave on a guitar string. If you pluck a guitar string, there are certain frequencies that it can jiggle at. It can jiggle in the fundamental or it can jiggle at twice the fundamental. There are various modes that the guitar string can vibrate in, but not all possibilities are open to you; you can't just change the frequency of the guitar string a little bit without re-tuning it, without changing the tension.

So de Broglie applied this idea to electrons in an atom and discovered that Mr. Bohr's model was explained. The quantum nature of electron orbits is explained if you believe that electrons are waves. So we're heading toward a theory. It's still not a fundamental theory; there's not a rigorous mathematical framework yet. In the next lecture, we're going to see the transition that was made in the 1920s from these primitive ideas of quanta to a more rigorous mathematical framework, which we now call quantum mechanics.

Lecture Five
Quantum Mechanics Gets Serious

Scope: What were the key early developments of quantum mechanics? What did it teach us about the world and why would anyone believe such a theory? This lecture serves as a qualitative introduction to the work of Schrödinger and Heisenberg, along with the Dirac equation (which marries quantum physics with relativity) to describe electrons.

Electrons are the first fundamental particle discovered: the carrier of electricity, constituent of all atoms, and the key to understanding chemistry! In this lecture, we highlight properties of the electron: what does it mean for a "pointlike object" to have properties, and what might they be? The lecture also introduces *spin* and tries to make sense of this purely quantum mechanical concept. We look at Dirac's equation, which predicted antimatter and turned out to be smarter than he was, and discuss the birth of *quantum electrodynamics*, or *QED*. (We'll talk about the words *quantum*, *electro*, and *dynamics* and put them together to get a sense of what QED tells us about.) We conclude with the experimental discovery of antimatter and the neutron and their significance for the developing story.

> Anyone who is not shocked by quantum theory has not understood it.
> —Niels Bohr

> I repeated to myself again and again the question: "Can nature possibly be as absurd as it seemed to us in these atomic experiments?"
> —Werner Heisenberg (*Physics and Philosophy*)

Outline

I. Quantum mechanics forms the conceptual underpinnings of particle physics. For our purposes, we need a qualitative sense of quantum mechanics.

A. Quantum mechanics was born when data contradicted the predictions of classical physics, which was a well-developed field in the late 1800s.

B. In the early 1920s, a German physicist, Erwin Schrödinger, was asked to give a colloquium on the ideas of Louis de Broglie about wave-particle duality.

    1. An audience member suggested that what was needed in the study of quantum waves was a wave equation, similar to what had been developed for sound waves and water waves.

    2. Schrödinger began to develop an equation that could describe the wave nature of an electron.

    3. The Schrödinger equation yielded quantitative results for certain questions, such as the energy of an electron, but the values were inexplicable by classical physics.

C. The remaining question, then and now, is, "What is waving?" The best interpretation that can be given came from another German physicist, Max Born.

    1. Born's interpretation of the wave is that it is a *probability wave*.

    2. To understand this term, think of a water wave. It is high in some places and low in others. In a probability wave, the high places represent locations where the electron is likely to be found, and the low places represent locations where it is unlikely to be found.

    3. This analogy does not mean that the electron is traveling in a wave-like pattern. It travels on some path, and from the wave equation, we see certain places where it is likely to be and certain places where it is not likely to be.

    4. This probabilistic interpretation of Schrödinger's equation was the key to making quantum mechanics "work" and for describing a range of other physical phenomena.

    5. Some physicists and philosophers to this day still debate and discuss the structure and meaning of these "interpretations."

D. Another German physicist, Heisenberg, also played a key role in developing quantum theory.

    1. Heisenberg was more interested in developing equations that described observable phenomena, such as the light that goes into, and is emitted from, radiation-testing apparatus.

2. Heisenberg's equations were *matrix equations* and were later found to be mathematically equivalent to Schrödinger's equation.

3. Heisenberg also derived the *uncertainty principle*: In certain well specified cases, if you know one quantity well, then you are obligated to lose information about another, related quantity.

E. Quantum mechanics is counterintuitive, but we should keep in mind that although it has some mysterious aspects, it is rigorous science; it does predict the outcomes of experiments.

II. Almost as soon as quantum mechanics was developed in the 1920s, it was observed that the new theory had a conceptual problem: Special relativity was not completely consistent with the early version of quantum mechanics.

A. The early version of quantum mechanics was excellent for describing small objects that moved slowly but not for objects that moved quickly, such as electrons in orbit in an atom. Special relativity was needed to describe the rapid movement of objects.

B. Paul Dirac, a British physicist, tackled the problem of marrying quantum mechanics and relativity.

C. The name for this theory is *quantum electrodynamics*, or *QED*.

1. *Quantum* applies because it's a quantum theory.

2. *Electro* applies because the description is of electrical phenomena.

3. *Dynamics* applies because the theory describes why electrons move the way they do.

4. QED looks at the question of why electrons behave the way they do in relation to quantum mechanics and relativity.

D. In 1927, Dirac published the *Dirac equation*, which was motivated purely by the beauty of the mathematics, not directly by data. Dirac believed that the formulas of quantum mechanics and relativity should match.

1. The Dirac equation is consistent with quantum mechanics and relativity, and it makes certain quantitative predictions. It is also, like Schrödinger's equation, a wave equation.

2. This theory predicts the behavior of an electron in certain situations, such as when it's running free, when it's bound to an atom, and so on.

3. This theory evokes as a hypothesis the idea that an electron is a point particle.

4. How do you describe a point particle? We can say that it has certain characteristics, such as mass and electric charge.

   a. Another property of electrons is *spin*. Think of a spinning ball; the "amount" that the ball is spinning, or its *angular momentum*, can be measured (in certain units).

   b. In the same way, electrons spin, all at the same rate. This rate, derived from Dirac's equation, is *spin 1/2*.

5. Another consequence of the Dirac equation is the following:

   a. If you solve the equation for the behavior of an electron traveling in a cathode ray tube, for example, you get another solution.

   b. The second solution also reveals the existence of a particle with a certain mass and a certain magnitude of charge, but the "sign" is the opposite. The particle is the opposite of the electron, with a positive electrical charge.

   c. We call this particle discovered in the second solution *antimatter*. In other words, Dirac's equation in the late 1920s predicted the existence of antimatter.

   d. Dirac himself did not understand or pursue this result of his equation, but in 1932, antimatter was discovered in photographic emulsions in the laboratory.

III. A final discovery in the transition from classical physics to quantum physics was the neutron, which fit in well with the then-current understanding of the atom.

   A. Protons had been known since the early 1900s. A positively charged, massive object, the proton is the fundamental building block of the nucleus.

   B. Chadwick discovered the neutron, which is like a partner for the proton.

      1. If we think about a simple nucleus, such as one with two protons, we can imagine that the protons should repel each other because they are both positively charged.

2. Some new force of nature must exist that binds these two protons together. That force is not electricity, nor is it gravity; it is the *nuclear force*, or the *strong force*.

   3. The data being collected in the 1930s on the nucleus suggested that something else must exist in the nucleus other than protons. Some electrically neutral particle is needed to help the whole system stick together. That is the role of the neutron.

C. At this point, physicists had begun to develop a coherent story of quantum physics, just as they had earlier developed a coherent story in classical physics.

   1. This worldview, based on the components of the atom, was reasonably satisfactory but not complete either.

   2. In the period from the 1930s to the 1950s, physicists began to see that just as the classical picture of atoms was not complete, neither was the quantum picture of the protons and neutrons.

   3. The next step was to understand the nuclear picture. Although the explanation of the electrons was satisfactory, the nuclear part of the story was a bit more complicated. In the next lecture, we will travel into the nucleus.

Essential Reading:

Barnett, Muhry, and Quinn, *The Charm of Strange Quarks*, chapter 2.6.

't Hooft, *In Search of the Ultimate Building Blocks*, chapter 3.

Recommended Reading:

Calle, *Superstrings and Other Things*, chapter 22.

Greene, *The Elegant Universe*, pp. 97–108, 112–116.

Lederman, *The God Particle*, chapter 5 (second half, starting from "The man who didn't know batteries").

Wilczek and Devine, *Longing for the Harmonies*, Fifth Theme.

Questions to Consider:

1. Why is quantum mechanics considered so puzzling? What single aspect of quantum theory do you find most counterintuitive or hard to believe or understand?

2. What practical uses can you think of for antimatter? (This is a "real-life" question, not a Star Trek question. You cannot assume that antimatter just appears from nowhere; it would have to be produced somehow.)

Lecture Five—Transcript
Quantum Mechanics Gets Serious

Quantum mechanics forms the conceptual underpinnings for particle physics. Certainly if you want to understand the mathematics of particle physics, you really have to develop a deep understanding of quantum physics. For the purposes of this course, for the purposes of just understanding the ideas of particle physics, it's okay just to have a sort of qualitative sense, a very crude sense, of what quantum mechanics is about. It's a very rich and deep field, absolutely fascinating physics, and I would encourage you to go learn more about quantum mechanics. It's got enormous applications throughout all of physics and technology, but I want to talk about the evolution of ideas that led us from classical physics to the quantum mechanics that was developed in the 1920s. That's going to be the topic of today's lecture.

Quantum mechanics was a theory born out of desperation. There was data, and it was in disagreement, clear disagreement, with classical physics predictions. Classical physics was well developed; people really thought, in the late 1800s, that we had a great understanding, a deep, core understanding of everything: the world, what the world is made of, how it works. Yet, there was this evidence that there was something, at some level (in particular, when you get down to the level of individual atoms) that wasn't quite right. It wasn't fully formulated correctly, so we were confronted with, in certain cases, disastrous experimental results that just disagreed with the theory and had to repair it.

So the early days, the days of Max Planck and Albert Einstein and Niels Bohr, were just tentative steps, ideas towards the formulation of quantum mechanics. The word "mechanics", to a physicist, just refers to a description of how things move, and why they move that way. So if you have a tennis ball, and you whack it with a tennis racket, mechanics, classical mechanics explains how it's going to behave as a function of time, after you hit it. The laws of classical mechanics work just fine when you're describing ordinary objects, but they seemed to be breaking down when we were talking about electrons or atoms, the constituents of atoms.

In the early 1920s, a German physicist named Erwin Schrödinger was asked to give a little colloquium on these crazy ideas of Louis de Broglie, the French prince who had first formulated an idea that perhaps electrons should be thought of not just as particles, but also as waves. So Schrödinger gave this little talk, and an elder physicist in the audience said that he thought all this talk of waves, with people waving their hands, was kind of silly. If you want to talk about waves, you need a wave equation. This is a physicist's way of thinking. In the 1800s, we had described sound waves, and water waves, mathematically, rigorously. So, let's come up with a wave equation which describes these quantum waves that people are talking about in such a casual fashion. So Mr. Schrödinger, who was a very smart young fellow, went off and started thinking about this. He was really trying to come up with an equation to describe the wave nature of an electron. The story goes that in the winter of 1926 he went off on a little winter holiday—didn't take his wife, he took his girlfriend. His goal was keep his girlfriend happy and work out a wave equation; apparently succeeded. Certainly he succeeded in getting a wave equation; he got the Nobel Prize for it in 1933, seven years later.

The Schrödinger equation is a equation. So physicists are excited; they can solve it, and what does solving it mean? Well, it's a sort of a funny business at first. They're not quite sure what this wave represents physically; it's got something to do with electrons. You could solve for an electron in a cathode-ray tube. You could solve for an electron buzzing around in an orbit around a proton, which would be a hydrogen atom. The equation yielded quantitative results for things like the energy of the electron, something that you can measure in the laboratory, and it gave correct values. These values were inexplicable by classical models. Niels Bohr had kind of built this, almost hip boot, quasi-descriptive quantum picture, but it wasn't very rigorous. However, now we had an equation, and to solve the equation, you come up with numbers, and they agree with the data. So it's looking good, and the question was, what is waving? It's not a water wave; it's not a sound wave. What is waving? And I don't really have a good answer for that, even today. But the interpretation, the best interpretation I can give, came from a fellow named Max Born, another German physicist. And Max Born, just a year after Schrödinger's equation, came up with what is still today accepted as the best possible interpretation of this wave: it's a probability wave. So what on earth does that mean?

Imagine. Let me draw a wave for you, so it's big in some places and small in other places. Like a water wave that's tall in some places and low in other places. Max Born says, "No, think of this wave abstractly. Where it's big, that's where the electron is likely to be, that's where it's probable to be found. When the wave is small, that's where the electron is not likely to be." So it's not that the electron is traveling along in a wave-like pattern; it's not going up and down. It's travelling along in some path, and what were saying from this wave equation is there are certain places where it's likely to be found and other places where it's not likely to be found. This probabilistic interpretation of Schrödinger's equation was the key ingredient, it was the necessary ingredient to make quantum mechanics work, and describe all sorts of other physical phenomena.

People didn't like it. People didn't like it then, and even today there are plenty of philosophers, and even physicists, who argue about, what it means. Does it mean that we've lost the rigorous, deterministic sense of science that Isaac Newton introduced us to in the 1600s? There are aspects of quantum mechanics that are very hard to grasp, very mysterious, and I would love to just continue to go into those, but it's really a detour from the course on particle physics, so I have to leave you to study quantum physics on your own. You're not alone if you find the ideas of quantum physics a little bit disturbing. Albert Einstein never really bought into this idea of probability. He thought that nature, at its core, had to be rigorous and deterministic. He said, "God does not play dice with the universe," and there were lots of people who felt that way. In fact, even Mr. Schrödinger, at a later time, said, "Had I known that we were not going to get rid of this damned quantum jumping, I never would have involved myself in this business." So even Schrödinger—who invented the equation—thought that it was some physical kind of wave and has people re-interpreting his work.

Schrödinger was not the only one developing a quantum theory; there was another German physicist named Werner Heisenberg. Heisenberg actually came first, chronologically; he was a year before Schrödinger, in his publication. Heisenberg had a rather different way of thinking. He was very mathematical, and he said, "Look, I don't want to ask questions about electrons that nobody can see. It's too small to look at an electron or an atom. So instead, let me write down equations that describe things we can se and we can measure, like the light that goes into our apparatus, and the light that comes back out again. That's something that I have dials and meters, and I can read that stuff. That is physics." So he wrote a mathematics which connected what goes in to what comes out, which after all is what science is supposed to do, and he succeeded.

It was very fancy mathematics; at the time, physicists weren't so well acquainted with what's now known as matrix mechanics, and they were very familiar with wave equations. So I think, even though Werner Heisenberg came first and was actually successful in his theory,

Schrödinger gets a lot of the credit because everybody else in the community was able to take his ideas and do the math more easily. Since then, people have gone back; and in fact nowadays, when we teach quantum physics in undergraduate and graduate school, we teach both the wave equations and the matrix equations. It was proven, shortly afterwards, by Schrödinger, that they were equivalent. They are mathematically equivalent ways of describing this rather difficult-to-picture behavior.

Mr. Schrödinger and Mr. Born left Germany during the early 1930s, as the Nazis came into power, but Werner Heisenberg stayed behind, and in, fact became very influential. He was the director of the German program to try to develop nuclear weapons during World War II. There were very interesting stories about these characters, as you can imagine. There's a fairly recent play, in fact, that talked about what did Mr. Heisenberg actually know. He claimed afterwards that he was trying, in some sense, to sabotage the Nazi war effort, which was why they didn't develop the nuclear bomb. It is an interesting story, but a little bit of a detour.

Quantum mechanics is very counter intuitive, and we're not going to go much further into the details, certainly not into the mathematics, but I do want to leave you with the following sense of quantum physics. Although there are some somewhat mysterious philosophical aspects about the interpretation, in terms of physics, in terms of predicting the results of experiments, this is rigorous science. You have equations, you do calculations, you check and they agree with the data, and it agrees well. In fact this theory, quantum mechanics, which was developed in the 1920s, is still used today. It describes semiconductors and superconductors and regular conductors; and just about any system that involves small objects is quantitatively well described in quantum mechanics. So even though Werner Heisenberg wrote down what is now called the Uncertainty Principle. The Uncertainty Principle isn't just some nebulous statement that, well, quantum mechanics is weird and everything is uncertain. It is in fact a rigorous mathematical formula, and it says, "If you know the following quantity very well, then you are obligated to lose information about another related quantity." It's a fairly bizarre statement, but it's accurate. It agrees with data, and there have been lots of experimental tests of the Uncertainty Principle. The Uncertainty Principle doesn't say everything is uncertain; you can calculate the energy of an electron in a hydrogen atom with extraordinary accuracy. It just says that knowing one thing often compromises your ability to know something else at the same time.

So quantum mechanics is developed in the early 1920s, and almost right away it was observed that there was one theoretical, or conceptual, problem with this new theory, and that was Special Relativity. Albert Einstein's reformulation of space and time is not completely consistent with this early version of quantum mechanics. There's some disconnect. The early version of quantum mechanics, as written by Schrödinger and by Werner Heisenberg, is basically describing objects that are moving slowly. It works great for objects that are moving slowly, but what about if you have a little teeny object—so you need quantum mechanics—and it's moving really fast? As for instance, an electron in orbit in an atom, it's both small and fast, and so you should really need quantum mechanics and relativity. Einstein's theory of relativity is most important, practically speaking, when things are going at speeds close to the speed of light.

The physicist who really put that story together for the first time was Paul Dirac, a British physicist, sort of famous as a man of few words. In fact, the story goes that when he was a child in England, his father, who was Swiss, insisted that everybody should speak French at the dinner table. I think he wanted to encourage his kids to learn his native language, but in fact, what that meant was the kids never said a word at dinner because they didn't speak French. So poor Dirac sort of grew up in this climate of not speaking a whole heck of a lot. So he's kind of famous. Later in his life, he would answer questions sort of yes, no, or I don't know. It's not completely fair, because his writings are very articulate. He was quite a brilliant fellow, and he tackled this problem of putting relativity and quantum mechanics together.

The buzzword, and we're going to talk about this theory in more and more detail as this course goes along, goes by the fancy name of quantum electrodynamics. "Quantum" because it's a quantum theory; "electro" because he's trying to describe electrons and their behavior. "Dynamics" is another one of those physics buzzwords that just generically means "why things move the way they do." Dynamics and mechanics are almost synonymous words. Quantum electrodynamics—why do electrons behave the way they do when you are worrying about relativity and quantum mechanics at the same time? It's abbreviated QED, and we'll come back and talk more about QED.

But in 1927, when Dirac first introduced this theory, he writes down an equationn and the equation was motivated purely by the beauty of the mathematics. He took the formulas from quantum mechanics and the formulas from special relativity and said, "These have to match." So he wasn't looking at data; he was just looking at the elegance of the formalism. This is a novel concept at that time, to just let the beauty of the mathematics guide the writing down of a theory, rather than just trying to describe data in some effective way. Nowadays, this is a very powerful guide to modern physicists who are trying to write down contemporary theories in particle physics.

Dirac was, in some sense, the first truly modern theorist, and he wrote down what we now call the Dirac equation. The Dirac equation is very simple to write down. It's consistent with relativity and with quantum mechanics, and it makes certain quantitative predictions. It's like the Schrödinger equation; it's a wave equation. It's just that it's consistent with Albert Einstein as well. When you write down this theory, you can predict the behavior of an electron in various situations—when it's running free, when it's bound by an atom. You can imagine describing all of chemistry, which basically amounts to electrons in this atom and electrons in that atom, communicating with one another via electrical forces. In principle, and in practice, you could describe that by either Schrödinger's equation, if things aren't moving too fast, or Dirac's equation, if the speeds become relativistic.

This theory invokes as a hypothesis that the electron is a point particle. That's a really weird idea, and I want to spend a minute talking about what it means to talk about a point particle. Of course, we're talking about particle physics in this course, and pretty soon everything we're going to be talking about are other point-like particles, and it's just such a hard concept to grasp physically. This theoretical physics stuff gets pretty abstract at times, and this is one of those abstractions that—I don't know if you can ever really picture a point. I think, "In my mind I have a little dot in my head, but's sort of like a little period at the end of a sentence." Even a period at the end of a sentence is itself an object which has a finite size, made of something else: little ink spots, little atoms inside that dot. Now I'm talking about a fundamental dot; an electron is truly a point. So I can't really tell you how to visualize it, except that this is an abstraction, and it's what went into Dirac's theory. Then you calculate, and you measure, and the calculations agree with the measurements. So this is evidence, scientific evidence, that your hypothesis—that the electron is a point—is in agreement with the data, so maybe you just have to accept it.

If you have a point particle, how do you describe it? Well it has certain characteristics. Even though it's a dot, that means it's not spread out over space, it still has mass. The electron is a massive particle, and you can measure that mass. Every electron in the universe, apparently, has the same mass. You can look up that number in kilograms, if you like; and every electron in the universe has an electric charge, bnd what does that mean? Well it means that if you put an electron into an electric field, as near a battery, it will move in a certain well described way. The strength of the interaction of an electron with a battery is characterized by a number, and that number is called its electric charge. It's a negative electric charge, and it has, again, a certain numerical value. Any object that has electric charge on it, you can compare with the electric charge on an electron. Now I say "on the electron," but remember it's a dot so it's just a property of that dot that it has mass and it has charge.

Fundamental particles have other properties which we associate with them; this point in space has a property which physicists called spin. Spin is a metaphor; it's really a mathematical property, but I think of spin in a kind of a classical sense. You have a ball ,so imagine this is my electron. Now obviously it's not an electron because it has a spatial extent, but try to imagine that this ball is really, really small. This ball has a mass, and if I rub it on my shirt, then it gains some electric charge; you know, as when you rub a balloon against your shirt, it gains static electricity and will stick against the wall. This could be a metaphor for an electron, and I can spin it. Now it has another property, it has a certain amount of spin, and you can quantify that. Physicists can measure what we call angular momentum, but it's really just a measure of how fast the thing is spinning.

Now this is a classical object. I can spin it fast, and I can spin it slowly; it's my choice. With an electron, it turns out you have no choice. If you take an electron and you measure its spin, all electrons spin at the same rate. That's a kind of a weird fact of nature; it's a quantum fact of nature. Some things come in chunks; electric charge comes in certain definite chunks and so does spin. Not at all a classical idea, it's very, very hard to picture why that should be. It's just a postulate of quantum mechanics, and we accept it as being consistent with the data.

Spin is measured, in physics, by a certain set of units, and there's a constant of nature called Planck's Constant, named after Max Planck. It's given a symbol, the letter H with a little slash through it; we call it H-bar. Physicists, quantum physicists are always talking about H-bar, and H-bar is a number which tells you how much spin something has. So you can measure spin in units of H-bar. It can have one H-bar or two H-bar; and we get tired of saying H-bar, so physicists just say you have one unit of spin or two units of spin. Niels Bohr, when he first wrote down a theory of the hydrogen atom, said, "I believe that the electron has either spin zero or one or two." So he was beginning to think of "quantizing" angular momentum.

What Mr. Dirac says is, "No, according to my formula, the electron is very special. It's unusual. It's got half a unit of H-bar." Spin one-half is what we say. So spin still comes in chunks, but the smallest chunk turns out not to be one, but one-half, and this isn't put into his equation; it comes out. That's one of the most amazing things about Dirac's equation.

In the early 1920s, there was some experimental evidence that electrons have an intrinsic spin; they seem to be rotating, even though they're points. It's hard to imagine a point rotating, but that's what we're trying to visualize. There was evidence that it was spin one-half, and now the mathematics of Dirac's equation says, "Yes indeed, spin one-half is required for mathematical consistency." That's a very compelling argument. When you write down a theory and then it makes a clear statement about something that's known experimentally—that you didn't put in by hand—it comes out. Then you think to yourself, "Gosh, this theory must have some intrinsic, deep, scientific truth to it."

Dirac's equation had more truths to it than that, and another one, which I think is one of the most amazing things about the Dirac equation, is the following. Dirac observed this himself. If you solve this equation, for example, just for an electron traveling along in a cathode-ray tube, just a beam of electrons, you solve the mathematics for the behavior of this electron. You discover that, when you look at the equations, there's another solution. Sometimes this happens in physics; you get a second solution. Dirac studied that second solution, and what did that solution say? It should be a particle; it should have the same mass as the electron. It should have the same charge—sorry, same magnitude of charge—but opposite sign. It's like an opposite particle that's solving the equation. It's got positive electric charge instead of negative, so what would you call this thing? You might call it a positron or nowadays we just call it an anti-electron. It's antimatter. So this equation is predicting the existence of antimatter.

Now this is the mid to late 1920s, and Dirac was faced with a solution to the equation that's predicting some essentially preposterous thing; it's science fiction, right, antimatter. So Dirac, he was kind of puzzling over this, and he wondered, "Maybe my equation is trying to describe a proton, which is another particle that sits in the nucleus, and it's positively charged. But no, no, the proton is not the antiparticle of the electron because they've got different masses. The proton is very massive; electron is very light." So Dirac is kind of scratching his head. He says, "I don't know, I guess, you know, it's just an accidental solution." He didn't really run with it like he probably should have, in retrospect, because just a few years later, in 1932, antimatter was discovered in the laboratory.

Antimatter was discovered in, basically, photographic emulsions. It's like a piece of photographic film where high-energy cosmic rays were traveling through the film leaving a little trail of dots behind. We'll talk more about the technology of detecting particles in upcoming lectures, and you could measure the properties of this little object that was just spontaneously generated by radiation in the atmosphere. It was the same mass as an electron, but positively charged, and when an electron and an anti-electron met in the laboratory, they annihilated one another. Antimatter is real.

It's, you know, too bad for Dirac; he was a smart guy. It's too bad he didn't make this prediction, it would have been very dramatic, and in later years, he made the comment that his equation was smarter than he is. I love that idea. He came up with this equation, and it had truth about nature in it, and even the guy who invented the equation, took a while before he could really appreciate it, that truth about nature. Antimatter is used today; it's a piece of technology that we use like many others. It's produced regularly in laboratories, for instance, PET scans. It's used as medical imaging technology; that's positrons, antimatter going into your brain and forming an image. So this stuff has gone from esoteric science, that a brilliant physicist was timid to propose, to something that's now becoming everyday scientific material.

I want to finish today's talk, which is the transition from classical physics to quantum physics and the development of modern science, with another discovery that was made in 1932. That was the year that antimatter was discovered. In Mr. Rutherford's lab in England, where people were intensely studying radiation and radioactive materials, a fellow named Chadwick measured a new particle called the neutron. What was the neutron? It fit in pretty well with people's understanding of atoms at that time. At that time, people were, visualizing atoms—the Bohr model, Mr. Rutherford's model— as a nucleus (very massive, with positive charge) and electrons in some sort of orbit. Now Mr. Bohr had been picturing little particles running around in planetary orbits; then Schrödinger and Heisenberg and Dirac were telling us, "Well you have to modify that mental image a little bit; it's more an electron wave that's present around the proton." But still, it's okay to think about it as a little particle because you can think of things both as particle and as wave at the same time in quantum physics.

But what's that nucleus, what's down there at the bottom? People had known about protons since the early 1900s: a positively charged, very massive object. The proton is the fundamental building block of the nucleus. You've got positive protons that are heavy down there and light electrons around it. What Chadwick had discovered, the neutron, was like a partner for the proton. Now it's not antimatter. The proton is positive; the neutron is not negative, it's neutral. So you have something with positive charge, something with zero electric charge, so an electron doesn't really care about the presence of neutrons.

Why are they down there? Well let's think about the second simplest nucleus in the world. The simplest nucleus in the world is a proton, the nucleus of a hydrogen atom. A hydrogen atom has one proton, the mass, and one electron, which balances the charge. The next nucleus that we know about in the periodic table, the next element, has two protons and two electrons; and those two protons are positively charged. Just as electric charges repel one another, opposite charges attract. Electrons are attracted to the protons; that's why they go around in a circle instead of running away, but the two positive protons want to repel one another. There must be some kind of new force of nature. It's not electricity. It's definitely not gravity; gravity is much too weak. There's some new force of nature which makes the two protons stick together. We call that the nuclear force or the strong force. Kind of a silly name, but it is a strong force, and that's what we call it, the strong force. It turned out that people were having a hard time matching data that they were collecting on nuclei with the idea that we have only protons down there, down in the nucleus. You need something else. Just something that's electrically neutral, that's not electrically repelling, that would help things stick together. That was the role of the neutron, and so Chadwick discovered this particle.

Now physicists are beginning to form a coherent story, so we've come from the 1800s, where we have a coherent story that involves atoms and classical physics. Then there's this period of tumult, as we learn about quantum mechanics, learn about relativity, learn that atoms have pieces; they've got electrons in them and protons at the middle and now neutrons in the middle. Once again, there was a period in the 1930s when people thought, "Okay, now we've got it. We understand what the world is made of." Electrons and protons and neutrons are the sort of holy trinity of particle physics at that time. The electrons orbit around the outside; they're very light. Protons and neutrons stick together with this strong force. The strong force is sort of like Velcro. Okay? Neutrons and protons don't even notice each other's presence until they get really close, and then whoomp, you get them close enough and they stick like two pieces of Velcro.

So this idea, this introduction of a new force of nature, the strong nuclear force, combined with electricity and magnetism and quantum electrodynamics—the primitive quantum electrodynamics that we had in the 1920s—was beginning to look like it was going to explain fundamental physics. It was going to explain how you build up atoms. So more complicated atoms were nothing more than just adding a couple more protons and neutrons into the nucleus and then balancing them by adding a couple of more electrons. As long as you keep the thing electrically neutral, you have a nice stable atom. Now you can begin to understand chemistry as simply the interaction of one atom with another, which boils down basically to the electrical interaction of the electrons with one another, obeying the laws of quantum mechanics. Perhaps, in certain cases, the Dirac equation, which has quantum mechanics and relativity put together.

This is a worldview which is reasonably satisfactory, but it's not complete either. It took people a while to figure this out, then, World War II occurred, which formed a hiatus. Iin the period between the 1930s and the 1950s, it was discovered that, just as the classical picture of atoms was not quite complete, there was still something underneath it that needed to be understood. It's the same story with protons and neutrons. The nuclear story is where we're going be headed next, because the electron part of the story, and the quantum electrodynamics, were really very satisfactory. The agreement with data was spectacular. The nuclear physics story was a little bit more complicated, and what people began to discover was there's more to it than just point protons and point neutrons. A point electron turns out, still today, to be an okay approximation or metaphor, but a point proton, and a point neutron, is not right. That's where we're going to be going next, into the nucleus.

Lecture Six
New Particles and New Technologies

Scope: This lecture looks at particle physics in the first half of the 20th century, including cosmic rays, and the discovery of the muon. We also see the dawn of nuclear physics with Yukawa's theory of nuclear force and ensuing "models" of nuclei. We discuss the discovery of the pion, along with bubble chambers and cloud chambers, the first modern tools to detect subatomic particles.

> Who ordered that?
> —I. I. Rabi (on the unexpected discovery
> of the muon)

Outline

I. Up to now, we have been talking about the transition from classical physics to modern physics. We have moved from a "mechanistic," clockwork worldview that enables quantitative predictions about all aspects of the world to a quantum worldview that still enables predictions, but not about everything, and encompasses some indeterminacy.

    A. Part of what drove this transition was the discovery of radioactivity and subatomic particles, including the electrons and the nucleus.

    B. The worldview in the 1930s used quantum mechanics as a framework and understood protons, electrons, and neutrons as building blocks.

    C. This worldview enabled scientists to study larger and larger subjects, such as chemistry, using these fundamental ideas.

    D. Of course, this view also enables us to look at deeper levels and try to find the ultimate building blocks of the world.

II. In the 1930s, the nucleus was still somewhat mysterious. It was seen as the heart of the atom, containing the bulk of matter.

    A. Scientists were beginning to discover how protons and neutrons in the nucleus hold together.

1. They knew that protons tend to fly apart because they have like charges, but some strong nuclear force binds protons together, and both protons and neutrons feel this force.

2. In the 1920s, experimentation with protons began. For example, a beam of protons could be directed at some material, then scattered to see what would happen.

3. The development of nuclear physics, followed by particle physics, began, then, as a mix of experimental work and new theories. The interplay between experiment and theory is always significant in physics, and this was especially true in the 1930s.

B. In the mid-1930s, a Japanese theorist named Yukawa hypothesized a mechanism that would mathematically describe the strong nuclear force, which was one of the central mysteries of physics at that time.

1. To understand Yukawa's model, we must examine *forces* in general, that is, the interactions between objects.

2. Think of a simple push-and-pull situation, such as two electrons coming near each other, feeling a repulsion because they are both negatively charged, and flying apart.

3. Newton called this type of situation *action at a distance*. The electrons feel the force even though they don't touch. The same principle keeps the earth in orbit around the sun; earth is attracted to the sun by the force of gravity.

4. A modern idea along similar lines is the idea of force as mediated by a particle.

5. In the same situation of two electrons approaching each other, imagine that one electron emits a photon, the fundamental quantum of electromagnetic radiation. The act of emitting this photon forces the original electron backward; the photon's interaction with the other electron also repels it, similar to two people tossing a medicine ball back and forth. The force between the two electrons is mediated by the photon.

6. This description is not to be taken literally; it is an analog for what the mathematics describes in the theory of quantum electrodynamics.

C. Yukawa wondered if this same idea of forces could be applied to the strong nuclear force.

1. The strong nuclear force is different from electricity in a variety of ways. It is, for example, at least 100 times stronger than the electrical force.

2. The strong nuclear force is also a contact force. If the protons and neutrons are far apart, they don't feel any force, but if they are close, they are bound together quite tightly.

D. Yukawa posited that the strong nuclear force might be mediated by a particle.

1. This particle would be similar to a photon, but unlike a photon, it would be massive. A photon, the transmitter of electricity, is itself massless.

2. Where would the energy come from to create this new particle? Classical physics would not allow energy to be created from nothing, but quantum mechanics, using Heisenberg's uncertainty principle, allows a loophole in the laws of classical physics.

   a. According to the uncertainty principle, energy can be created out of nothing, but only for a very short time.

   b. The *virtual particle* of the strong force, then, is created for a very short time, travels from one proton to the next, then disappears.

   c. The uncertainty principle is quantitative; it states that if a certain amount of energy is "borrowed" from nature, only a certain amount of time for the existence of the particle is allowed.

   d. The particles can travel no faster than the speed of light and only for a certain maximum distance, which explains why protons and neutrons must be close to each other to feel this force.

3. Yukawa also deduced what the mass of his new particle would be. His conclusion was that the particle must be less massive than a proton but more massive than an electron. Because the mass of this particle is in the middle, Yukawa called it a *meson*.

E. For the first time, a theorist, Yukawa, had predicted the existence of a radical new particle of nature. Later, many other mesons were discovered, and Yukawa's particle became known as the *pi meson*, or *pion*.

F. With this idea of mesons, a new era was introduced. Scientists could begin doing calculations in the realm of nuclear physics.

III. Experimentation continued, especially in Europe, up until the hiatus of World War II.

   A. Experimentalists were eager to create a pion and document its existence with real, direct evidence.

   B. To achieve this goal, a source of energy was needed, and scientists looked to *cosmic radiation*, which is natural radiation from the stars, the sun, and other sources.

   C. Alpha and beta rays are, in effect, electrically charged particles. When such a particle passes through matter, its electric charge attracts and repels all the electric charges in the atoms and tears the atoms apart. When a particle passes through matter in this way, it leaves an *ion trail*, a trail of atoms that have been temporarily torn apart.

   D. Particle detectors were developed to provide evidence of the passage of these submicroscopic particles. For example, a charged particle could be sent through a photographic plate. The film could then be developed to reveal an ion trail.

   E. Another form of particle detector was the cloud chamber.

   1. Imagine a container filled with gas. If you were to pull a piston on the chamber very rapidly, the gas would cool and would tend to condense.

   2. The gas can condense along the walls of the chamber or along an ion trail.

   3. In other words, an electrically charged particle is sent through the chamber, the piston is pulled, and a chain of droplets forms along the ion trail, verifying the existence of the particle.

   F. The bubble chamber is similar to the cloud chamber, but it uses a superheated liquid that is on the verge of boiling. Again, bubbles from boiling liquid can form along the edges of a container or along an ion trail.

   G. The next step is to analyze the somewhat puzzling picture provided by the particle detectors to identify the details of the particles that passed through.

IV. Recall that all these experiments were aimed at finding evidence for Yukawa's pion.

    A. Scientists were looking for a trail with certain well-defined characteristics, including a certain mass, electric charge, and so on.

    B. In 1937, such a trail was seen, but it was not clear whether the observed particle was Yukawa's predicted pion or not.

    C. Pions were "invented" to explain the strong nuclear force; therefore, they should be strongly absorbed by nuclei. Yet, the results from various particle detectors revealed that the particles travel from one end of the detector to the other, hardly interacting at all.

    D. These mysterious new particles, which were not behaving as pions should, were called *muons*.

        1. The muon was similar to a heavy electron.

        2. Muons with both positive and negative charges existed.

        3. The mass of the muons was in between the mass of protons and that of electrons.

        4. Finally, muons were radioactive. Every now and then, a muon would be seen decaying into an electron at the end of its track. It was radioactively unstable; it could transform itself into an electron.

        5. Even today, the muon remains something of a mystery; its purpose is unclear.

    E. It was another 10 years before Yukawa's pion was discovered, but the discovery of the muon was, in a sense, the birth of particle physics.

Essential Reading:

Barnett, Muhry, and Quinn, *The Charm of Strange Quarks*, chapters 2.5 and 3.1 (to p. 47).

Schwarz, *A Tour of the Subatomic Zoo*, chapter 3 (up to pp. 30) and chapter 8.

Recommended Reading:

Calle, *Superstrings and Other Things*, chapter 8 and the first half of chapter 24 (up through "Pions").

Riordan, *The Hunting of the Quark*, chapter 2.

Questions to Consider:

1. If the pion were suddenly significantly more massive than it currently is, what would happen to the range of the strong nuclear force (that is, would protons "feel" each other's presence when they are farther apart—at long ranges—or would they need to be even closer together still—at short ranges)?

2. What happens to muons as they travel downward through the atmosphere? (What happens to their overall number and their average energy?) What happens to an individual muon when it decays? Where does all the energy from cosmic rays ultimately go?

3. How much, and what types of, background radiation are you exposed to daily? Where does it come from? To what extent do you think it is a significant health hazard?

4. Colorado has a much higher flux of cosmic radiation than sea-level states (because it has much less atmosphere for protection), but overall cancer rates are not noticeably higher in Colorado than in sea-level states. What conclusions might you draw?

Lecture Six—Transcript
New Particles and New Technologies

Up to now we've been talking about the transition from classical physics to modern physics. We've been going from a mechanistic, kind of clockwork image of how the world works—where you have certain fundamental particles, certain laws of physics and can make direct, concrete predictions about everything and anything that happens, a classical worldview, to a quantum-mechanical worldview, where you can still make definite, numerical, quantitative predictions but not about everything, and not about all aspects of one particular system). There is now some quantum indeterminacy in the universe, which we just have to accept. This seems to be the way the world works. Einstein's theory of relativity tells us that there's also a kind of a new way of thinking about space and time; that these things depend on the observer.

So physics has been going through this transition period. In the meantime, part of what drove this was the discovery of radioactivity, sub-atomic particles, the electrons, which are tiny particles that make up atoms; and the nucleus, which itself is now understood, in the 1930s, lets say, by itself being composed of protons and neutrons. So you have a worldview that is beginning to form that involves quantum mechanics as a framework for understanding how things work. The particles involved, the protons, and the neutrons, and the electrons, are the building blocks. That's a nice, kind of coherent picture, and people began trying to understand various phenomena.

One of the directions that you can go is up to larger and larger-scale things. For instance, you could try to understand chemistry based on these fundamental ideas, and that was an extremely successful program. People developed mathematics which described the known chemical laws, derived them, expanded upon them and allowed us to make predictions of new types of chemical reactions and physical reactions. That was going up in the scale of complexity of the world, and what I want to do, what we want to talk about in this course, is going in the other direction—looking deeper, trying to find the true, ultimate, miniscule building blocks, the real fundamentals of this story. The place where this gets especially interesting is when you look down into the nucleus.

The nucleus, in the 1930s, was still a little bit of a mysterious place. It's the heart of the atom. It's where almost all the mass is. It's where the bulk of matter exists, in a certain sense. The electrons are very, very important. They are responsible for chemistry, and chemical reactions, and therefore biology, but down there in the nucleus, in the heart, how exactly are those protons and neutrons holding together? We know that the protons want to fly apart, because they are electrically charged particles, and like charges want to move apart from one another, so there's something, there's some strong nuclear force that binds them together. Protons and neutrons both seem to feel this nuclear force, so people began doing experiments, beginning in the 1920s and the 1930s, and working their way forward, performing all sorts of experiments, as you can imagine. For example, you might just take a beam of protons. Mr. Rutherford had come up with a beam of alpha particles; you could use those as well. Mr. Chadwick, in 1932, had discovered neutrons, and fairly quickly afterwards people could produce beams of neutrons, and you could scatter—bounce these things off of some chunk of material—and watch what happened. The development of nuclear physics, and then of what we call particle physics, began, in this era, in the 1930s. It was a mix of experimental work—of novel devices to look and see what was going on, and also some novel theories to try to understand and explain what was going on.

This interplay between experiment and theory is always a big part of physics. The theorists sometimes are trying to explain the data, which is already out there, and other times they're kind of off imagining how the world works, and coming up with new ideas, which the experimentalists want to test. There's a give and take, at all times in history, and in the 1930s it was especially pronounced.

For example, there was a Japanese theorist whose name was Hideki Yukawa. In the mid 1930s, Mr. Yukawa published a paper in which he hypothesized a mechanism which would not exactly explain, but would describe, in some mathematical detail, the strong nuclear force, which was one of the central mysteries of the nuclear story at that time. In order to understand Mr. Yukawa's model, let me take a step backwards and talk about forces in general. Physicists talk about forces. As with the way of describing pushes and pulls, it's the interaction between objects. Let's think of a simple push-and-pull situation, as with two charged particles, maybe two electrons, which come near one another and feel this electrical repulsion, because they have the same charge, and then they fly apart. Now how are we supposed to visualize that?

Isaac Newton had been thinking about forces; he was thinking more about gravity than he was thinking about electricity, but he could have extended his ideas easily to electrical phenomena. He called what was going on there "action at a distance." Here's an electron over here, there's an electron over there; they come closer together, and the closer they get, the stronger the force between them, and then they fly apart. They feel the force, even though they don't touch, so it's some kind of action at a distance. This is a very classical idea; Isaac Newton invented this idea. How is it that the earth knows the sun is there? We feel gravity. Why doesn't the earth just go flying off into outer space? It's attracted to the sun by action at a distance, by the force of gravity. That's a classical idea, and that idea has been modified and improved over the years. In future lectures we're going to talk about a modification of that idea that refers to what we call force fields.

Let me talk about a very modern idea now, which is the idea of force as mediated by a particle. What does that mean? When I think of this electron over here, and another electron, separated by some distance, and they want to communicate with each other, they want to let each other know that they're there. How do they do it? The modern way of thinking about it, at least one modern way of thinking about it, is: imagine that the electron, one of them, spits out a photon. What's a photon? A photon is that fundamental quantum of electromagnetic radiation. It's a bundle of electrical energy, if you like. So the electron actually produces a little particle, a particle of light, a photon, which travels along, and it interacts with the other electron.

So think about, you know, you and me trying to communicate with one another, and I take a big medicine ball, and I toss it at you. I pull it out of my pocket; it expands to a big heavy medicine ball, and I toss it at you. So when I toss the medicine ball, I go recoiling backward. I feel a force. It's local, it's a contact; I'm pushing on this ball. Then it gets to you, and you catch it, and you feel this heavy thing whack you, and you also feel a force. So now we're understanding the force between you and me as mediated by this particle: a medicine ball in our case, or a photon in the case of two electrons. It's a neat way of thinking about forces. It's a kind of invention. What I'm really describing shouldn't be taken too literally. I'm really trying to create for you a metaphor, an analog, of what the mathematics is describing in the theory of quantum electrodynamics, the real detailed mathematical description of how it is that particles interact with one another.

Mr. Yukawa had been thinking about this idea of creating a particle, tossing it; that's how you feel forces. He wondered, "Can we make this same idea work for the strong nuclear force?" Now the strong nuclear force is very different from electricity. It's different in a variety of ways. First of all, it's different because the strong nuclear force is very strong. It's at least a hundred times stronger than the electrical force, when protons and neutrons are close together. It's also what we call a contact force; it's like Velcro. The protons and neutrons don't feel anything if they're far apart; they have to be right next to each other, and then they really stick big time. So you have to understand, how could this force be short-range? Electricity is long-range. You feel electricity, even when you're far away. Gravity is long-range. We feel the sun, even though we're really far away.

Mr. Yukawa was thinking about this, and he said, "Okay, I've got an idea. Suppose the strong nuclear force is also mediated by a particle, kind of like a photon, except there's going to be one big difference. Number one, this new particle, this medicine ball that I'm tossing back and forth between protons and neutrons, is going to be very strong; it's going to be massive. What do I mean? I'm really talking about creating a particle, and sending that particle between a proton and a neutron. A photon, you can think of as a particle, but it's a very unusual kind of particle. The transmitter of electricity is itself without mass. Every particle has a mass: the electron has mass, the proton has a mass. The photon, when thought of as a particle, has no mass. It's always traveling at the speed of light, it can't sit still; it has no rest mass. That's a feature of the electrical force." Then Mr. Yukawa said, "I believe my new force carrier is going to be massive; it's really going to be an honest-to-goodness particle, just like other particles."

Now how is this going to work? If you've got a proton here, and a, say another proton nearby, and they want to attract one another, this proton has to create a particle, out of nothing, toss it; and the other particle has to catch it. Now you stop and say, "Wait a minute. How can I create a particle out of nothing?" Albert Einstein says $E = mc^2$; energy and mass are intimately connected. If you create a particle with mass, that takes energy. Where's the energy going to come from? I'm just talking about two protons sitting still; they just attract one another. It's as though you are standing here on planet Earth, being attracted by the planet; there's no energy being used up. Planet Earth will happily hold me on its surface forever. Nobody is using any energy: I'm not, the planet's not. Same thing with two protons; they're just sitting there. So where is the energy coming from to create this little massive particle that's going to travel from one to the other?

This is a dilemma for classical physics. Isaac Newton could never have come up with this idea, because to Isaac Newton, energy conservation, the fact that you can't just pull energy out of nowhere, is inviolate. But quantum mechanics, this strange theory of microscopic particles, allows us for the following loophole. Heisenberg's Uncertainty Principle tells us many things; one of the things it says is, yes, you can create energy out of nothing, as long as it's for a very short time. This probably doesn't make any sense, and it really can't make any classical sense. It's a completely non-classical idea. You can borrow energy from nowhere, from the vacuum, from empty space, as long as it's for a short time. It's kind of like, I don't know, a bank, that's willing to lend you a dollar, but just for a second, and then you have to give it back. They don't really care; it doesn't do you any good. You get a little bit of satisfaction that your bank account went up; you might even be able to use this to your advantage in some way. This bank might be willing to lend you one dollar for a second; they might even be willing to lend you a million dollars, but maybe only for a millionth of a second. So, you know, again, you might get some satisfaction out of this, but there's not a whole lot of practical benefit, but you do get something out of it; you get the sense that, "For a microsecond, I was a millionaire." The protons get something out of borrowing energy, creating an intermediate particle, for just a brief moment. It's only got a flash of an existence, it's called a virtual particle, because it doesn't appear and then run away. It appears, travels, and then disappears again, very, very quickly.

Heisenberg's Uncertainty Principle is quantitative: it says, if you want to borrow this much energy, you've got this much time. It tells you there's a formula; it tells you exactly how much time you've got. If a particle is traveling along for a certain amount of time, the fastest it can possibly go is the speed of light, because nothing can go faster than the speed of light, so it's got a certain maximum distance that it can travel. That's the explanation, in Mr. Yukawa's mind, of why it is that protons and neutrons have to be pretty close to one another, in order to feel this force. Because if you're going to borrow energy from the universe, it can only last for a certain distance before it's got to be reabsorbed, so that's the distance of the interaction between protons and neutrons. If they're any further apart than that, they feel nothing. Nature cannot produce these virtual particles and have them extend over large distances.

Mr. Yukawa looked at the data, and he knew how far apart protons and neutrons have to be before they start feeling a strong nuclear force, and so he did a little calculation. How massive would his new particle have to be? He discovered it was kind of an intermediate sort of number. It was less massive than a proton itself, by quite a bit, but more massive than an electron, by quite a bit. It was right in the middle. So he looked up an old Latin word for "in the middle", which was *mesos*, and he called this new particle a meson. This is, I think, the first time when somebody has really predicted a new, a radically new, particle of nature, the meson. This was a theorist talking, and he predicted that such a particle was real, that it exists. In fact (although it's virtual inside the nucleus), once you predict that this particle is part of the world (if I give you that much energy, if I'm not trying to borrow it from the vacuum, if I just pour energy into an experiment), you should actually be able to make these mesons.

In later years, Mr. Yukawa's meson needed an adjective, because it turned out people discovered many different kinds of mesons, many different particles of this nature, so Mr. Yukawa's meson became a pi meson. Remember, we're using Greek letters to name particles; pi is a Greek letter, not apple pie, but the Greek letter. So a pi meson later got shortened to pion. In the present era we refer to the pion as this intermediate particle. Meson is a nice word though, because it reminds you of "in the middle." Not only is it in the middle mass-wise, but it's also in the middle; it's between the proton and the neutron, or the proton and the proton. It's what's holding them together; it's the force carrier.

Mr. Yukawa's theory really introduced us to a new era, because now that we had this idea, people could start doing calculations about what would happen if you smashed two nuclei together, what would be the lifetime of this unstable nucleus. People could actually start doing nuclear-physics calculations, and the presence of these pions really could help to understand, and consolidate, and explain a lot of data. It's just another one of those situations where a theorist's idea is invoked for one particular situation. Then it's very generalized, and one can begin to explain a large amount of data, from a whole variety of situations, all in terms of this one idea, the virtual particles. Mr. Yukawa's proposal was in the mid 1930s, and that was kind of towards the end of the time when the Japanese and American scientists and the European scientists were able to collaborate easily on scientific issues; that is, until after World War II. So there was a long hiatus in the development of these ideas, until the late 1940s, early 1950s. In the meantime though, especially in Europe, there was a lot of continuing experiment.

Here's this idea of Mr. Yukawa's, that there is a new particle of nature, and, of course, everybody wants to go and look for one. They want to try and find a situation where a pion is not just virtual, going in between the proton and neutron, because you're never going to see that. What you want to do is actually pour energy in, make a pion, and have it leave a little track behind that you can photograph it and have some real, direct evidence. That's where the experimentalists sort of took over, and experimental physics sort of dominated the story for quite some time.

Where are you going to find this kind of situation, where you've got lots of energy? You might just take some uranium salts, and hope that there's enough energy in the radiation that comes out that you could do something interesting. But it turns out those are fairly low-energy radiations, low-energy particles, that are coming out of the uranium salts. There's another source of radioactivity that was available to physicists in the 1930s. This was before the era of big particle accelerators, where we actually take particles and speed them up ourselves with electric fields, and smash them into targets. That's where we're going to be heading in future lectures. But in the 19 0s and 1930s, that technology didn't exist. They had to look just to nature for radiation.

As it turns out, there's natural radiation raining down on us from the sky all the time. It's called cosmic radiation, because it's coming from stars, and the sun and the galactic center. In fact, today it's still active research to determine what exactly is the source of a great deal of cosmic radiation. Most of it is in the form of high-energy protons, which strike the Earth's upper atmosphere, interact with nuclei up there and create showers of particles. It's present wherever you live. It's part of daily life. It has nothing to do with human beings and our artificially created radioactivity.

We're exposed to radiation all the time. All right, your body has evolved to deal with exposure to radiation, your DNA can repair itself; and part of the reason for that is because it must because there's cosmic radiation which travels through you all the time; and when radiation goes through matter, think about what it does. There are various forms of radiation: alpha, beta, and gamma. Two of those, alpha and beta rays, are themselves electrically charged particles. When an electrically charged particle travels through matter, it's a little, teeny, electric charge which is attracting and repelling all the electric charges in the atoms. It's ripping apart those atoms, so as a particle goes through matter, it leaves a little trail behind it. It's called an ion trail. An ion is just a fancy word for an atom that's been temporarily ripped apart, so it's left behind with some electric charge.

If you take a piece of photographic film and expose it to sunlight, (the sunlight itself is not charged, nonetheless it can interact with electrons) the sunlight can expose that film, ripping apart, making little ions. Then you develop the film, and it looks black where it's been ionized. So people developed particle detectors, and a particle detector is any device that you can think of that is capable of showing you evidence of the passage of an invisible. Remember, these are sub-microscopic particles. The way you could do that, for example, would be to take a photographic plate, and send a charged particle goes running though it, as when light goes through it, and it will leave behind a little trail of ions. You develop that film, and you'll see a little black line. In fact, that little black line, the intensity of that line will tell you information about the energy and the character of the particle. A more highly charged particle will leave a thicker line. You can learn about the particles by looking at these developed plates. These plates were called emulsions. They were like photographic film, a little bit thicker. The technology of emulsions developed in the 1920s and 1930s and was used for many years as particle detectors.

There were other technologies that were developed, for instance, the bubble chamber. The bubble chamber was a little bit more of a mechanical device. Imagine taking some gas in a chamber; there are bubble chambers and there are cloud chambers. Let me talk about cloud chambers first. In a cloud chamber, you take some vapor, and you pull a piston out very quickly. This gas is present in the chamber. It has cooled because you expanded the piston, and cool gas really wants to condense. Now, it turns out that when you really want to condense, it's nice to have some place to condense. You might condense along the walls; or, if there's an ion trail in there, that's where you're going to condense. So here's the idea. An electrically charged particle goes flying through, perhaps a cosmic ray. You pull the piston; that little trail of ions is going to form a little chain of water droplets—water, or whatever the material is in your cloud chamber. Those water droplets, then, are visible. They are not microscopic, even though the particle that made them was. So this is the way you now take a picture, and you see a little line, just like the photographic emulsion. There are various advantages and disadvantages to both of these. The cloud chamber is three-dimensional and can be quite large, unlike the emulsions, which are fairly flat and tend to be small.

The bubble chamber is like the cloud chamber, except instead of water condensing where the ions are (imagine that you had something liquid already that was super heated), it really wants to boil, so bubbles are going to form. Where are they going to form first? They always form on impurities, or edges; so, if there are ions in there, they're going to form right around the ions and once again you get a little line. It's like when you look up in the sky and you see this trail of a jet airplane. The jet airplane may be too small for you to see, but behind it is a trail of exhaust, The clouds, well, there are no clouds up there yet; but there is water vapor. It's cold up there, and the water vapor really wants to condense and form a cloud. It will tend to do so when there is a place for it to do it, which is on the exhaust trail. So you look up in the sky and you see this beautiful line, which tells you where the airplane has been, and if you're really skilled, you might even figure out what kind of an airplane it was by looking at the characteristics of that line. It's the same thing with the cloud chamber, which was invented in 1911, and the bubble chamber, which really wasn't developed into technological usefulness until the 1950s.

You're getting photographic images, and what you get are images which look something like this. Here is a picture of a bubble chamber, and you can see trails of particles. It looks like a big nest because there was a whole bunch of particles interacting and knocking into nuclei. Now you've got to try to take that picture, piece it apart and figure out, okay, what was in there. By looking at the density of bubbles, I can figure out how electrically charged it was, and how energetic it was. If you immerse this whole thing in a big magnet, if you just put a magnet next door, then the magnetic field will cause electrically charged lines to bend. The amount of bending teaches you something about the particle's properties. The more energy and momentum it has, the less it bends, right, because it's basically just plowing on through in a straight line, and the lower its energy, the more easily the magnetic field makes it go in a circle. You'll see some low-energy particles that form little corkscrews in these pictures, so you can begin to piece together, in a bubble chamber or a cloud chamber, or an emulsion, the details of the process that you saw.

This was the experimental world of the 1920s and 1930s. They might take a big balloon, and launch some emulsions up, way up high in the air so that the cosmic rays hit them; or they might take some radioactive material and put it next to a bubble chamber, or a cloud chamber, in the earlier days. One of the things, one of the most important things that people were trying to look for was Mr. Yukawa's pion. I mean, they were looking for a trail with certain very well defined characteristics. They thought they knew its mass, they thought they knew its electric charge. Mr. Yukawa predicted, in his original paper, that his virtual particle could be positive or negative or neutral. There were three different kinds of pion; we call them the pi plus, the pi minus, and the pi zero, for obvious reasons, for shorthand. This was a prediction of Yukawa's theory, and we should be able to see a trail in a bubble-chamber picture; I should say "cloud chamber" because bubble chambers didn't come until the 1950s. That's what you're looking for.

In the 1930s, around 1937, I believe, such a trail was seen. It was a charged particle. It had the mass predicted by Mr. Yukawa. It was intermediate in size, a little lighter than a proton and a little heavier than an electron, but this particle didn't quite make sense. People looked at these trails, and in some respects it seemed like this predicted particle of nature, but in other respects it didn't. The most important thing, the glaring issue was, pions were invented to explain the strong nuclear force. Pions should be strongly absorbed by nuclei. That's what they're for; that's why they were proposed. Yet, in these bubble-chamber—cloud-chamber pictures, the particles that you saw began at one end of the chamber, and went all the way through. Actually, the original pictures of the muons came from emulsions; in fact, they started at one end and went all the way through to the other end, hardly interacting at all. What did they look like? They looked like electrons. Electrons also just cruise on through. They're electrically charged, so they leave a little trail of bubbles behind, but that's about it. They don't hit a nucleus and cause some explosion, or reaction of any kind.

So what they were seeing, this new particle, didn't appear to be behaving the way Mr. Yukawa's pion behaved. So people scratched their heads, and they started collecting more and more data, and they said, "Look, we can't call this thing the pion, because it's not Mr. Yukawa's particle. Let's give it a new name." It's a "something –on", so we come up with another Greek letter. We're just going through the alphabet now; it's a muon." I have no idea why they picked the Greek letter *mu* for this thing; it's just a muon, that's its name. The muon is a big mystery particle. It's discovered in the emulsions, but why is it there, what role does it play? It's like a heavy electron. They came, it turned out, in both positive and negative charges. There was no evidence of an electrically neutral one, only positive and negative muons. The mass was this intermediate mass, in between the proton's mass and the electron's mass, and they were radioactive, in the following sense. Every now and again, you would see a muon decaying, at the end of its track, into an electron. So this muon was radioactively unstable, it could transform itself into an electron; and this was all just a big puzzle.

It was the first time that a brand new particle, a sub-atomic, fundamental particle had been discovered, and nobody had a clue about why it was there, what role it was playing, what it meant. In a certain sense, this is how particle physics began to appear, more and more and more, over the next couple of decades. People were finding particles, and they were mystery particles. We're going to come back in this course and revisit the mystery of the muon, to try to understand what this thing is and what role it plays. I have to confess that even today, although we know all about muons, and we use them in technology, muons are constantly present. They're present in cosmic rays, they are present in accelerators, but we don't really have a completely deep understanding of why the muon is there. A famous physicist, I.I. Rabi, said, "Who ordered that?" It was a quote, you know, like you're in the restaurant, and a bunch of people are there, and this dish comes, the mystery dish, "Who ordered that?" That's the muon.

It was another ten years before Mr. Yukawa's pion—which made sense, it was predicted, it was expected— was indeed discovered. It turned out that the reason we hadn't seen it is they're produced very easily, very strongly, in the upper atmosphere, and then they hit nuclei very quickly because they're strongly interacting particles. They tend not to make it down to our laboratories, so we needed higher energies, and particle accelerators, before we were able to actually see those pions, the particles that Mr. Yukawa had predicted.

So at this point, we have a developing scenario in the world of nuclear physics. We've got quantum mechanics, which is a mathematical framework that allows us to describe these phenomena quantitatively. We have protons, and neutrons, and electrons; and now we've got pions, that are expected, but not yet seen until the late 1940s, early 1950s; and the muons, which are seen, but not expected. So the world of particle physics is beginning to develop. There are new species of particles that are mysterious and unexplained, and people are getting really curious. What are these things? What are their properties? I would say the discovery of the muon is, in a real sense, the birth of particle physics. It's a transition from just studying nuclei, and their forces, to saying, "Hey, there's other stuff going on. There are actually new particles in the world, and we've got to understand them."

Lecture Seven
Weak Interactions and the Neutrino

Scope: What is an *interaction*, and how does that differ from the concept of a force? What is a weak interaction, and how is it connected to radioactivity? How weak is weak? In this lecture, we address these questions and look at the importance of weak interactions in the sun. We also examine the carriers of weak forces, W and Z particles, and the modern idea of transmutation. We close with the birth and early story of the neutrino. What are these ghostlike particles, and how can we speak of a particle with no mass?

> I have done a terrible thing. I have postulated a particle that cannot be detected.
> —Wolfgang Pauli

Outline

I. One of the most interesting developments in the history of particle physics was the discovery of a new force of nature, the weak force, and a particle that is associated with that force, the neutrino.

   A. In the 1930s, scientists were concerned with fundamental forces of nature, such as electrical and magnetic forces, gravity, and the strong nuclear force.

   B. Today, instead of *force*, physicists usually use the term *interaction*. The act of two electrons repelling each other can be thought of as an electrical interaction.

      1. As we said in the last lecture, on a microscopic level, we might think of one electron generating a photon and transmitting it to another electron. This photon causes the recoil between the two electrons.

      2. Instead of an electric force between the two electrons, what happens is more of an interaction between an electron and the photon it creates. On the transmitting end, one electron transforms itself into an electron and a photon, and on the receiving end, an electron and a photon transform into an electron.

C. In the 1930s, scientists were studying various types of radiation, particularly beta radiation because it is unusual.

    1. In beta radiation, an electron flies out, and when we look at what's left behind, we see that a transformation has occurred.

    2. The simplest beta decay we can think of is as follows: After a period of time, a neutron in the lab radioactively decays, spontaneously transforming into an electron and a proton.

    3. What causes this transformation? We now explain it with the term the *weak force*, a weak interaction of nature. This is not an electromagnetic interaction, not the strong nuclear force, but a transformative force.

    4. This way of thinking about a force was new in the 1930s because the weak force was not a push or a pull, but a force of nature that causes radioactive decays. We now know that the weak force does exert more traditional pushes and pulls, but at the time, it was thought of only as transformative.

II. Remember that when we think about forces of nature now, we visualize the involvement of a force carrier.

    A. Electromagnetism, for example, involves the photon. The Japanese physicist Yukawa proposed that the strong nuclear force involves the pion as a force carrier.

    B. What kind of force carrier is involved with the weak force? We now call this force carrier the *W particle* (or *W boson*) for "weak particle." Because the W particle can have either positive or negative charge or zero charge, we speak in terms of *W plus*, *W minus*, and *Z* (for "zero charge"). These are three possible force carriers for the weak force.

    C. Why is this interaction termed "weak"? If a neutron decays and a proton and an electron are generated, that interaction does not seem weak.

        1. The reason the interaction is called weak is that it is improbable.

        2. The interaction seems improbable because the W particle is extremely massive. As we learned in discussing the pion, if a particle is created from nothing—essentially "borrowed" from nature—then the more massive it is, the shorter the amount of time is allowed for it to exist.

3. The W particle is so massive that the amount of time it can be borrowed from nature is extraordinarily short. That is the origin of the improbability of the weak force.

D. Why are we concerned with the weak force?
1. The weak force plays some role in nature. For example, a number of radioactive materials decay in this way, and weak decays usually mean long lifetimes.
2. In medical technology, the long lifetime of certain materials is important to allow doctors to see the images.
3. A more important role of the weak interaction takes place inside the sun. The main power generation in the sun is the strong force, but the sun would not work with only the strong force.
   a. Protons interact with other protons in the sun, but the strong force alone will not bind them into a stable nucleus.
   b. The weak force is needed to transform one of those protons into a neutron to form a stable nucleus.
   c. We need the strong force as the source of the sun's energy, but we also need the weak transformation to stabilize the final product.
4. The weak force also plays a role in building larger and larger nuclei, as in carbon, nitrogen, oxygen, and so on.

E. How do we see the weak force? Beta decays are one of the ways that we have observed this force in nature. In studying the weak interaction, physicists have also come to see the significant role of the *neutrino*.
1. A neutrino is a particle of nature, similar to an electron or a proton, but it feels only the weak force. Because it has no electric charge, it doesn't feel electricity; it has no mass; and it has no strong interaction with protons and neutrons.
2. For this reason, the neutrino can emit and absorb only W bosons or Z bosons, which is a highly improbable event.

III. The "inventor" of the neutrino was Wolfgang Pauli, a member of the generation that was developing quantum mechanics.

A. Pauli formulated the *Pauli exclusion principle*, which states that two identical electrons cannot be in the same place at the same time. This idea is at the heart of chemistry.

B. Pauli also formulated tentative ideas in studying beta decays.

    1. If we again think of a simple beta decay, we can measure both the original energy of the system and the final energy of the system. The result is that the system has more energy to start with than it has in the end.

    2. This disappearance violates the principle of conservation of energy and was, to Pauli, an unacceptable possibility.

    3. Momentum is also not conserved in beta decay. The electron and proton created in beta decay often travel off in a similar direction, rather than traveling in opposite directions, as we would expect.

    4. Pauli imagined that a third particle must be involved in this interaction to explain the contradictions in conservation of energy and conservation of momentum. Later, this third particle was named the *neutrino*.

    5. Pauli shied away from his own proposition of a particle that could not be detected, saying, "I have done a terrible thing."

C. Another physicist, Enrico Fermi, developed the idea of the neutrino into a mathematical framework. Fermi's mathematics, which assumed the existence of the neutrino, enabled correct calculations of many other beta decays.

D. In 1956, Cowan and Reines set up a giant detector outside a nuclear reactor to increase the chances of occurrence of this low-probability event, and they found direct evidence of neutrinos.

IV. How do we make sense of a particle with no mass and no charge?

A. The answer lies in the fact that the prediction of the neutrino consolidates a great deal of physical data in a simple way.

B. This particle belongs to the class of particle called *leptons*, the same class as the electron and the muon.

C. Like the electron, the neutrino has spin 1/2. As we will see, all neutrinos spin in the same way, and anti-neutrinos spin in the opposite way.

D. When a neutrino is created as a result of a beta decay, it shares some properties of the electron that was also created. Thus, it is an *electron-flavored neutrino*. *Muon-flavored neutrinos* can also be created.

E. In technology, neutrinos are just beginning to become a wonderful astronomic tool to give us a picture of the inside of the sun, which would otherwise be impossible to see.

Essential Reading:

Barnett, Muhry, and Quinn, *The Charm of Strange Quarks*, chapter 2.10 and pp. 48–55.

Schwarz, *A Tour of the Subatomic Zoo*, chapter 1 (pp. 10–11), chapter 2.

Recommended Reading:

Lederman, *The God Particle*, chapter 7 ("The Weak Force").

Wilczek and Devine, *Longing for the Harmonies*, Third Theme.

Questions to Consider:

1. Why *specifically* did Wolfgang Pauli feel compelled to postulate the existence of a nearly invisible particle?

2. In your entire lifetime, probably only one solar neutrino will interact with an atom somewhere in your body. Using this fact, roughly how big a tank of water would you need if you wanted to observe a neutrino interaction 10 times a day? (After you "guesstimate" your answer, you might go on the Web to look up the size of the neutrino detector at SuperKamiokande in Japan.)

Lecture Seven—Transcript
Weak Interactions and the Neutrino

One of the most curious and intriguing developments in the really early history of nuclear and particle physics was the discovery of a brand new force of nature. In this lecture we're going to be talking about this new force, and a very unusual particle, called the neutrino, that's associated with that force. In the 1930s, people were thinking about fundamental forces of nature. A force is a push or a pull, as when you tug or push on somebody's hand, or you grab a rope; those are ordinary, everyday forces. Microscopically, you can ask, if a rope is pulling you, what's really going on is the atoms in the rope are connecting in some electrical way to the atoms in your hand, but ordinary, everyday forces really boil down to electrical or magnetic in nature. Then of course, there's gravity, which is a fundamental force that we experience every day. By the 1930s people knew about the strong nuclear force, which is this very short-range force that holds protons and neutrons together; and there was another one that was discovered at this time, and at first it seemed very mysterious. It didn't quite act the same as the other forces. It required a little bit of a rethinking about what you mean by force.

Nowadays, instead of using the word force, particle physicists usually talk about interactions. It's kind of the same thing. You talk about, say, an electron and another electron, and they're going to bounce off of one another because they electrically repel. You can think of that as an electrical force, or you can talk about it as the electrical interaction between those particles. If you think really microscopically about that interaction, what's going on? Here's an electron cruising along, and the kind of quantum way of thinking about the force is to say, this electron generates, or creates, a little photon. A little bundle of electromagnetic energy which travels across, hits the other electron, so it's like tossing a heavy ball between you and me. You toss the ball and you recoil, and then the other person receives the ball and recoils, so that's what makes the recoil. Microscopically I'm thinking about electric force not, anymore, in terms of the force between two electrons that are far apart, but in the interaction of the electron and a little photon that it creates. That's really the electromagnetic interaction in nature. It boils down to an electron, which transforms itself into an electron plus a photon; or, on the other side, an electron and a photon coming together, which transform themselves into an electron. So there's this transformative aspect of interactions which is now kind of an integral part of how we think about forces of nature.

In the 1930s, people were looking at radioactive decays; there were various types: alpha, beta, and gamma radiation. One of the types, beta radiation, was very unusual. Well, first of all, beta radiation means that an electron is flying out; but when you look at what's left behind, a transformation occurred. The sort of simplest beta decay that you could imagine is if you take a neutron, and it's just sitting there in your laboratory. It sits there for a while, and then it radioactively decays. It lasts for a long time, and then it spontaneously transforms into a proton and an electron, and, actually, something else, a neutrino; but that was not visible in the 1930s. So as far as the experimentalists were concerned, the neutron simply turned into a proton plus an electron. So what caused that to happen? We now use the phrase "the weak force". It was the weak interaction of nature. It was not electromagnetic. It was not some strong nuclear force. It was a transformation. So this is a, sort of a, novel way of thinking about a force that's not really a push or a pull. It's a force of nature that causes radioactive decays. Now, since then the weak force has also been demonstrated to cause pushes and pulls. Neutrinos get pushed and pulled by the weak force. But the origin of this force was more transformative than it was pushing and pulling.

When you have a force of nature, now, we're thinking about it as involving a force carrier. So electromagnetism involves a photon, which is the particle that's transferred that causes that force. Yukawa, Mr. Yukawa, in the early 1930s, mid 1930s, proposed that the strong nuclear force that binds proton and neutron should be thought of as involving the exchange of a pion, a pi meson, so again, it's a little particle that's transferred. The pion was really a particle, an honest particle. It had mass, and it could have electric charge. In later years, you could find little tracks of these pions in bubble chambers because you could actually create one of these things. What about the weak force? This is a new force of nature. What kind of a force carrier does it have? Now, this idea really didn't develop in the 1930s. The idea of the force carrier for the weak force was developed more in the 1960s and even later, where it was refined. We now call it the W, for "weak"; it's the W particle. Actually there are two W particles, because they can be electrically charged. You can have a W positive or a W negative; we call it the W plus and the W minus. We have a third one now, Z, because it's got zero charge. So there are really three possible force carriers.

We call it the weak interaction, and why is it weak? If a neutron decays, and a proton and an electron go flying out, that doesn't really look all that weak. They've got lots of energy and they leave good tracks; it's really easy to detect and measure. It seems like a fairly dramatic decay. The reason we call it weak is because it's very improbable. A neutron lives for a long, long time, so it's not so much that when it happens it's wimpy; it's that it's just unlikely to happen. Why is it unlikely to happen? Because this new force carrier, the W—sometimes called the W boson, and we'll learn what boson means later—this W particle is extremely massive. Remember when I was talking about the strong force? I said, here's a proton, here's another proton; why do they stick together? Because one of them creates a pi meson, and the pion is massive. Now, you're actually asking nature to create a particle out of nothing, and that seems like it violates conservation of energy, which is a pretty crazy idea. You can only get away with this because of quantum mechanics. Quantum mechanics says, "Yeah, you can create energy out of nothing, but just for a really short time," so the pi meson can live for a really short time. It has to be eaten up by another proton or neutron right away, and the more massive the particle is, the shorter of a time you're allowed to borrow it from nature. That's the Heisenberg Uncertainty Principle. So the W particle is so incredibly massive that the amount of time that you can borrow it from nowhere is extraordinarily short, and that's the origin of the improbability of the weak force, the weak interaction. You're just asking nature to give you this huge bundle of energy. The mass of a W boson is something like 80 times more than the mass of a proton, so it's huge. In general, in nature, when weak interactions are present, they are very improbable. So it's kind of like the lottery. The odds for any individual to win are very low, but when it happens it can be quite dramatic, and the evidence is very clear once somebody wins the lottery.

Why do we care about the weak force? You've never felt it. You know, this push, this pull, it's not weak. Even when it is weak, it's not the weak force. It's not a part of our everyday life. Neither is the strong force, but the strong force is incredibly important. The strong force is responsible for nuclear energy, and nuclear weapons, and solar energy, right? That's what's powering the sun. The weak force is a little more obscure. As we go on in this course, we'll be finding more and more obscure particles, and interactions, that somehow seem less connected with our life. The weak force does play some roles. For example, beta decays, radioactive decays are quite common. Lots of radioactive materials decay this way, and typically, weak decays mean long lifetime. So for instance, in medical technologies, if you want to image the inside of somebody, you might inject them with some radioactive material, and typically, you'll use a material with a long lifetime. You want something that might beta decay so that it doesn't just radiate away right away; you need some time in order to make the images. It has some practical applications in medicine. Some street signs—not street signs, but lighted signs in buildings—have tritium in them, which beta decays. It lives for a long time and glows a little bit, and actually you can make the decaying beta particles make some red signs light up and say "Exit".

There's actually a much more important role for the weak interaction that is critical to our lives. Inside the sun, the main power generation is the strong force. A proton from hydrogen and another proton from hydrogen—it's very, very hot—slam together. They stick by the strong force, and that sticking actually ends up releasing some energy which is the power supply for the sun. The problem with that is that a proton and a proton sticking together would form something like two protons, and that's not a stable nucleus. So they would actually just come together and fly back apart again, and the sun wouldn't work if there were only the strong force. So what do you need? You need the proton and neutron to come together—sorry, proton and proton to come together—and then one of the protons has to transform into a neutron because a proton-neutron system is stable. That's a perfectly content nucleus. You need both the strong interaction, to release the energy, that's the supplier of the energy; but the weak interaction's absolutely critical. You need that transformation in order to stabilize the final product, so the weak interaction plays this critical role in making the sun work.

Even more important than that, you might argue, is that if you want to build up more complicated nuclei, if you want to take two of those things and slam them together, and then two of those things and slam them together—which is what happens in stars, especially towards the end of their lifetime—that's the way you build larger and larger nuclei. All right? That's how you build carbon, and nitrogen, and oxygen. I don't know if you ever thought about this, but your body is filled with atoms that have nuclei, carbon and nitrogen and oxygen and so on, that are all very big. Where did they come from? I mean, where did you come from? You started off little and you've grown this big; where did all those carbon atoms come from? Now, you might say, "I ate them." But where did the food carbon atoms come from? Well, you might say they came from planet Earth, but where did the planet Earth come from? Ultimately, all of the stuff, all of the heavy atoms, everything beyond hydrogen that we are made of, must have come from some nuclear process in a star. Not our star, because our star hasn't gone supernova yet. So it must have been actually a previous generation of star, which underwent this nucleosynthesis and built up the larger nuclei, for which the weak interaction was absolutely critical to do the transformation from one nucleus to another. So the weak force is in some sense responsible for all of the stuff we're made of.

When you want to talk about the weak force, you have to ask, how are you going to see it? Beta decays are one of the ways that we have observed this force of nature, and there's an associated process, that in the early days was extremely speculative, and nowadays it's becoming a kind of a big part of the story of studying the weak interaction. It's the story of the neutrino. Neutrino is a particle of nature, like an electron, or proton; it's another one of those fundamental particles. This particle is completely bizarre, for the simple reason that it only feels the weak force. Okay? Imagine a particle with no electric charge, so it doesn't feel electricity. It has no mass. Okay? That's a very weird concept. I'm talking about a particle with no mass. How can that be? Well, we've met a particle with no mass already; it's the photon. The photon, the carrier of electricity and magnetism, is itself a particle with no mass. We've seen it before; it's kind of hard to visualize, but there you go. It's what we have to postulate in order to make sense of the experiments. So it has no mass, and no charge, and it has no interaction with protons and neutrons, no strong interaction. It can just go whizzing right through the middle of a proton, and it doesn't care because it only feels the weak force. That means that it can only emit and absorb W or Z bosons, which is a highly improbable event, so most of the time neutrinos just cruise through matter and don't interact.

Let me tell you a little bit about the, sort of, origins of the neutrino. How on earth could we have come up with this crazy idea for a particle that nobody can see? Of course, nowadays we can see them, but in the early days we couldn't. The inventor of this particle, and I use that word sort of loosely. I mean, you don't invent a particle; nature has the particles. I don't know— you discover the particle—like Wolfgang Pauli. Pauli was a remarkable physicist. He was of a generation; he was of the generation that was developing quantum mechanics. Mr. Schrödinger and Mr. Heisenberg had invented the theory of quantum mechanics, but Wolfgang Pauli was really one of the big, key players to turn the theory from a fledgling idea into a robust, well-developed mathematical framework. Mr. Pauli was the one who figured out what we now call the Pauli Exclusion Principle that says that two electrons can't be in the same place at the same time. It may seem obvious; after all, how could you put two particles on top of each other, but this was a very important idea in the development of quantum mechanics. In fact, that idea, that two electrons can't be in the same place, is what's responsible for chemistry.

So Wolfgang Pauli's work (really, that's what he got his Nobel Prize for, this understanding of the chemical properties of the elements) boils down to this: If you start adding electrons to some nuclear system, so the electrons go into orbit, the first one goes down into a nice, low, stable orbit. Remember electrons have spin, so an electron can be spinning clockwise, or it could be spinning counterclockwise. Quantum mechanically, those are two different states of matter. Clockwise spin, counterclockwise spin are distinguishable, different states. So Mr. Pauli says, you can actually put two electrons right next to each other if their spins are in opposite directions, because they're different from one another. You can put different things together; you just can't put identical spin one-half particles together. A helium atom has two electrons, and that's nice and stable. They're both down in this low orbit, but what about lithium, which has three electrons? You add the third electron, and it's not allowed to go down into that same orbit because it's got spin one-half. It could be clockwise or counterclockwise, but it can't be either down where the other ones are. It has to go into some more loosely bound orbit, which means that lithium has a loosely bound electron; it's very chemically reactive. So this is the way that we begin to understand chemistry, from the underlying physical principles.

Mr. Pauli was responsible for lots of amazing developments in quantum mechanics, and one of his ideas, perhaps his most tentative—he was a little bit afraid of this idea—was in response to looking carefully at beta decays. So, imagine the simplest beta decay—it wasn't actually the one that Pauli was considering—but think of a neutron sitting in the laboratory. You could watch it. You might have a little emulsion or bubble chamber image of a neutron just sitting there, and then poof!; it decays, and turns into a proton and an electron. Now, if you're measuring data like that, you can very carefully measure the original energy of your system and the final energy of your system. You have to use Einstein's $E = mc^2$ to figure out the initial energy, and in the end, you've got $E = mc^2$; you've got rest-mass energy of the proton and the electron. You've also got some kinetic energy, energy of motion. You have to add it all together very carefully, and when you do, you discover that there is more energy to start with than you have in the end. That's kind of a scary idea. To Wolfgang Pauli, it was a completely unacceptable possibility. Conservation of energy is a deep principle of physics, and so he's asking himself, "How can this be? How can energy just disappear from nature?"

Momentum is also not conserved. Momentum is a fancy word that means oomph, so if you start off with a neutron, and it has no oomph, it's just sitting there. It has energy, but it has no motion; now it creates a proton and an electron, and oftentimes they go off together. You know, one goes in one direction, and the electron might go in a similar direction. This is really weird. You would expect, if the proton goes one way, the electron would have to go back to back, recoiling against it to keep the total momentum conserved, but again, it's not seen in the experiments. So this is the usual story; you've got some experiment that doesn't agree with some fundamental principles of physics. You had a choice in the 1930s—actually it was 1930 when Pauli came up with this idea—when you could abandon conservation of energy and conservation of momentum, but a slightly less radical proposal would have been to imagine that there was another particle involved, a third particle. He didn't actually name it; it was named later This particle was the neutrino. It could run away as well, carrying energy and momentum, and that would preserve these conservation laws. It's just another particle. Nowadays, you know, we propose new particles all the time. It's not a big deal in this generation. But in 1930 (this was 1930), it was before the discovery of antimatter, before the discovery of the muon, and even before the proposition of the pion. This would really be the first new particle proposed,. and it was just too scary.

Wolfgang Pauli didn't publish a paper about it; he wrote a letter to his colleagues at a conference. He began the letter, "Dear Radioactive Ladies and Gentlemen." He had a good sense of humor. Pauli was an amazing fellow. People sometimes call him the conscience of physics because he was so smart, and so sharp—not mean, just sharp. When people would relate ideas, he was very quick to criticize but in a scientific manner. There's a famous quote of him at a meeting; he says, "What Professor Einstein has just said is not so stupid." This is sort of typical Wolfgang Pauli. He had a big ego, and, it's sort of interesting to speculate as to why didn't he publish his idea about the neutrino. I think it's because he was so critical of other wild ideas that he was a little bit timid about proposing his own wild idea.

In fact, here's another quote from Pauli: "I have done a terrible thing. I have postulated a particle that cannot be detected." Okay? Think about why that's such a terrible thing. This is a physicist. Physics is about measuring stuff. Okay? We can talk all we want, until we're blue in the face, about elegant theories and beautiful mathematics, and nowadays that actually carries some weight. But even nowadays, ultimately, it doesn't matter how beautiful your theory is. I don't care how elegant, and formal, and lovely, and consistent the mathematics is, it's only physics if you can go in the lab and verify it. Right? That's the essence of physics. You have to be able to check these things. You want to see little tracks in a bubble chamber that prove that your mathematical proposition is correct. What Pauli was proposing was a particle of nature with no electric charge, so it didn't create ions in a bubble chamber; it had no mass, it had extraordinarily weak interactions. Once it was produced, which was a low-probability event, it just cruised through matter, and left no trace of its self. So this was a particle that was extraordinarily difficult but not impossible. All right? It had low probability, but not zero probability, so in principle he could detect a neutrino. But in practice, Pauli knew that this would be essentially impossible with 1930s technology.

So the idea was out there in the literature, and another physicist, Enrico Fermi, took the idea just a few years later. He thought about it and developed it into a mathematical framework—not just the idea of a particle, but really a consistent mathematics. He described the interaction of the neutrino with the neutron, with the proton, and with the electron. With that we have a mathematical framework, and we can start calculating the probability of various different kinds of beta decays. There are lots in nature; every nucleus, in principle, can beta decay. Now we have a theory, so we can do calculations, and they work. They work like a champ. So Mr. Fermi's theory is telling us that yeah, you have to assume a neutrino for his theory to work, and then you start to agree with lots of data, not just the original one. There were other events that didn't involve neutrons and protons. For example, remember those crazy muons, "Who ordered that?" Right? The muon was this particle observed in cosmic rays. It appeared to be some sort of a heavy electron, which would occasionally decay into an electron. It took a long time. It was a weak decay, an improbable decay. Fermi's theory actually allowed you to calculate the lifetime of the muon, and it worked. It was consistent with all the other data; so in fact, according to Mr. Fermi's theory, when a muon decays, not only does it produce an electron, it produces two neutrinos.

The story fits in consistently and mathematically, and this, you know, allows people to believe in the reality of neutrinos. But still, I think deep in our hearts we wanted to see one, and it took over 20 years, 25 years. It was in 1956 that Cowan and Reines, two physicists working outside of a nuclear reactor, were able to do so. What they did, basically, was they stood outside of the nuclear reactor because nuclear reactors produce lots and lots of neutrinos, just as the sun does. A nuclear reactor is primarily strong occur that emit neutrinos. When the reactor was on, it was just spewing neutrinos like crazy, which just went flying off, interacting with nothing, most of the time. Cowan and Reines set up a giant detector outside of the facility, right outside, really, really big. So there were lots and lots of target atoms, target nuclei; and so, you know, they had a low-probability event with lots of chances. It was like rolling dice over and over and over and over again, and once or twice an hour they would see an event. It was called Project Poltergeist, because they were looking for these little ghostlike particles. They found them, and when the reactor was off, they got nothing. When the reactor was on, they would get two events an hour. They won the Nobel Prize for this work. There was direct evidence. All right? You see, it was sort of the opposite process. Instead of producing a neutrino, the neutrino came in, whacked into a neutron and created a particle. Actually, it whacked into a proton, and created a particle.

How do you make sense of a particle with no mass and no charge? It's hard to picture, and I think you just have to look at the mathematics and at the calculations, and you realize that the prediction of this particle consolidates an enormous amount of physical data in a simple way. You take one prediction, one theory of Mr. Fermi's, and then you can essentially make an enormous number of generalizations to a whole bunch of different circumstances. This particle belongs to the same class of particles as electrons. We call those leptons. That's just a fancy word which means light particles, although later on we've discovered leptons that aren't so light anymore, but in those days we were classifying particles and trying to organize this world of sub-atomic particles. The electron, the muon, and the neutrino are kind of all in a family together. They do not interact with protons and neutrons in a strong way. The electron has electric charge, so it can interact with a proton electrically. The neutrino feels only the weak force, and they're somehow connected. As we progress in this course, we're going to talk more about the connections among muons, electrons, and neutrinos. Neutrinos have spin, just like the electron did. They have spin one-half, which means they have a certain amount of angular momentum. In principle, that means they could spin clockwise or counterclockwise; and it's an interesting fact of nature, which we'll be again talking about in a future lecture, that every neutrino that we've ever found always spins the same way. Anti-neutrinos, which also exist, spin the other way. (According to Mr. Dirac, every particle should have its anti-particle.) We're going to have to talk about that story because that's definitely not the same as an electron, by any means.

It's also the case that neutrinos are rather rich, in the following sense. If you have a beta decay that produces an electron and a neutrino, that neutrino has some property, some character of the electron in it. We call it an electron-flavored neutrino because it was created in partnership with an electron, and as it cruises along, it retains that property, that characteristic of being an electron neutrino. It's one of the ways that you describe a particle. Do you have electron number? Even though it's not an electron, it's associated with an electron. What that really means is that if it whacks into a nucleus later (a low probability event) but if it does interact weakly, then what it will produce is an electron, or maybe an anti-electron, depending on whether you have a neutrino or an anti-neutrino. It will never produce a muon. There are other neutrinos that have been discovered in more recent years that are muon-flavored neutrinos. So for instance, when the muon decays and turns into an electron, it produces two neutrinos. One of them is an electron-flavored neutrino, and one of them is a muon-flavored neutrino. So you can actually create beams, nowadays, of muon-flavored neutrinos or electron-flavored neutrinos. If you have a muon-flavored neutrino beam and it whacks into a target (again, a low probability, but if it does have this weak interaction), it will always produce a muon, never an electron. So it has this character to it, and nowadays, one of the puzzles about neutrinos is whether there's any vague possibility of an electron-flavored neutrino turning into a muon-flavored neutrino. It's a completely weird idea, and we'll come back in a future lecture which will be devoted to the topic of neutrinos.

They're fascinating particles. I don't know why I find them so interesting. You know, there's a saying that it's like love and antiques. If something's hard to get, then it's interesting, and neutrinos are hard to get. They're hard to measure and they're hard to detect, but it is possible, and we're getting better at it. Maybe it's because my dad called be Steverino when I was a little kid, I don't know. These particles are, in some sense, the most esoteric, wispy, ghostlike particle I could think of. But nonetheless, we have actually come up with applications for these things. People have thrown around ideas and asked if there any technological uses for neutrinos. Kind of a—at first it was a whimsical question. I mean, people were making jokes about communication from submarines because the neutrinos would travel, you know, all the way through planet Earth without interacting with anything, and you'd have some giant detector. But that's a fairly ridiculous idea, at least for the present day. What we have done with neutrinos, already, is…they're produced copiously in the sun. Every nuclear reaction in the sun produces a neutrino, so the number of neutrinos coming out of the sun is astronomical.

There are about 100 billion neutrinos per square centimeter (That's you know, the size of your thumb, or your eyeball.) passing through every square centimeter of you every second. Most of them are coming from the sun. Some of them are coming from stars, and even from the early universe, and they're just passing through your body. During the daytime, they're coming through the roof of the building, if you're inside of a building and in the direction of the sun, most of them passing through you, going through the Earth. The people who are on the other side of the earth are sleeping right now, because it's nighttime over there, and the neutrinos are coming up from underneath the floor, through their bed, through them, and off out into outer space. The same number: 100 billion per square centimeter per second. These solar neutrinos, these astronomical neutrinos are very low energy, and stupendously difficult to detect, but we have detected them, and that's going to be a future story as well when we ask how we do that. Now we have actually seen an image of the sun, if you like. It's like a photograph of the sun; but not in sunlight, but in neutrino light. So we have actually seen the sun by detecting the neutrinos that come from it, and we have even seen a supernova which occurred a long time ago, hundreds of thousands of years ago. It was in 1987 that we saw the light from that supernova, and at the same time we saw a few of the neutrinos from that supernova.

Now, why is that interesting? Well, neutrinos come from the center of the sun, where the nuclear reactions occur. The light from the sun comes from the outside edge. We have no way of seeing the inside of the sun. There's no way anybody could imagine of getting inside the center of the sun, except through neutrinos. So they're like this microscope that looks inside, to the center of the sun, or supernova, or the center of our galaxy, a wonderful astronomical tool using these exotic particles.

Lecture Eight
Accelerators and the Particle Explosion

Scope: World War II spawned a fresh burst of scientific energy and achievements that lasted for decades. Particle accelerators were born. What were these machines? How did they work and what did they do? In some respects, this era saw the origin of "big science" in the United States. With the birth of accelerators came a steady stream of new discoveries. Fundamental particles began to appear by the handful. What were they like, and how could they be organized to allow physicists to make sense of what they were detecting? We include a quantitative discussion of energy, setting numerical scales to make sense of the particle discoveries.

> As experimental techniques have grown from the top of a laboratory bench to the large accelerators of today, the basic components have changed vastly in scale but only little in basic function. More important, the motivation of those engaged in this type of experimentation has hardly changed at all.
>
> —Wolfgang Panofsky (*Contemporary Physics*)

Outline

I.   Research in nuclear and particle physics slowed considerably during World War II, but the war had a significant influence on some aspects of science.

    A. In the United States, a number of bright physicists were brought together in the Manhattan Project to develop the atomic bomb. This project was more about technology and engineering than physics, but the exchange of ideas among physicists on the project brought about an explosion of ideas in theoretical and experimental physics.

    B. Politically, physicists were seen as valuable to the U.S. government for national security purposes. A certain amount of prestige became attached to the science of physics, which led to funding and public support for projects.

II. Throughout the 1910s–1930s, physicists were using natural sources of radioactivity, such as uranium or cosmic radiation, in their experiments in nuclear physics, but natural radiation does not offer very high energy.

    A. One reason that physicists were looking for higher energy sources stems from quantum mechanics.

        1. The uncertainty principle states that if we want to examine very small objects, we need higher levels of energy.

        2. To understand this idea, think of water waves and their wavelengths, which are defined as the distance from one peak of a wave to the next. As water waves propagate, the wavelength can be changed. Thus, if you create a wave in a pond by moving your hand in the water and you increase the rapidity of the movement of your hand, the waves tend to get closer together.

        3. In general, then, if you want to look at a small object, you need a small wavelength of light, which is created with more energy.

    B. Einstein's $E = mc^2$ also led physicists to seek higher energy events in order to observe new phenomena.

        1. Energy is required to produce a particle that has a mass; the amount of energy required is the mass × the speed of light^2.

        2. The speed of light^2 is a large number; in other words, we need a good deal of energy to produce a new particle.

        3. Further, if we're looking for more exotic particles of higher and higher mass, we must have more energy to produce them.

    C. Energy can be measured in various unit systems. Particle physics uses a unit called the *electron volt*, or *eV*, which for our purposes, is just a scale for measuring energies.

        1. The measure 1 eV is a rather small energy, which might be the typical kinetic energy of an electron in an ordinary atom.

        2. The typical energy of a proton inside a nucleus is 1 million eV, or an *MeV*.

    D. In the 1920s and 1930s, scientists were trying to build particle accelerators. These are mechanical devices used to create small particles, such as an electron or a proton, using a lot of energy. The accelerator then smashes the particle into a target to enable scientists to learn about other interactions and to see what other particles are generated.

1. One such device was the Van de Graaff accelerator, which could generate up to a million volts and allowed the study of protons and neutrons. To look at the constituents of protons and neutrons, however, requires much more than a million volts.

2. Ernest Lawrence, a Midwestern physicist, made a breakthrough in the development of a particle accelerator with his *cyclotron*.

3. The idea of the cyclotron is as follows: If you want to speed up an electron or a proton, you need some voltage, such as from a battery. If you set up a high-voltage battery and produce sparking from one terminal to the other, that voltage difference accelerates the charged particles.

4. Lawrence's idea was to direct the sparks from the high-voltage battery into a region with a large magnet. The charged particle in the magnetized region will travel in a curved path. When the particle reaches a certain point in the curve, the poles on the battery are reversed, and the particle is sent back through the device. Each time the particle is pushed back through the battery, it gains energy.

5. The resulting beam of protons can be used in various ways, for example, aimed at a target, much as Rutherford had been doing.

E. With this innovation, funding for physics increased, and the United States began building centralized national labs with particle accelerators. The energies produced increased steadily; right after the war, the energy reached 200 million eV.

1. Recalling Yukawa's work with the pion, physicists knew that about 100–150 million eV was required to produce one pion. Ultimately, these particles were created in the cyclotron.

2. By the 1950s, the energies of the accelerators reached 1 billion or more eV, which is termed *GeV* for giga-electron volt.

III. At the same time that the particle accelerators were developing, the particle detectors were improving.

A. In the early 1900s, Rutherford had used students to count particles by eye as they hit a phosphorescent screen. Later, the cloud chamber and bubble chamber were used to detect and identify particles.

B. The invention of the bubble chamber in 1952 improved the ability to observe high-energy particles because of the high density of the liquid inside it. The higher the density, the greater the likelihood of observing interactions.

IV. At this time, the need for automating the detection process grew, although computers were not yet in common use.

A. As the devices for acceleration and detection improved, hundreds of new particles were discovered. Physics experienced an explosion of information but lacked an organizing principle to characterize these discoveries.

B. Physicists wanted to learn the properties of these new particles, including their mass, electric charges, lifetimes, and spin. Physicists also discovered that almost all these new particles interact strongly, or have a high probability of interaction, and have very short lifetimes.

C. A classification system began to emerge for these particles, based on their interactions. The terminology is as follows:

1. *Hadrons*—From the Greek for "thick," strongly interacting particles, such as the proton, neutron, or pion.

2. *Leptons*—From the Greek for "light," not strongly interacting, such as the electron, muon, or neutrino.

D. Hadrons were subcategorized as follows:

1. *Baryon*—From the Greek for "heavy"; what one might think of as truly a particle, such as a proton or neutron.

2. *Meson*—The intermediary particles, such as the pion.

V. In the 1950s, a great deal of overwhelming information was being collected. The situation in the physics community was similar to what it had been in the field of chemistry in the 1800s.

A. The physicists in the 1950s needed some kind of organizing principle, similar to the periodic table for the chemists, to begin to classify their data.

B. In the next lecture, we discuss the first attempts by theorists to organize these data. Ultimately, we will discover that this seemingly complex story is really quite simple.

Essential Reading:

Barnett, Muhry, and Quinn, *The Charm of Strange Quarks*, chapter 3.2.

Recommended Reading:

Lederman, *The God Particle*, chapter 6.

Riordan, *The Hunting of the Quark*, chapter 3.

Questions to Consider:

1. In a cyclotron, the amount of time it takes a non-relativistic particle to swing around a circle in the magnetic field is *independent* of the energy of the particle. (In other words, all particles go around at the same frequency, no matter what their energy.) Why would Einstein's theory of relativity contradict this principle at super-high energies?

2. Under what circumstances do you suppose a cloud chamber would be more suitable as a particle detector, and when might one rather have a bubble chamber? Can you think of advantages and disadvantages for each? (You might want to go web-hunting at some of the major particle physics sites to see what kind of detectors are being used today. Detector technology is a field that continues to make great strides all the time!)

3. Energy can be measured in many units (just as length can be measured in feet, or miles, or meters). Particle physicists often use "eV" but a more common measure of energy is the "Joule" (J). You can figure out one from the other (just as you can figure your height in cm—if you know it in inches—as long as you know how many cm there are in one inch!) The conversion you need is $1 \text{ eV} = 1.6 \times 10^{-19}$ J. Now, Einstein's $E = mc^2$ (mass × speed of light^2) tells you what a particle's "rest energy" is. If you plug in the mass in kg (kilograms), and use $c = 3 \times 10^8$ m/s (meters/sec), the energy automatically comes out in Joules, not eV! Given that the proton mass is 1.7×10^{-27} kg, see if you can find out the "rest energy" of a proton, in eVs. (To check yourself, notice that I said that by the 1950s, energies had reached 1 billion eV, and that was, not coincidentally, the era that the anti-proton was finally observed.)

Lecture Eight—Transcript
Accelerators and the Particle Explosion

We've been following an historical progression, learning about the developments of modern physics starting from the late 1800s, working our way up now to the 1930s. And then came World War II was kind of a hiatus, not a hiatus in scientific development, or technological development, but physicists were busy with the war effort, by and large, and so, pretty much, nuclear and particle physics research, *per se*, especially theoretical research, shut down. The war had a big influence in a lot of ways. First of all, a lot of very bright physicists were brought together in the United States for the Manhattan Project, which was the development of the atomic bomb, and that was really a project in technology and engineering; it really wasn't a project to develop any understanding of physics. There were a bunch of physicists there who were helping in the project. One of the things that happened was that these people, who normally were separated geographically, were together a lot of time, and they were talking and brewing ideas so that after the war there was a real explosion of both theoretical and experimental development.

There were other aspects of the war. For example, radar was developed, and the technology and devices that were used for radar ended up being very applicable to making experimental physics apparatus that was much more accurate and useful than what they had had before. On a political level, I think the physicists had demonstrated a certain value to the U.S. government for national security purposes. There was an element of prestige also associated with the science that was developed during the war that led to a lot of funding, and a lot of public support for the physics community for, and it lasted a long time after World War II. I think today the situation is somewhat different, and in future lectures we'll talk a little bit about the shifting role of particle physics in the world of science and science funding. But certainly, right after World War II it became big science. Lots of money became available from the U.S. government; large institutions and laboratories were constructed, which, of course, had a big impact on the kinds of developments that were possible when people were trying to continue their study of the nucleus and the sub-nuclear particles.

Natural radioactivity is what people had been pretty much using through the 1910s, 1920s, and 1930s, to study what was going on in nuclear physics. So you take some sample, like uranium, and you hold it next to a detector; or perhaps you go up in a balloon and you use comic radiation, naturally occurring radiation. Naturally occurring radiation was good enough in the 1920s and 1930s, but it was not high energy, and people wanted more energy. Why was that? There were various reasons. One of the reasons came from quantum mechanics itself. The laws of quantum mechanics, Heisenberg's Uncertainty Principle in particular, says that if you want to look at something very small, you need more energy. There's a tradeoff; the smaller you want to get, the more energy you need. It's not at all an obvious fact of life. It's not a classical idea, but you might visualize it in the following way. If you think about water waves, water waves have a certain wavelength associated with them, it's the distance from one peak to the next peak of the wave. As water waves propagate, you can change the wavelength of the light. For instance, if you are jiggling on a pond, and you make a little wave, if you jiggle with higher frequency, with more energy, and the wave carries more energy, the waves tend to scrunch up. They have a smaller wavelength. In general, if you want to look at a small object, you need a small wavelength of light. The smaller the wavelength, the smaller the object you can detect. So waves with a big wavelength will go past a duck, and the duck will just bob up and down, but the wave won't be perturbed. But if you have a little, short wavelength, then it will bend and refract, and you'll see some shadow image of the duck behind it. It is the same way with particles. The higher the energy that you put into your waves (and, of course, quantum mechanics says everything's both wave and particle at the same time) the higher the energy. Consequently, the higher the energy, the better the microscope you have. So if you're looking to understand sub-atomic particles, if you want to look at smaller and smaller objects, you need higher and higher energies.

There's another argument, a physics argument, for why you would like a high-energy collision, or event, in order to look at new things. It's Albert Einstein's $E = mc^2$. If you want to produce a particle with a mass, it requires some energy, the mass times the speed of light squared. That's how much energy you need. The speed of light is a big number; you square it, it's a huge number. You need a lot of energy to produce a new particle of nature. So the more energy you have, the more particles you can produce. If you're looking for more and more exotic particles, of higher and higher mass, again, you'll need higher energies. I'm going to be talking about energies quantitatively, although we won't be doing any math with them, but it's just nice to compare energy scales.

Energy can be measured in various different unit systems. Particle physicists tend to use a number, a unit called the electron volt, an eV. And never mind what an electron volt is; it doesn't really matter. It's just a scale for measuring energies. One electron volt is a rather small energy. It's the kind of energy that an electron might have if it's sort of sitting in a molecule, and that's the kinetic energy that it has. It's sort of a typical, chemical kind of energy. That's one eV. A million eVs—an MeV is how we abbreviate that, also a mega eV—that is the sort of more of a typical energy of a proton sitting inside of a nucleus. It has a million electron volts of energy just rattling around in there, inside of a carbon nucleus or something.

The rest-mass energy of an electron is 500,000 electron volts, a half a million electron volts. So that means $E = mc^2$; if you wanted to produce an electron, then you would need a half a million electron volts of energy. Unfortunately, you can't just produce an electron by itself because you can't just produce one negative charge in nature. Charge is conserved. So if you want to produce a negative charge, you also have to produce a positive charge; you produce an electron and an anti-electron together. This is the remarkable idea that led us to the creation of particles in machinery in the post-war era. If you pour energy into a small region of space—by slamming a particle into a target or some other, similar, mechanism—if you put energy into a small place, $E = mc^2$ says you can create particles out of that energy. Energy can be converted into mass. It's like matter-anti-matter annihilation run backwards. When matter and anti-matter annihilate, they produce energy, or, photons; but if you bring photons in, you can create an electron and a positron, and they'll go flying off. You can detect them. They'll leave tracks in your detector, and that's how you could create anti-matter in the laboratory.

In the 1920s and 1930s people were trying to create what is now called a particle accelerator. It was a mechanical device to create a small particle, like a proton or an electron, with a lot of energy. You could smash that into a target, and then see what came out, what kind of particles have you made. There was some success even in the pre-war era; there were various devices that people came up with. For example, Mr. Van de Graaff invented the Van de Graaff accelerator. You may have seen this in a science museum, or even sometimes in high school physics classes. It usually has a tall post with a rubber belt running around it, and there's a big metal ball at the top. You charge it up with just static electricity, and if you touch the thing, your hair stands on its end. If you really charge it up big time, you can get a million volts on a Van de Graaff, and it'll start to spark. What's a spark? It's really a lot of electrons traveling and jumping off that thing, and those electrons have about a million electron volts of energy. If you have a million volts, and one electron jumping off, you get a million eVs, one MeV. So the Van de Graaff accelerator was actually designed not for science demos, but to accelerate particles.

A million electron volts is enough, if you pour that energy into a nucleus you can sort of excite the nucleus. So if you're studying nuclear physics, and trying to understand how protons and neutrons fit together, it's a useful kind of energy. But if you really want to go deeper than the protons and neutrons themselves, if you want to look at the fundamental particles, the constituents, if you want to see what the proton and neutron are made of, you need much more than a million volts. The Van de Graaff technology kind of got stuck at a million volts, because it starts to spark at that point, and it's really hard to prevent the sparking. It's hard to get to higher energies.

There was a big breakthrough made by an American physicist named Ernest Lawrence. Ernest Lawrence was a big, tall, Midwestern physicist. You know, usually when I introduce some famous physicist who gets a Nobel Prize, I always say, a really smart guy. Ernest Lawrence was a fairly smart guy, but not one usually characterized as a genius, by any means. He was just a go-getter who was politically savvy and had big ideas. He could organize folks and get money, and he had some good physics ideas. He invented the idea of a particle accelerator, but really, I don't think he did the work. I think his graduate students did most of the day-to-day, practical development work, but Ernest O. Lawrence did win the Nobel Prize, just before World War II, for his development of a device called the cyclotron. The cyclotron was a sort of prototype particle accelerator, which has been advanced and improved. Nowadays, we have devices that are kind of descendents of the cyclotron, that we still use to do particle physics research. Here's the idea of a cyclotron.

If you want to speed up an electric charge, such as a proton or an electron, what you need is some voltage, such as a battery. So you take a battery, (Don't try this at home.), and if you were to take a screwdriver and drop it from one end of a car battery to the other, you'd get a spark. It's a very dangerous spark, because it could cause a lot of damage. You'll get about 12 electron volts of energy, because it's a twelve-volt battery. So imagine setting up a big, high-voltage battery, and you're sparking across it; that accelerates the charged particles. Now, here's the trick that Mr. Lawrence came up with. Suppose you have some high-voltage battery, and you get sparks across it; and then, when the spark reaches the far side, you have a little hole for it to go through, and it enters a region where there's a big magnet. What we knew already, from the previous 100 years of physics research in electricity and magnetism, is that when you have a charged particle and a magnet, it goes around in a curved path; it goes in a circle. So you've accelerated this particle, and then it leaves the battery region and it goes into a magnet region, and it goes around in a half-circle, and then it comes back. Now the trick is to reverse the battery poles; flip positive and negative. So you accelerate it on its way back again. It's as though you're whacking this charged particle every time it reaches the central region, and then it reaches the magnet, and goes in a semi-circle and comes back.

It's kind of like a kid on a swing. Right? You give him a little tap, and he takes a swing, and you just wait for them to come back. Then you give him another little tap, each time he comes back you give tap him again, and each time he gets more and more and more energy. He swings further and further, and more and more energy is created. That's exactly the idea of the cyclotron. It's a cyclotron, because these things are going around and around in cycles, getting a little bit more energy each time, so you don't need super high voltage anymore that's going to spark all over the place. All you need is a bunch of little taps in resonance. You have to tune the frequency of the taps, just like with the kind of the swing. If you just shake them, they don't go faster and faster; you have to tap them at just the right natural frequency.

The only potential problem is, just as with the kid on the swing, each time around he makes a bigger swing. At a certain point, your magnet runs out of steam, and it leaves. Basically, you are limited by the size and strength of your magnet. In the early days, in 1920, these things were the size of your fist. In the post-World War II era, when there was some big money involved, these things began to be housed in giant laboratories. There were gigantic magnets, and the energy just kept getting bigger and bigger. I have a quote from Mr. Livingston, a graduate student of Mr. Lawrence, "Dr. Livingston has asked me to advise you that he has obtained 1,100,000-volt protons. He also suggested that I add, whoopee!" This is a telegram from 1931 in the Nobel Prize-winning work that first developed the cyclotron. So these were protons: they were accelerated up to one million electron-volts and then they would go flying out the end, and you'd have a beam of protons, which you can do whatever you want with. You can aim them at a target, like Mr. Rutherford had been doing with his naturally produced radiation, and study the atoms and what makes them.

So after World War II, Ernest Lawrence was able to get a tremendous amount of funding. This was the beginning of big science in the United States. It was a government financed research laboratory. Everybody wanted an accelerator, but not every university could afford one, so the idea was to centralize, to build a couple of national laboratories, and people from all sorts of different places in the U.S. could go, do their research, and use the accelerator. There would be a dedicated staff to run the machine, so each physicist didn't have to learn how to design and build a cyclotron on his own. Then he could collect his own data with his own experiment and go home and analyze it. There were various laboratories that started cropping up in the United States. Mr. Lawrence's laboratory was called the Radiation Lab, or the Rad Lab. It was most sophisticated and had more and more funding.

Right after the war, the energy reached 200,000,000 electron volts. Now, that's a very special number, because of the pion. Remember Mr. Yukawa, who had predicted the existence of a new particle of nature, way back in the mid-1930s? He had even estimated the mass of the pion, which was, according to his calculations, about 100, or 150 million electron volts, when you take the mass and multiply by c^2. You need about 100, or 150 million electron volts to produce one pion, so if you've got 200, you have more than enough. You take your protons, smash them into the target, and you should be able to get some pions coming out. That's how we discovered pions; basically, not through random cosmic radiation, but through the actual creation of these particles in a machine.

They were smashing a proton into an atom, and so these things were called atom smashers. I don't think physicists really like that word very much; it's not all that descriptive. First of all, you're not really smashing the atom; you're really just adding energy to a nucleus, creating particles and seeing what comes out. You might argue that what physicists are doing is—you know, I hand you a beautiful Swiss watch, and I say, figure out how this thing works. So you look at it for a second, and then you smash it against the wall, and you look at the gears that come flying out, and you say, "Okay, I know it's made of gears. Give me another one." That's sort of the idea here. You're trying to figure out what is inside of protons and neutrons, what kind of fundamental particles are there, by smashing things together, getting lots of energy and looking to see what comes flying out.

At the same time the particle accelerator was developing, the particle detectors were improving. Think back to the early 1900s, when Mr. Rutherford had two graduate students—actually one of them was an undergraduate—sitting in front of a little piece of phosphorescent screen in the dark, counting flashes with their eyes, very painful work. Later the cloud chamber was developed, and then in the early 1950s, the bubble chamber, which is denser because it's a liquid. Particles go through this very hot liquid, and as a charged particle goes through, leaving a little ion trail, it also creates a bunch of little bubbles because this liquid is superheated. It really wants to boil, and the little charged particles leave a track where it can more easily boil. You take a picture, and you see an image of the tracks of the particles. If they happen to hit a nucleus, you'll see all the spray that comes out, and you can begin to identify all the particles that come out. You look for the density of dots, which teaches you about the energy and the charge of the particle. You stick it in a big magnetic field, so that the tracks are curved, and you can learn about the energy and the momentum. You can identify particles and learn about their properties, which, of course, is really what you want to do, if you want to understand new particles like a pion. You want to know its mass, you want to know its charge, you want to know its spin; you want to know everything there is to know. How long does it live? How long is the track before it decays, in turn?

The energies increased. By the 1950s Berkley had what they called the Bevatron. The Bevatron was a super cyclotron which reached energies of billions of electron volts. In the metric conventions, that would be called a GeV, a giga is the metric prefix for billion. So a giga is like a gigabyte in your computer. That's a billion units of memory, and a billion electron volts is a lot. In fact, the Bevatron could reach six billion, six GeVs, and how much energy is that? Well, a proton has a rest-mass energy of approximately one billion electron volts, so six is more than enough to produce a proton and an anti-proton. Mr. Dirac, back in the 1920s, had an equation that predicted that every particle should have an anti-particle, and we had discovered anti-electrons early on, naturally, in cosmic rays, because electrons are so light. It doesn't take so much energy to make an electron-anti-electron pair. To make an anti-proton requires billions of electron volts, and this facility was actually designed specifically to have enough energy to do that; and surely enough, the anti-proton was there right where you expected it to be.

The bubble chamber invention, in 1952, really improved our ability to look at high-energy particles because of the high density of the liquid. The higher the density, the more likely it is that you'll see some interactions. There are more opportunities for the high-energy particle that's cruising through to interact. The story goes, by the way, that the bubble chamber was invented by a physicist named Glaser, who was sitting in a bar looking at his beer glass, watching the bubbles rising and sort of thinking about what was making those bubbles form. This gave him his inspiration for the bubble chamber. It's kind of a nice, a nice idea for future physics research.

As people built higher and higher energy machines and better and better detectors, the need for automating the detection process grew. Computers weren't really in big use at that early stage, but they quickly came into use. For example, instead of using a bubble chamber, where one needs to take a picture and develop it, then make measurements with a ruler to look at the image and try to categorize what particles he has seen, we now have devices where it's all completely automated. Instead of a bubble chamber, we have a device with wires, and as a charged particle goes through, producing a little track, the little ions will drift over—in a drift chamber—to those wires. Then there'll be an electrical signal which can be fed into a computer. Essentially the process became more automated, but fundamentally was still the same thing that was happening. It's just a little bit more sophisticated, and human beings aren't directly involved in the interpretation at the first stages.

So as we're discovering new particles, and we immediately started to discover new particles, not only the ones that we were expecting, (Mr. Yukawa's pion and the anti-proton were the two that we expected, but we started finding new particles. Now the naming convention was Greek letters; so there was the delta particle, the rho and the omega. There were first tens, then dozens, and later hundreds of new particles being discovered, sort of an explosion of information and factual data. At first there was no organizing principle. It was kind of an exciting time, but also a little bit overwhelming. There were all these new particles, so what do you do? Well, you carefully measure their electric charge, their mass, their lifetime, their spin, and their angular momentum. You try to characterize each particle. You try to characterize, how strongly it interacts, and you discover that almost all the particles we're talking about here interact very strongly, just as protons and neutrons interact very strongly. The pion, the carrier of the strong force, interacts very strongly.

When I say interacts very strongly, remember what that means in quantum mechanics. It means there's a high probability of interaction. If you send a pion into a chamber, it's going to interact just right away, and it's going to produce other particles. They're going to interact right away; and if they decay, they're going to decay right away. So these strongly interacting particles tended to be radioactive, which means they decayed, they turned into other particles, very quickly. These particles had very, very short lifetimes. This strong interaction, leading to a short lifetime, allows us to classify all of these particles as strongly interacting, so people started coming up with nomenclature.

Now, I don't want to burden you with nomenclature, and it's not so important, but it's useful, when you're talking about the developing story, to at least have a little bit of a framework of words to use to describe different categories of particles. It's like, when I go for a walk in the woods, I can't identify very many trees, but at least I can say these are deciduous, and these are coniferous. It's nice to have some words, just so that I can separate in my mind these different categories of trees.

For example, we're going to describe particles as either strongly interacting or not. Strongly interacting particles will be called hadrons, from the Greek word "hadros" which means heavy, or thick—actually not heavy, but thick. Hadrons is any particle—a proton, a neutron, a pion, anything that is being produced in these particle accelerators—that interacts strongly. The other, flip side is that you're either a hadron or you're a lepton. A lepton comes from the Greek word "leptos", which means light, and that would be like the electron, the neutrino or the muon. Those particles do not interact strongly. They tend to just go right on through, for instance, your bubble chamber, leaving a little track behind, if it's an electron or a muon, because they're electrically charged, but not really doing much. These are very rare interactions with the nucleus. So you've got hadrons and leptons. Now, as we studied the hadrons—that's mostly what people are studying in the late '40s through the '50s and onwards, that's what the particle accelerators were really focusing on—you begin to subcategorize those strongly interacting particles, and there were two big categories, two main categories of hadrons.

You're either a baryon or a meson. Every particle that's found is either a baryon or a meson, if it's strongly interacting. What's the difference? The baryons are kind of what I think of as particles. Baryon, from the Greek word "baryos," is heavy. Like the proton and the neutron, it's really a particle. If you measured its spin, it had spin one-half. Spin one-half is very characteristic of anything that you think of as a particle, as an object that makes up other objects. They are the baryons. The meson was like Mr. Yukawa's particle. Remember Mr. Yukawa invented this intermediary. It was called a meson, "mesos" for in the middle. It was middle in mass; it was in between the protons and neutrons. It has a different nature. When Mr. Yukawa proposed it, it was like a force carrier; not a particle, but the force carrier between the particles. It's still a particle, and so, as we begin to discover more and more mesons, it becomes a little bit difficult to just think of them as force carriers. We don't think of them that way at all anymore. That's sort of an old 1930s idea, which helped us to develop our more modern ideas that we're going to learn about in the future.

So what characterizes a meson? Well, the first ones that were discovered were intermediate in weight; at first, they tended to be lighter than the protons and neutrons. The pion, for example, has spin zero. It doesn't spin. It has no angular degree of freedom. You have a pion, it's just there. It has electric charge; it might be positive, negative or zero, but it's not rotating in any sense. It has no spin degree of freedom, is the fancy way of saying that, and you can produce pions like crazy. There are no conservation laws for pions. Remember, I said if you create a proton, you're going to have to create an anti-proton. You can't just create a positive electric charge out of nowhere. You have to create proton and anti-proton pairs together, but you can just make a single pion, especially if it's a—well, it has to be a neutral pion, because you can't produce electric charge. Pions, just as in making photons—if you jiggle an electric charge, it produces light. What's light? It's just a bunch of photons flying off. You can produce as many photons as you like. There's no conservation law for them. So mesons are these particles that are kind of like photons, except they're real; they're massive. They have charge; they're kind of like heavy versions of the pion.

Now, let's think a little bit more about what I'm saying, a heavy version of a particle. So you have the proton and the neutron. Those are the simplest, lightest baryons. But then you've got these new particles that are being discovered, and they have names such as the delta and the lambda, and the sigma, right, all these Greek letters. What's a delta? Well, in a certain sense, you can think of it as a big, heavy proton. If you produce one, it will just fall apart fairly quickly; it's radioactive, and it will turn into a proton and some other stuff. Some pions, for example, might go flying out to carry away the extra energy. So you have heavy versions, making a kind of a tower of more and more and more particles that are all baryons. Then you have the pion, which is the lightest kind of meson, and you have this tower of more and more and more other kinds of mesons, which themselves decay, usually into lighter mesons, and then ultimately into electrons and neutrinos and so on.

So this was the story that was developing in the 1950s. We were beginning to collect data, information, and it was pretty overwhelming, and people didn't quite know what to make of it. Think of the situation 100 years earlier, in the 1800s. People—chemists were the big players back then—were collecting data on all the elements: hydrogen, helium, carbon and nitrogen, but they were just collecting data. What was the mass, and what were the chemical properties? It's kind of an analogous story, and at first, the chemists didn't realize that there was a pattern to it. In the late 1800s, Mendeleyev discovered that the chemical elements actually fit into a beautiful periodic table. They were ordered. They were ordered because atoms are really simple things; they were built up of protons and neutrons, with electrons flying around them. There is a picture that we have nowadays that simplifies our understanding of those maybe hundred-odd elements that were being discovered.

So in the 1950s people were hunting for something similar. They were looking for a new periodic table to organize this zoo of particles so that life wouldn't be so complicated. And it, it's a little bit frustrating to think that these particles…they seem to be fundamental particles, and there are so many of them. I mean, it could be that nature's just complicated; we don't know. Maybe nature just has 100 or 200 fundamental particles. Now, you might say to me, "Well, I think the proton and neutron are fundamental, and these new ones like the delta, the lambda, and all those are radioactive. They're not really fundamental," and I would argue, "There's not really much of a difference. One particle can transform into another. A neutron can transform into a proton; a proton can transform into a neutron. They all, in a certain sense, are equally valid, equally fundamental particles. There's no real way that you can say this one is less fundamental than that one. They're just all there." They really do exist in nature, although most of the time you don't see these particles in cosmic rays. Every now and again you do, but you usually don't have enough energy in the cosmic rays. So mostly these are artificial creations; but they're still fundamental particles. We've created artificial, radioactive elements, such as plutonium, and they are real. You could argue that we've created them; they not usually present in nature. Nonetheless, I would argue that they are elements like any other element.

So we've got this big "zoo" of particles, and in the next lecture, I'm going to talk about the first attempts by the theorists to try to organize this story. Ultimately we're going to discover that the story is really very simple. Nature is, at its core, beautiful and simple, and all this zoo can be understood, once again, in terms of just three fundamental, underlying particles. It took another ten years for people to figure that out, so that's where we're headed, down into the sub-structure of these fundamental particles.

Lecture Nine
The Particle "Zoo"

Scope: As the accelerators churned out new particles, many were easy to understand and categorize, but some appeared especially strange, with a curious mix of strong and weak properties that seemed, at first, to be mutually contradictory. The proper description of these *strange particles* was a turning point in our developing understanding of the zoo of particles. How many particles exist? How can we keep track of, and make sense of, this zoo of particles? New words were coined for new particles and concepts, including *mesons* and *baryons*, *hadrons* and *leptons*, *bosons* and *fermions*, *flavor* and *families*. The words are alien and unfamiliar, but they help organize our thinking about the world. We'll talk about these words and try to connect them with the stories and particles we've already learned about to give them form and context. What happened to our theme of "simplicity"? What is the meaning of "fundamental" if there are hundreds of fundamental particles? We will also see how the idea of a *conservation law* helped to organize these particles, along with the various properties, often referred to as *quantum numbers*, of the individual new particles.

> Physics is simple, but subtle.
> —Attributed to P. Ehrenfest

Outline

I. In the 1950s–1960s, particle physics was experimentally driven.

   A. The advent of particle accelerators enabled physicists to produce whole families of new particles and observe their tracks in bubble chambers and more sophisticated detectors.

   B. This lecture addresses these families of particles and the first steps toward organizing the wide range of data that were being collected at the time.

II. In the 1950s, some unusual tracks began to appear in bubble chamber images. The particles that left these tracks were called *strange* *particles*, because scientists could make no sense of some aspects of the tracks.

A. First, some V-shaped tracks appeared in the images, caused by what was then termed a *V particle.*

1. The V particle seemed to appear out of nowhere, but it could have been generated by the decay of an electrically neutral particle that left no track in the bubble chamber.

2. More puzzling was the question of why these tracks took so long to appear. Producing particles in a bubble chamber usually results in a strong interaction; producing a neutral particle should result in a rapid decay. Why, then, is the V particle so widely separated from anything else?

3. Another strange aspect of these V tracks was that they almost always appeared in pairs. This phenomenon is called *associated production.*

B. The first part of the explanation for these strange particles came from a young theorist named Murray Gell-Mann.

1. Gell-Mann proposed that fundamental particles might have another property besides the ones that had already been considered, which were mass, charge, and spin. Note that the measures for all these properties are called *quantum numbers.*

2. Gell-Mann proposed that there was one more quantum number associated with these particles, which he called *strangeness.*

3. For example, strangeness 0 meant that the particle was "normal," such as a proton, neutron, or pion. The V particle, an exotic new particle produced with higher energies, has strangeness 1.

C. How did Gell-Mann's proposition explain the associated pairs of the V tracks?

1. When a strong interaction takes place, such as a pion hitting a proton, and an event occurs, such as a transformation or the creation of a new particle, the strong force preserves total strangeness. In other words, if an interaction begins with strangeness 0, it must end with strangeness 0.

2. This conservation of strangeness is an idea that is easy to accept because it follows similar ideas of conservation. We know, for example, that energy is conserved in an interaction, as is momentum and electric charge.

3. This concept explains the associated pairs of the V particle. One particle will have a strangeness number of +1, and one will have strangeness −1.

D. How did Gell-Mann's idea explain why these particles have such long lifetimes?

    1. The strong interaction of nature conserves strangeness, but the weak force does not. The weak force transforms a neutron into a proton; it can also transform a strange particle into a non-strange particle.

    2. This explains why the particles last so long. Once a strange particle has been produced, it cannot decay strongly if it doesn't have enough energy to produce other, lighter particles with nonzero strangeness. (The strong force must conserve strangeness.) The particle must "wait," then, for the improbable weak decay to occur, which takes a longer time.

E. Throughout the 1950s, Gell-Mann's theory was tested and shown to fit with a wealth of data. As a conservation theory, it was useful in trying to understand complex phenomena.

    1. Under the principle of conservation of electric charge, if a neutron strikes a proton, the resulting charge is $0 + 1 = 1$. That event may also produce some new particles, but the production is limited by conservation of charge; the *total* charge produced must equal 1, which it started with.

    2. In the same way, the conservation of strangeness limited the possible outcomes of these accelerator experiments.

III. Let's briefly review the naming conventions for these particles.

A. Strongly interacting particles are called *hadrons*, including protons, neutrons, pions, and strange particles.

B. Weakly interacting particles are called *leptons*, including electrons, muons, and neutrinos.

C. Hadrons are further subdivided into *baryons* or *mesons*.

    1. Baryons are particle-like. For example, the proton and neutron, which are baryons, serve as building blocks of nuclei. They also have half-integer spin.

    2. Mesons were proposed by Yukawa as force carriers. After the pion came heavier versions of pions, such as kaons and rho particles. All mesons have integer amounts of spin.

D. Each of these particles can still have various other quantum numbers. For example, some mesons may have strangeness 0, some may have strangeness +1, and so on.

E. Another way to categorize these particles is based solely on spin. Although this concept is very complex, for our purposes, we only need to realize that every particle of nature has either half-integer or integer spin. This fact is significant.

   1. Particles with half-integer spin are called *fermions*. Again, fermions are what we tend to think of as particles, such as electrons and muons.

   2. Particles with integer spin are called *bosons*, named after an Indian physicist of the early 20th century, Satyendranath Bose.

   a. This physicist argued that any particles that have integer spin, even two identical particles, can exist right on top of each other.

   b. The most famous example of a boson is the photon, the "particle" of light. We can easily visualize light sitting on top of other light; the result is just brighter light.

   c. Fermions are the opposite; they cannot be made to exist in the same place at the same time.

IV. The goal in the 1950s was to make some sense of the data from all these different particles. In fact, scientists were beginning to observe deeper patterns, which served as hints that some part of the story of physics was still missing.

A. For example, consider the proton and neutron, which are electrically different but, in every other respect, almost identical. We call such particle "partners" *doublets*.

B. Later, we learned that there are many particles with similar partnerships. For example, the three kinds of pions, pi+, pi–, and pi 0, form a triplet.

C. When we look at particles with strangeness, we discover a strange meson, called the *kaon*.

   1. The kaon can have various strangeness numbers. In fact, there are four different kaons, a positive one, a negative one, and two neutral ones.

   2. What's the different between the two neutral kaons? One of them has positive strangeness and one has negative strangeness. The two neutral kaons are a kaon and an anti-kaon.

D. How can we explain these doublets, triplets, quadruplets, and so on? Gell-Mann created an organizational scheme to assign numbers and categorize these particles. He called his scheme the *eight-fold way*, a tongue-in-cheek reference to Eastern philosophical writings, but it was still a description of the data rather than an explanation.

E. Later, Gell-Mann came up with an idea to explain what seemed like the overwhelming complexity of these particles. He suggested that all these strongly interacting particles, such as the kaons, pions, and protons, are constructed of three simple building blocks, which he called *quarks* after a phrase in James Joyce's *Finnegan's Wake*.

F. As we get further into this course, we will discuss the theory and phenomenology of quarks, which are the ultimate building blocks.

G. Before we begin our discussion of quarks, however, we must leave the world of particles briefly for a look at forces. In the next lecture, we will further develop the idea of force as mediated by a particle.

Essential Reading:

Barnett, Muhry, and Quinn, *The Charm of Strange Quarks*, chapter 4.2.

Schwarz, *A Tour of the Subatomic Zoo*, chapter 4.

'l Hooft, *In Search of the Ultimate Building Blocks*, chapter 5.

Recommended Reading:

Calle, *Superstrings and Other Things*, chapter 24, sections on "Particle Classifications," "Conservation Laws," and "Strange Particles,"

Lederman, *The God Particle*, chapter 7 ("The Strong Force").

Riordan, *The Hunting of the Quark*, chapter 4.

Questions to Consider:

1. The following nuclear reaction is observed to occur frequently (it is a *strong* interaction):

   pi– + p => K+ + Sigma–.

What this means in words is that an incoming pion with charge –1 (that's the pi–) can strike a proton (charge +1), and the result is a positively charged kaon (the K+) and a negatively charged Sigma particle.

(a) Verify explicitly that this reaction is not forbidden by charge conservation.

(b) The K+ meson has strangeness number +1. What can you conclude about the strangeness number of a Sigma– particle?

(c) Later, the K+ meson is seen to decay into a pi+ and a pi0 (one positively charged pion and one neutral pion). Could this reaction be a "strong" decay, which happens quickly (meaning that the K+ will have a very short track; the decay will happen immediately), or is it a "weak" decay, which means that it takes a long time (and, thus, the K+ meson should leave a long track)? Why and how do you know?

2. Consider the following four particles: a neutrino, a positive pion, a proton, and a positive kaon, or K+ (see question 1 for more information about the K+). Which one(s) is (are) a hadron? a lepton? a meson? a baryon? a fermion? a boson?

3. Consider the following nuclear reactions. Based only on the principle of charge conservation, which one(s) are allowed and which are forbidden to occur? Are any that are allowed by conservation forbidden based on the principle of baryon number conservation?

Note: *n* means neutron, *p* means proton, *e* means electron, *pi* means pion. Protons have charge +1, electrons have charge –1, neutrinos are neutral. For all other particles, the electric charge is written as part of the name. For example, *pi+* means a pion with charge +1, *pi–* means a pion with charge –1, and *pi0* means a neutral pion, that is, with charge 0.

    (a)   e + p => neutrino + n
    (b)   e + n => neutrino + n + pi–
    (c)   e + n => anti-proton + n
    (d)   e + p => anti-proton + proton
    (e)   e + p => pi+ + pi0

Lecture Nine—Transcript
The Particle "Zoo"

Particle physics in the 1950s and the early 1960s was definitely experiment driven. There were these new accelerators which were being built, mostly by the U.S. government, and some were being built in Europe, where people were producing new particles. You would see tracks in bubble chamber pictures or more sophisticated detectors. People were beginning to discover that there was more to the world than just protons, neutrons, and electrons, even more than adding in the muon and the pion. There were whole families of new particles which were more massive and which required a great deal of energy to produce. They were radioactive and would decay quite rapidly; people were a little bit overwhelmed. What were these things? So they started just to categorize. It was, in a certain sense, analogous to what was going on when Tycho Brahe, in the 1500s, was looking at the motion of the planets, just trying to collect the data, hoping that somebody like Johannes Kepler might come along later and start to explain what was going on.

In today's lecture we're going to talk about the "zoo of particles" and the first steps towards organizing and understanding what was going on. We're going to talk about a puzzle that was solved, which helped us to organize these particles. That puzzle was kind of another, "Who ordered that?" sort of situation, something that had not been expected. There wasn't any obvious reason for it, but it was a principle that allowed us to begin to make some sense of this really overwhelmingly complicated zoo of particles. These heavy things that decayed, and turned into protons, and neutrons and pions, and electrons—the kind of ordinary stuff that the regular world is made of.

Around the 1950s, some rather unusual tracks started showing up in the bubble chamber pictures. They were called strange particles, because at first there were some just curious aspects of these tracks that didn't really make sense. There were a couple of curious aspects. First of all, these particle tracks, and here's a bubble chamber picture where you can see nothing, and then a V which just appears. This was called a V particle at first, and how can you explain that, some particles just appearing out of nowhere? Well, that's not so hard to explain. Most likely, there was some electrically neutral particle that had been produced earlier, and was traveling through your bubble chamber, but since was electrically neutral, maybe it wasn't interacting, wasn't ionizing; then it fell apart. It radioactively decayed.

What's puzzling is, why did it take so long? When you produce a particle in a bubble chamber, it's usually a strong interaction of nature. It's protons whacking into other protons, and neutrons, and pions. These are all very strong forces of nature, highly probable events; and what you expect, if you produce a neutral particle, is that it should decay very quickly. So what's unusual is that that V is so widely separated from anything else in the picture. Apparently you have some long-lived particle, and that seems a little strange. If you produce a particle strongly, it should decay strongly because it's a strongly interacting particle. That was the idea. That was the worldview that people had; that was the puzzle that we were facing. There was an additional strange element of this story, which was that when you see one V track, you almost always see another V somewhere else in your bubble chamber image. Sometimes it's off to the side of the image, but usually they come in pairs. This was called associated production. So if you're making one strange event, you usually make two of them, and that also was a puzzle that really just couldn't be explained in any obvious way.

The solution to this puzzle, at least the first solution, came from a young theorist named Murray Gell-Mann. Murray Gell-Mann turned out to be one of the big players in this story. As events progressed, he was the one who really helped us make sense of this zoo of particles, and here was his proposition. He said, "Suppose that fundamental particles have another characteristic, another property, besides all the ones we've considered. What have we considered so far? We have mass, we have electric charge, we have spin. How much angular momentum do they have?" Those are properties that we can tabulate. We can write down the numbers. Every particle has a certain amount of all of those things. We call those things quantum numbers, because they're numbers that describe and identify a particle. Murray Gell-Mann proposed that there was one more quantum number associated with every particle that we'd never thought of before, and he gave it the lovely name of "strangeness," because these were strange particles. So he said, "You can have a certain amount of strangeness as a particle." You could have strangeness zero, and that means you're normal. That's like a proton, or a neutron or a pion, or any of the particles that had been known up to this point. They have zero strangeness. But then, he proposed that there was a new particle being produced. Because we've gone up to higher energies, and we could make new exotic particles, that was strangeness one. It's a quantum number, so it's going to come in chunks.

This was Murray Gell-Mann's hypothesis, and then you had to see whether it matched the data or not.

Now, how did this explain what was going on? Here is his proposal. When you have a strong interaction, like pions whacking into protons, and something happens—there might be a transformation, you might produce a new object, but it's happening because of the strong force—his proposal says, the strong force preserves total strangeness. If you start with zero, you have to end with zero. Okay? This conservation of strangeness was a very easily acceptable idea at that time, because we knew that energy was conserved. The total energy coming in had to equal the total energy going out. We knew that total momentum was conserved; we know that total electric charge, another quantum number, was conserved. If you start off with no charge, you might produce a whole bunch of particles, some pluses, some minuses; but if you add them all up, it has to add up to what you started with. So, since all those things were conserved, maybe this new quantum number was also conserved by the strong force. So if you're going to produce a strange particle, you're going to also have to produce, at the same time, a particle with strangeness negative one. That's why you're getting this associated production; you always have to make them in pairs. One will be strangeness plus one and one will be strangeness minus one; they add up to zero, so you preserve total strangeness. So that explains why the particles would have to come in pairs; but now why do they live so long? Why were those V tracks so isolated? Here's Murray Gell-Mann's answer.

Even though the strong interaction conserves strangeness, maybe the weak force of nature does not. Okay? Different forces can act differently. After all, the weak force was already known to be kind of a bizarre force. The weak force transforms a neutron into a proton. That's a pretty radical change in character of a particle, and that's only caused by the weak force of nature. So maybe the weak force can also change a strange particle into a non-strange particle. It can transform strangeness into non, or non into strangeness, so that explains why it lasts so long. Because once you've produced this strange particle, how's it going to decay? Okay, it's massive, it wants to radioactively decay; but it can't decay strongly, because if it's a strong decay, and you have strangeness one, you have to produce particles, and at least one of them has to have strangeness. There are no other light particles that have strangeness. All right? We are producing the first, very lightest particles with strangeness here. There is nothing lighter that carries strangeness, so you're kind of stuck. You're just traveling along. You want to decay strongly, but you can't, because you can't get rid of your strangeness. So you have to wait until this improbable weak decay occurs. Improbable means you live a long time.

Now, you can get quantitative with this idea, because with weak lifetimes, you can actually do some estimates. You go all the way back to the theory of weak interactions written down by Enrico Fermi in the 1930s. Remember, that was the theory that involved neutrinos, and it predicted quantitatively what kinds of weak decays can happen, and how strongly or how weakly, they can occur. So you can plug the numbers into that theory, and surely enough, the lifetime that you're observing in these tracks is what we would call a weak lifetime. It corresponds very nicely, so this picture is coherent, and it matches up with these so-called associated production of strange particles.

Now, you've produced various different strange particles. Apparently, one of them had strangeness one, lived a long time and then decayed. Another had strangeness negative one, lived for a while, and decayed. So now, whenever we're tabulating particles, we have to tabulate not just charge, mass, and spin but also strangeness number, and make sure that this story is coherent. All right? You make up a story like this, and it explains one or two pieces of data, but you're going to throw it away if that's all it does. All right? Especially if it's such a radical proposition, a whole new quantum number. You're only going to keep this idea if you can start to explain, not just these events, but all events. So the search was on. As the energies went up and new particles were discovered, you would look to see, does this particle get created strongly and then decay weakly? Then it's a candidate for having strangeness. Look for the interactions of those particles with other particles, and make sure that strangeness is always preserved in all strong interactions, those that occur with high probability, and that it's violated only in those that occur weakly. It worked like a champ. It worked quantitatively. It began to assimilate a vast wealth of data. Throughout the 1950s, there were many accelerators with lots and lots of particles being discovered. Hundreds, by now, had been discovered, and this story, this idea of strangeness, was really fitting in well. So it looked as though an organizing principle had been added to the story.

Conservation laws are very useful when you're trying to understand complex phenomena. Let's just think about that for a second. If you have a principle such as conservation of electric charge, it says that if a neutron strikes a proton—so you've got total charge zero plus one is one—you might produce a whole bunch of particles. At first you throw up your hands and say, "Well, anything can happen. This is quantum mechanics. If it can happen, it will." All right? Particles happen, and the conservation law says, "Well, sorry, not anything can happen." You might produce some pions, but you have to make sure that the sum total charge is equal to zero, so you're limiting the possibilities of nature by a conservation law. This new conservation law is making some simplicity, some order, out of the complexity of everything that could happen, only a more limited number of events can occur. So that's nice.

As people began to look further, they discovered a few new particles, very massive, which had two units of strangeness. Strangeness two, or minus two, and again, they fit in beautifully. If a strangeness two particle was created strongly, it would have to be involved in a process where you either had two strange particles coming in, or two strange particles going out; strangeness was conserved no matter what. So the principle continued to work. The weak force is getting a little bit weirder. All right? The weak force can now violate a conservation law, and we will see, as we go on in future lectures, the weak force has a history of being weird. It violates lots of conservation laws. It's just an unusual force of nature. It's the force associated with neutrinos, which are also kind of unusual particles, very exotic.

All these particles…the naming conventions are a little confusing. I want once again to reiterate some of the various naming conventions. It's not that you necessarily have to memorize these words, but I will, from time to time, be talking about the types of particles we have, and it's nice to just have a sense of the scheme. So we've got strongly interacting particles, or not. The strongly interacting particles are going to be called the hadrons; and the weakly, or not strongly, interacting particles are going to be called leptons. The leptons include an electron, or a muon or a neutrino. Those don't interact with the nuclei very strongly. Everything else, all the stuff you're seeing in bubble chambers, and all the strongly interacting particles: protons, neutrons and pions, etc., including all these strange particles, are all hadrons.

Hadrons are subdivided into two categories. Hadrons can either be baryons or mesons. Baryons are, kind of, like particles. I think of them as particle-like, the proton and the neutron. They build stuff. Baryons have spin one-half, or spin three-halves. They have what we call a half-integer of spin. That might seem like a funny coincidence, but it turns out there are some deep theoretical reasons why this is the case. But in the 1950s, it was just an observation. These things, like protons, and neutrons, and delta particles, the ones that look like heavy protons and neutrons and decay into protons and neutrons, they all have spin one-half, or spin three-halves, or maybe even five-halves, but never anything that's not an odd number divided by two. So that's just a, sort of a weird fact at the time, which we'll come back to later.

The mesons were first invented by a theorist, Mr. Yukawa, in the 1930s. He proposed the existence of mesons as a way to explain forces. They were like the force carrier of the strong force. Well, at first, we only had the pion, but then there came all these heavy versions of pions. The kaons, and the rho particles, and a whole bunch of them that were like heavy versions that could decay into other mesons or other particles. All of these mesons had either spin zero, like the pion; some of them had spin one, or spin two. They were always integers, pure integers. All mesons seemed to have integer amounts of spin. So there was some order in this chaos. There were certain patterns beginning to appear, and in the 1950s, it was still a little bit overwhelming, but people were hunting. They were trying to find some organizing schemes, anything so that we didn't just have to have a random collection of data in a table that was the only way of understanding the world. That would really have been a shame.

When you're organizing particles, each particle can still have quantum numbers. So, for example, you have various kinds of mesons. Some of them might have strangeness zero, some of them might have strangeness plus one and some of them might have strangeness minus one. They're still all potentially mesons, if they have integer spin, if they can be produced in any numbers; there are various characteristics that define what is a meson or a baryon. So you can talk about the strange mesons or the non-strange mesons, and the strange baryons or the non-strange baryons. So we're beginning to categorize in this way.

There's another way of categorizing that people think about, based on this idea of spin. At first, it wasn't clear exactly why spin was so important. I'm never really going to be able to explain to you deeply, mathematically why spin is so important, because it's pretty formal mathematics that connects spin to the properties of particles. But it was observed that every particle in nature, every particle, either had half-integer spin—one-half, three-halves, five-halves—or integer spin of zero, one, two or three. That's a different way of categorizing particles. If you have spin one-half, then you're called a fermion, named after Mr. Fermi, in honor of Mr. Fermi. Any particle of nature that has spin one-half or three-halves is called a fermion—whether you interact strongly or weakly, an electron or a proton, they're all fermions. I generally think of fermions as particles. Electrons, muons, protons, neutrons all have spin one-half, and I think of them as little fundamental particles.

If you're not a fermion, you're a boson. I love that word. It's a little bit of a silly name, although it's named in honor of a great physicist whose name is hard to pronounce, Satyendranath Bose. He was an Indian physicist in the early part of the century. It was in the 1920s, I believe, when Bose was did his theoretical work. He argued, mathematically, that particles with spin zero or one or two, integer spin, should behave very, very differently than particles with spin one-half. He said that if you have two particles of integer spin, even identical particles, that they can sit right on top of each other, and they're not only perfectly happy, they actually like it. It's more probable for bosons, any particle with integer spin, to sit on top of identical bosons. That's a crazy idea. We're talking particles here. How can two particles sit on top of each other? Well, bosons are unusual particles. The most famous example, my favorite example of a boson, is the photon, the particle of light. And, yes, it's a particle but it's also a wave. I think it's not so hard to visualize light sitting on top of other light; it's just brighter light. In fact, that's exactly what a laser beam is, a bunch of photons, which are identical, which sit right on top of each other, in space and in time. You can pile on more and more and more. The more intense your laser beam is, the more bosons you have put together. So photons, it turns out, have spin one; photons do carry angular momentum. And because it's an integer, they can sit on top of each other. This is what Mr. Bose discovered, and hence the name "bosons". The name was given to these particles by Mr. Dirac, who was also thinking about a theory of bosons and fermions in later years.

Fermions, spin one-half, are the exact opposite. They're the antisocial ones. They don't like to sit on top of each other. In fact, if they're truly identical, you cannot make them sit in the same place at the same time. That's Mr. Pauli's Exclusion Principle, and it's a very deep, mathematical property of quantum mechanics. So I think that's why I think of spin one-half, or fermions, as "particle-y", because they don't want to sit on top of each other. That's my image of particles; that they can't be on top of each other. Bosons are more like photons, so I think of them more as force carriers; except, you know, is the pion, and its exotic heavy partners. Is it a particle or is it a wave? Well, it's a boson. It doesn't make sense to ask is it a particle or is it a wave in quantum mechanics. Everything is both. Stuff—you know, it hurts your brain a little bit. It's hard to visualize, but at least there is some mathematical order and organizational scheme to all of this stuff.

So what about simplicity? All right, we began arguing that the goal, in the 1950s, was to make some sense of this pattern of particles. At first was not even a pattern; there were just masses and charges. It's just numbers. It's like collecting data, we were beginning to see some organizational schemes. We were kind of, the physicists of the 1950s, acting a little bit like Johannes Kepler, who figured out from the data that particles move in elliptical orbits. We were learning that particles had new quantum numbers, such as strangeness; they could be bosons or fermions. There were categorizations.

In fact, there were some deeper patterns that people were beginning to observe, and these deeper patterns were really hints that there was some part of the story that was still missing. For example, you had the proton and the neutron. These two particles which are, in almost every respect, identical. Okay? They do have different electric charge, so they're electrically different. But as far as these strong interactions are concerned, a proton attracts a proton, or a proton attracts a neutron in basically the same way, in the strength of their interaction, their properties and their masses; everything's almost nearly identical. So we say they're a doublet; two particles that are intimate partners. You start looking around in nature, and you discover there are many such partners, for instance, Mr. Yukawa's pion. Mr. Yukawa predicted three pions: a pi plus, positively charged, a pi minus, negatively charged, and a pi zero, electrically neutral. They're three particles, a triplet of particles which are, in almost every respect, identical. They have almost exactly the same mass, the same spin and the same basic strong interaction properties, but they are distinguishable because of their different charges. This pattern continued.

When they looked at these particles with strangeness, they discovered that, for example, that there was a strange meson called the kaon. It was called the kaon, not a Greek letter, because it was discovered in an American laboratory, and somebody noticed that many times this particle would leave a track in the bubble chamber. Where one particle was came in, three particles went out, and they looked like the letter "K". So, right? You get to name the particle if you discover it, so they called it the kaon. People were getting a little bit whimsical, and it will only get worse as time goes on. The kaon is a meson; it's kind of like a heavy version of a pion, and this kaon has strangeness number. Well, it can have various different strangeness numbers. In fact, it turns out that there are four different kaons: a positive kaon, a negative kaon, a neutral kaon. Now you're asking, what on Earth could the fourth one be, and the answer is, another neutral one. So what's different between the two neutral kaons? One of them has positive strangeness; one of them has negative strangeness. It's a kaon and an anti-kaon! Okay? If you have a particle and its anti-particle, they have opposite electric charges; they will also have opposite strangeness number. So, in this way, people were discovering a quadruplet of particles that were intimately connected. They were very close in mass and in properties. So what was going on? How could they explain these triplets and quadruplets? In fact, people discovered octets of particles, and decaplets, so there was, there's some order coming out.

Murray Gell-Mann came up with sort of an organizational scheme, in which at least you can assign numbers and categorize particles as to whether they belong in a doublet or a decaplet. He called his scheme the "eight-fold way." It's kind of a tongue-in-cheek reference to some old Eastern philosophical writings of the eight-fold way, and I think he later came to regret this use of a mystical term for physics because people jumped on it. Right? This was in the late 1950s and early '60s. I think people began to connect particle physics with mysticism in the popular literature, in a way that I think pretty much annoyed the particle physicists because they were just collecting data. There's nothing mystical or magical about what they were collecting. There were tracks in a bubble chamber, and they were just organizing them, but it just seemed like a cool name to him at the time. Murray Gell-Mann's eight-fold way is still not an explanation of anything. It's kind of like Johannes Kepler's, observing that certain organizing principles, such as planets, move in ellipses or that the radius of the planet is related to the period; the radius, meaning the distance to the sun, is related to the amount of time it takes to go around. He was just noticing order in the complexity of the data. Still, nobody has an underlying story, an explanation for why are there four kaons or why there all these particles.

Ultimately it was Murray Gell-Mann, actually not alone, who came up with an explanation, an underlying picture. There were lots of ideas. There were people, for example, who argued, that it was just nature, and nature is complicated. There was a program of theoretical research called Nuclear Democracy, where every fundamental particle that we had discovered was equal in every respect; it was a democracy. It meant that no particle was better than any other particle, that all particles could be thought of as part of this web of fundamental particles. It was a neat idea, right? There's always some connection between the scientific ideas that you have, and the politics and worldview of the scientists who are trying to be creative, who are trying to understand what's going on. That idea ultimately failed, and Murray Gell-Mann's idea, which was more "reductionistic", said that we can explain this complexity in terms of something underneath, and that ultimately has proven to be a much more valid and descriptive way of understanding the zoo.

Murray Gell-Mann suggested that, perhaps, all of these particles—the kaons, and the pions, and the protons, all the strongly interacting particles—were built of three simple building blocks. He gave them the name "quark." It's a whimsical name, and I'm not quite sure why he picked that name. He says that he thought they were kind of quirky, to propose something brand new, something smaller than anything that had ever been seen. He was a fairly literate man and was reading James Joyce's *Finnegan's Wake*. There's a passage in *Finnegan's Wake* that says, "three quarks for Muster Mark," and there were three quarks in his model; anyway, that's the term that he picked. By the way, if you're a literature fan, and you look at that passage, you might argue that it should be pronounced "kwark", because it rhymes with "Mark"; but Murray Gell-Mann invented the particle, and he insists that it's "quork". He invented it, so he gets to tell you how to pronounce it, and that's what we call it.

The quarks, then, are the ultimate building blocks. In our next lecture, we're going to really talk about the theory of quarks and how we make sense of the zoo by postulating the existence of three particles. They're going to be strongly interacting. You're going to put quarks together, and a proton, for example, is going to be a combination of some quarks. Just as an atom can be understood as being built of simpler things, a nucleus and some electrons. So that's a lovely organizing and simplifying principle. In fact, one of these three quarks, or rather, the sort of exotic one, is going to be called a strange quark. And so what is strangeness number? It's how many strange quarks are inside of you. So if you have a particle that's observed, and it was already tabulated to have strangeness one, we will understand that what that really means is that one of its quarks is an "S" quark, or a strange quark. If you've got strangeness two, it means two of your quarks are strange quarks. So the strong interaction can create a strange quark and an anti-strange quark, and that's how it is that we are going to conserve strangeness numbers. We're going to understand all the phenomenology, all this orderliness that was beginning to be developed in the 1950s, by proposing this simplifying idea.

In order to really understand quarks, and the strong force that binds them together, we really have to take a break, for a lecture, from all these particles, and talk about forces. Now, I've been talking about forces in a rather casual way so far. I've introduced the idea of force as being mediated by a particle. You create a particle, you toss it over, and the other particle absorbs it. That's a way of thinking about forces, and it's useful to try to develop that a little bit more formally.

In the 1950s, a lot of physicists were working on this idea, and they were, in particular, trying to understand the simplest force at that time, which was electromagnetism. Electricity and magnetism was sort of the oldest and dearest force to everybody, well understood already in the 1800s. So this theory of fundamental forces developed, and there were a lot of trials and tribulations. We really need to talk about how that all works, and then we'll put everything together—the nature of the forces and the nature of the particles—and talk about the quark model of the strongly interacting particles.

Lecture Ten
Fields and Forces

Scope: This lecture examines the concept of a field, leading to the story of modern QED, the fully quantum theory of light and particles. We look at the early trials and tribulations, the idea of a "field of particles," the 1947 Shelter Island Conference, and the concept of *renormalization*. We will also examine Feynman diagrams, the most widely used conceptual tool of particle physics. QED is perhaps the greatest theory mankind has ever produced, vindicated by quantitative tests and stunning accuracy.

In 1968, one of the participants of the Shelter Island Conference, theoretician Abraham Pais, described the state of particle physics as:

> A state not unlike the one in a symphony hall a while before the start of the concert. On the podium one will see some but not all of the musicians. They are tuning up. Short brilliant passages are heard on some instruments; improvisations elsewhere; some wrong notes too. There is a sense of anticipation for the moment when the symphony starts.

> What I am going to tell you about is what we teach our physics students in the third or fourth year of graduate school … It is my task to convince you not to turn away because you don't understand it. You see my physics students don't understand it … That is because I don't understand it. Nobody does.
> —Richard Feynman (*QED*, p. 9)

Outline

I. In this lecture, we focus on the *theory* of particle physics, how we understand interactions between objects in the world. In particular, we will examine a way of thinking about forces that is between Isaac Newton's somewhat naïve idea of action at a distance and the idea that force is an exchange of virtual particles.

    A. The intermediate step between these two ideas was very important and, in fact, has been integrated into our quantum picture today.

This idea is called the concept of a *field*, also known as *field theory*.

B. In the 1800s, electricity was explained as being similar to gravity: Two objects are far apart and, somehow, they feel an electrical force, action at a distance.

   1. In trying to visualize the phenomena of electricity and magnetism, Michael Faraday, a British physicist of the mid-1800s, drew pictures—field lines—of the way he imagined the forces would manifest themselves.

   2. You may have seen similar pictures or performed similar experiments in elementary school science. For example, you place a piece of paper on top of a bar magnet, then sprinkle some iron filings on top of the paper. The iron filings spread out and form curved lines that go from the south pole to the north pole of the magnet. This experiment helps you visualize lines of magnetic force, spreading out through space.

   3. If you were to go off to the side of your work space with a sensitive compass needle, you could detect the compass needle deflecting because of the existence of this magnetic force field.

   4. Faraday thought about field lines as being real. He imagined that if he held up a magnet, magnetic fields created by the magnet truly exist in the world, everywhere in space.

   5. This idea of a force field explains how a charge that is far away can feel the force. The first charge makes the field, which exists in empty space, whether or not a second charge is present. If a second charge is added, it will react to the field, either attracting or repelling.

C. Later theorists wondered whether this idea was generalizable: Could it be incorporated into the theory of quantum mechanics?

   1. Dirac's equation, which married relativity and quantum mechanics, was the description of electrons in the presence of electric fields. It was the first primitive attempt at quantum electrodynamics, or a quantum theory of electricity and why things move under the influence of electromagnetic forces.

   2. Yet Dirac's equation didn't incorporate the field theory. Dirac was thinking more about the electron as a particle experiencing some electrical forces. Even so, Dirac's theory worked well without incorporating the concept of a field.

D. What happens if we take into account field theory?

    1. Using Dirac's theory alone, we can calculate the energy of light, or the *spectrum*, of radiation that is emitted by a hydrogen atom.

    2. If we include the idea of the field in these calculations, we must imagine that the space between the proton and the electron in the hydrogen atom is no longer empty. It's filled with an electric field.

    3. Now that we're thinking quantum mechanically, we realize that electric fields carry an energy density and, thus, we can borrow a bit of that energy to create, for example, a virtual electron and an anti-electron. Heisenberg's uncertainty principle says that we can pull a particle/anti-particle pair out of the vacuum (if only briefly).

    4. The problem that we originally thought was pretty simple, involving one electron and one proton, becomes more complicated because we've added an electron and an anti-electron.

    5. Further, if we can produce one particle/anti-particle pair in the problem, how about two or 10 or a billion? All of a sudden, the vacuum inside and around a hydrogen atom is a very complicated place. We call these virtual particles *vacuum fluctuations*.

    6. What are the consequences of the presence of an electron-positron pair? Imagine a positive charge in the middle and a negative charge that's attracted to it on the outside. In the middle, we have a (virtual) + and a (virtual) − charge. The plus will tend to shift toward the electron, and the minus will tend to shift toward the proton in the middle. The force the electron feels will change, because the electron is reacting to a different configuration of charges nearby.

    7. To determine the energy of that electron, we have to make this correction to our earlier estimate. The correction is numerically small, but we must account for an infinite number of small corrections.

E. That infinite number of corrections was a disaster for physicists in the late 1920s–1930s. When physicists tried to calculate the net total effect of all those vacuum fluctuations of the quantum fields, they discovered that the total effect on the electron, this infinite sum of progressively smaller and smaller effects, grew to be infinitely large. There was a sense that quantum field theory might not be the answer for describing nature.

II. After World War II, at the Shelter Island Conference of 1947, a group of physicists got together to share the problems they were working on and decide what they should look at in the future.

  A. An experimentalist at the conference named Willis Lamb presented an experiment he had done using radar technology from the war to measure the light emitted by hydrogen. Lamb made an unexpected discovery: One of the colored lines that comes out of hydrogen is, in fact, two colors that are very closely spaced, two frequencies that are very nearly the same. This phenomenon is called *Lamb splitting*.

  B. Another physicist at the conference, Hans Bethe, had heard in advance of Lamb's experimental work and began thinking of it in conjunction with vacuum fluctuations. Bethe acknowledged that he couldn't deal with an infinite number of vacuum fluctuations, but he could deal with just a few. What if he had an e+/e− pair that was produced in the vacuum? What would that do to modify the answer from the calculations about the light emitted from hydrogen? As you might guess, it splits the line.

  C. The outcome of the conference was Lamb's radical result, which was in contradiction with quantum mechanics, and Bethe's hint that a quantum field theory was needed to explain the data. Many scientists started thinking about this problem of how to make quantum electrodynamics a full-fledged quantum field theory.

  D. Among the other participants at the conference were Richard Feynman and Julian Schwinger. Together with Tomonaga, a Japanese physicist, they were awarded the Nobel Prize in 1965 for their work on the development of a successful quantum field theory.

III. We will try to understand the breakthrough that these scientists made— if not the details, then at least the basic premise.

  A. Here's the original problem: We start with a proton in the middle. It creates a field, and that field is modified by the presence of virtual fluctuations, e+/e− pairs popping up and disappearing again. We must add up all possible contributions for one pair, two pairs, 10 pairs, and so on, but when we add them up, we get infinity.

B. The inventors of QED thought that the problem was being approached from the wrong angle. If we start from a proton with an electric charge, we can measure that charge, but we don't know the bare charge of a proton, because we can't observe a bare proton. What we observe is the proton plus all the surrounding vacuum fluctuations.

C. Instead, we should start with some unknown original charge, then add the infinite number of corrections. The final answer should be the observed experimental charge of the proton. The term for this approach is *renormalization*. The idea is to fix up the starting point of the theory so that in the end, the calculation makes some sense.

D. Renormalization seems almost like cheating. We start off with something that we don't know and we fix it up. In fact, in our problem, renormalization is worse than cheating. In order to make the calculations work, the original charge of the proton must be akin to negative infinity, because we're adding an infinite number of corrections to it, and we end up with the right answer, which we measured in the first place.

E. Why do physicists accept this approach? The reason stems from the fact that we can calculate an infinite number of things about electrons and protons, such as energies, magnetic field strength, their reactions in an electric field, and so on. We have an infinite number of observables. But we have to cheat about only two things.

   1. First, we must know what the charge is going to work out to be; we can't predict that.

   2. Second, we have to know what the mass is going to be. Einstein's work tells us that if we borrow energy from a vacuum, we're shifting the energy or mass of the system.

   3. These are the only two quantities we have to measure. Every other physical quantity can now be calculated in this theory.

F. When scientists began doing the calculation for the Lamb shift, they found that it worked exactly. To the best ability of the experimentalist who measured the splitting, quantum electrodynamics, QED, gave the correct number.

G. The same thing occurred for every other observable that we measured, such as the magnetic strength of the electron and the energies of other orbits in the hydrogen atom. The theory was successful. In fact, for estimating the magnetic strength of an electron, QED was successful to 12 digits of accuracy.

IV. Let's examine Feynman's contribution to QED in a bit more detail.

A. Part of Feynman's genius was his ability to examine complicated problems from a different viewpoint to make them seem simple.

B. Feynman summarized his mathematical calculations in little pictures called *Feynman diagrams*. Each Feynman diagram is a symbolic representation of a mathematical equation.

C. To draw a Feynman diagram, we must use certain symbols. For particles, we draw straight lines. For example, imagine an electron moving along; if we draw a graph of its position on the vertical axis versus time on the horizontal axis, we have a straight line. The graph shows the electron moving forward in time and forward in position.

D. Now suppose that this electron has an electromagnetic interaction. The quantum view of electromagnetic forces is that a photon is spontaneously produced. Feynman's diagram for a photon is a wiggly line that leaves the electron and goes off on its own. The spot where the electron and the photon were touching for one brief instant is the interaction spot. It's the vertex where they interact.

1. What explains this occurrence? The interaction spot occurs if the photon is capable of being reabsorbed very quickly. When energy is borrowed from the vacuum to create the photon, it must be returned.

2. Perhaps there's a proton nearby, and it, too, interacts with the photon, then it recoils because it absorbs something. Both particles recoil. That's the force that the electron and the proton feel because of each other.

3. Each line, each vertex, each wiggle in a Feynman diagram represents some mathematical operation. The beauty of this approach is that you don't have to be a brilliant physicist to understand the representation.

4. This graph represents a number, and if we do the calculations to find that number, it is always between 0 and 1. The

Feynman diagram rules are constructed so that this number is the probability that this process occurred.

5. Quantum mechanics tells us about probabilities of the occurrence of events. The Feynman diagram predicts the likelihood that an electron and a proton with certain given initial properties will result in certain measured final properties.

E. How does this apply to quantum field theory?

1. We might naively think of the photon traveling between the electron and proton as the electric field. That electric field could split and create an electron and a positron pair briefly, and a Feynman diagram could be drawn to represent that occurrence.

2. At a certain place in the middle of the drawing, the photon turns into a positron that runs backward and an electron that runs forward. They re-annihilate—create a photon again— because they can fluctuate for only a brief moment in time, then the photon continues on its way. This is the quantum fluctuation that we mentioned earlier.

3. This diagram could be calculated just as the first one could be. The numerical value for this diagram is much smaller than for the previous one, because there are more vertices and each vertex introduces a numerical factor that decreases the result. In other words, this is a low-probability event.

4. The game with Feynman diagrams is to start adding diagrams. We can imagine drawing more and more complicated versions of our first diagram, but those diagrams always come out smaller. The more complicated the diagram gets, the smaller its value.

F. This diagram is universal. Using this kind of diagram, we can calculate any kind of observable. We add all possible intermediate diagrams to our initial scenario, and the answer we get is the numerical estimate that we're interested in.

V. This incredibly powerful theory of QED was developed in the 1950s and truly ties everything together. We now have a theory that accounts for electricity, magnetism, light, electrons, quantum mechanics, relativity, and fields—a truly unified theory of electricity and magnetism.

A. Next, physicists started to wonder if this idea could be used to understand all the forces of nature. What about the strong nuclear force? Could we understand the interactions among protons and neutrons and pions using an idea like this?

B. When scientists had first tried to understand Yukawa's ideas about pions with a theory similar to QED, they failed. They could not make the theory work.

C. It was impossible for us to create a relativistic quantum field theory with the ingredients that we had in the 1930s–1950s, but we now have the ingredients, which we will examine in the next few lectures.

Essential Reading:

Barnett, Muhry, and Quinn, *The Charm of Strange Quarks*, chapter 4.1.

Kane, *The Particle Garden*, chapter 4, pp. 69–70, and Appendix A.

't Hooft, *In Search of the Ultimate Building Blocks*, chapter 9.

Weinberg, *Dreams of a Final Theory*, chapter 5, pp. 107–116 only.

Recommended Reading:

Feynman, *QED*. (You may want to take your time reading the whole book. It's a great attempt to explain the theory of QED, in surprising detail but without any math, to the general public. Worth a look if you want to dive deeper.)

Lederman, *The God Particle*, chapter 7 (first half, up to "The Weak Force").

Wilczek and Devine, *Longing for the Harmonies*, Sixth Theme.

Questions to Consider:

1. Try to draw a simple Feynman diagram schematically showing Rutherford's experiment (where a positively charged alpha particle scatters, or bounces electrically, off a positively charged nucleus). Note that the symbol for any charged particle is generally the same—a straight line with an arrow showing the direction of travel.

2. Classically, there is energy in the fields that surround a charged particle. Also, the strength of the electric field near a charge is proportional to 1 divided by the square of the distance from the charge. Qualitatively, why might postulating a point-like particle, such as an

electron, lead to infinite energy in a classical calculation? (This is the technical issue underlying the need for renormalization.)

3. Richard Feynman once said he believed that renormalization was like sweeping dirt under the rug. What do you suppose he meant?

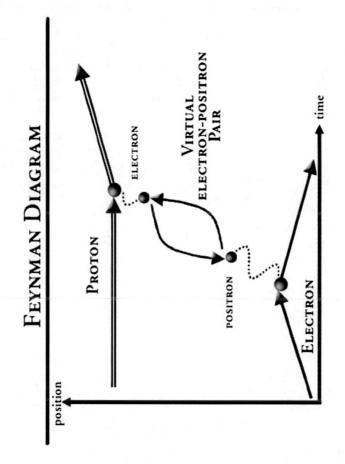

Feynman Diagram

position

Proton

ELECTRON

Virtual
electron-positron
Pair

POSITRON

Electron

time

163

Lecture Ten—Transcript
Fields and Forces

Particle physics involves some pretty hairball math, and I really want to avoid the math, but I don't want to avoid talking about the concepts behind the math because it's interesting and useful. So today we're going to leave behind the kind of content of particle physics that I've been talking about, the particles themselves and their properties. Instead I just want to focus on the theory: how we understand interactions between objects in the world.

This story goes back to classical physics, way before quantum mechanics was developed. What I want to talk about is a way of thinking about forces that's intermediate between Isaac Newton's somewhat naïve action at a distance, whereby he has a planet like the Earth over here, the sun over there, and they communicate. They interact with one another by gravity, and that's almost magical. Action at a distance, how do we know that the sun is over there? How do we, why do we, feel the gravity of the sun? Isaac Newton didn't really have an explanation for that; he just said we do. He wrote down a formula and described it, but he didn't really explain it. Nowadays, in quantum mechanics, we talk about exchange of virtual particles as explaining the force; but there was an intermediate step that was very important. In fact it's been integrated into our quantum picture today, and it's called "the concept of a field," also known as "field theory." We can talk about quantum field theory, or we can talk classical field theory.

So let's go back to the 1800s, when people were trying to understand electricity and magnetism. Electricity—you might think of static electricity between two charges as being a lot like gravity—the objects are far apart, and somehow, magically, they feel this force, action, at a distance. We could write down a formula, and we did, to describe what was going on. There was a British physicist, Michael Faraday, in the middle 1800s; this guy has an interesting history. He's an unusual physicist, in that he wasn't trained as a physicist. He was a bookbinder by profession, and as he was binding some books on science, he read them and became really interested. He became so interested, he went off and studied some more on his own. Finally, he went to a local university and asked if he could be a researcher. He started working in the lab, and he worked his way up until he was a professional research physicist. But he never really had the kind of rigorous mathematical training that most physicists have, and for that reason, it was wonderful. In order to visualize what was going on with electricity and magnetism, he would draw pictures, sketches, of the way he was imagining the forces would interact. He drew field lines. You've seen pictures like this in, I don't know, elementary school science, if you took a bar magnet, put a piece of paper on top of it, and sprinkled some iron filings on top. The iron filings spread out and begin to form these lovely curved lines that go from the south pole to the north pole of the magnet. They spread out all over the paper. You can visualize lines of magnetic force, spreading out through space, and if you were to go, I don't know, somewhere off to the side, with a little sensitive compass needle, you could detect the compass needle deflecting because there's this magnetic force field out there.

Faraday drew these pictures, and he thought about field lines as being real. He imagined that if he were to hold up a magnet, there really would be, in the world, magnetic fields everywhere in space, created by the magnet. So when you talk about electric forces, or magnetic forces, if there's a charge over here, it creates some electric force field; and that explains how it is that a charge far away can feel it. It's not a magical action at a distance. What it's really feeling is the field over there where it is. The field is there whether or not I put a charge at that other spot. You have charge number one that makes the field; whether or not charge number two is present, you still think of a field as being there in empty space. If you put a charge there, it will feel the field, and start to attract or repel. So this is a lovely way of thinking about action at a distance, the field concept. As time progressed, people wondered whether this idea was generalized. Could it be incorporated into the theory of quantum mechanics? Could we make a quantum field theory?

When Mr. Dirac wrote down his equation, the Dirac Equation, as it's now known, which married relativity and quantum mechanics, it was a description of electrons in the presence of electric fields. It was the first primitive attempt at quantum electrodynamics, a quantum theory of electricity, and why things move under the presence of electromagnetic forces. He hadn't yet incorporated the field idea, really; he was thinking more about the electron as a particle experiencing some electrical forces, and that theory was very successful. So the quantum theory, without really incorporating the concept of a field, worked pretty well. For example, you take that electron and put in orbit around a proton (a hydrogen atom) then start calculating, and you can predict the energy of the light that is emitted by this atom, the spectrum, as we call it, of the radiation that comes out of a hydrogen atom. If you use Dirac's theory, which was a kind of a primitive quantum-mechanical theory at that time, you get very good results. By today's standards you could do better, but, at that time, parts-in-ten-thousand kind of accuracy, that was really quite impressive.

But what if you want to include the idea of the field? So now what you're imagining is that the space between the proton and the electron is no longer empty; it's filled with an electric field. Now that you're thinking quantum mechanically, you realize that electric fields carry an energy density; and if they carry an energy density, you can borrow a little bit of that energy. We've talked about this idea before. You might be able to create, for example, a virtual electron and anti-electron. You can borrow energy from the field for a very short amount of time. Heisenberg's Uncertainty Principle says that you can pull a particle-anti-particle pair out of nothingness, out of the vacuum. Even though it's not quite a vacuum; there's an electric field there. You can create this particle-anti-particle pair if you just do it for a short time. It's as if you're borrowing energy from the universe, and the Uncertainty Principle allows you to do this.

That means that the problem you thought was pretty simple, called two-body problem, one electron and one proton., is really more complicated because the full problem is what you started with. Plus you have a proton and an electron, and in between a little electron and a positron, an anti-electron. So it's really a four-body problem. And shoot, if you can produce one particle-anti-particle pair, how about two, or ten, or a billion? All of a sudden, the vacuum inside of a hydrogen atom and around it, is a very complicated place. We call these virtual particles "vacuum fluctuations," and if you try to write down a quantum field theory, you have to accept they're a logical consequence of the theory that you've written down.

Now what is the consequence of the presence of an electron-positron pair? Well, the mathematics is, as I said, pretty hairball, but the physical consequence is pretty easy to picture. You have a positive charge in the middle, and a negative charge that's attracted to it on the outside, and in the middle you've got one plus and one minus. So the plus is going to tend to gravitate towards the electron, and the minus is going to tend to gravitate towards the proton in the middle. So, actually, you're going to end up changing the force that that electron feels, a little bit, because it sees a different configuration of charges nearby it. So if you want to figure out the energy of that electron, you're going to have to correct your earlier estimate by including this small extra correction. It is small, numerically small but you have to add an infinite number of small corrections. And that was the disaster that was experienced in the late 1920s through the 1930s. Any time people tried to calculate the net total effect of all these vacuum fluctuations of the quantum fields, they discovered that the grand total effect on the electron, this infinite sum of progressively smaller and smaller little effects, grew to be infinitely large. That's a disaster in a mathematical theory. If you get infinity as the answer, it's not physics; it's not the real world.

People were beginning to despair. I think there was a sense that quantum field theory might just not be the way we were going to describe nature; and, look, we were doing welll enough, even without worrying about these fluctuations. We could predict all sorts of properties, not just the energies. For example, an electron is actually a little tiny magnet. It's a spinning electric charge, like a kitchen magnet. It would stick to your refrigerator magnetically, and you could estimate the strength of the magnetism of an electron using Mr. Dirac's theory. That, also, was an accurate calculation that you could do, and it worked quite well, so there were a number of areas where his theory was succeeded.

That was pre-World War II, and during the war, the problem was pretty much left behind. After World War II, there was a conference. It was quite a famous conference, called the Shelter Island Conference of 1947. I think a bunch of physicists were getting together for the first time after World War II, just trying to decide what were the big problems of the day. Okay? After this long hiatus, they wondered where should they be looking, what should they be working on and worrying about? That was before all those particle accelerators were built, so they didn't have zoo of particles to think about yet. There was an experimentalist at that conference named Willis Lamb, and he presented an experiment that he had done. He was actually using radar technology from the World War II era to measure very, very accurately the light emitted by hydrogen, and what he discovered was something that was unexpected. It had not been a prediction of Mr. Schrödinger or Mr. Heisenberg, or even Mr. Dirac. One of the colored lines that came out of hydrogen was, in fact, two colors that were very, very closely spaced, two frequencies that were very, very nearly the same. There was a small splitting, so that one line was really two, and that splitting was a big mystery. It was not explained.

There was another physicist at that conference whose name was Hans Bethe. Hans Bethe is a wonderful character. He was a great physicist in 1947, and he's still a great physicist today. He's in his 90s. I heard him give a talk on physics just a couple of years ago; he'a an amazing fellow. Hans Bethe was the one who figured out how the sun works. He figured out the nuclear reaction chain that powers the sun. He had heard, in advance of this experimental work, of Mr. Lamb, and on the train ride on the way out to this conference, he had been sort of thinking about vacuum fluctuations. He thought, "I don't know how to deal with this infinite number of vacuum fluctuations, but let me just deal with a couple. What if I just had one, an e+/e− pair that was produced in the vacuum—what would that do to modify the answer?" And darned if it didn't do the right thing! It splits the line, and even though order of magnitude is roughly correct, it doesn't give you the exact experimental number, but it's pretty close. So at this conference, now, there's this radical result that's in contradiction with quantum mechanics, and this hint that you really need a quantum field theory to explain the data. So this is a big push. Lots of people at that conference started thinking about this problem of how to make quantum electrodynamics a real, full-fledged quantum field theory.

Richard Feynman was one of the people at that conference. Julian Schwinger was another young physicist at that conference, and they, and many other people, went off and started to work on this. There was another physicist, Tomonaga, in Japan, who had been working on this, but his work wasn't really communicated well to the Americans because it was just shortly after World War II. In the end, in 1965, the Nobel Prize was awarded to those three people, Tomonaga, Feynman and Schwinger, for their work on the development of what ultimately was a successful quantum field theory.

Now, we're really getting into some deep mathematics here, and I want to try to explain to you the fundamental idea, the breakthrough, that these folks came up with. If you get lost in the details, don't worry about it; I think you can understand the premise, at least. So look, here's the problem. You start with a proton in the middle. It creates a field, and that field is modified by the presence of virtual fluctuations, e plus-e minus pairs just popping up and then re-disappearing again. You have to add up all possible combinations. One pair, two pair, ten pair, and when you add them up, you get infinity. So, the inventors of QED said, "Look, maybe we're starting from the wrong place. We started from a proton with an electric charge. How much electric charge? Well, we measured that, in the laboratory. It's a certain number. Maybe we should sort of throw up our hands and say, I'm not sure how much charge a proton really has. I don't know what the bare charge of a proton is, because you can't see a bare proton. Even if you have just a proton, all by itself, and you're an observer off on the side, what you see is the proton—plus all the little vacuum fluctuations around it. So what you should do is start with some unknown, original charge, and then add all those infinite number of corrections; and then your final answer should be the observed experimental charge of the proton." The fancy word for this is "renormalization;" we're going to renormalize the original charge of the proton. Renormalize—if you can drop this word at a cocktail party, you know, you're a physics god. This is a very technical word, but the idea is, fix up the starting point of your theory, so that, in the end, the calculation makes some sense.

Now when I first learned about renormalization—this is a sort of an advanced, graduate topic in physics—I thought to myself, this is cheating. This is nuts. It's as though you take a student in freshman physics, and give him a homework problem with the answers in the back of the book. I do this, when I teach freshman physics, and I know every single student will get the right answer when I do it. How do they get the right answer? Well, if they know the physics, they just get it. If they don't know the physics, I don't know. They start with some stuff, and then they write some other stuff, and they always end up with the correct answer in a box at the end. It seems as though that's what renormalization does. We start off with something that we don't know, and we fix it. In fact, it's worse than that because to make things work out, the original charge of the proton has to be kind of like negative infinity because you're adding an infinite number of corrections to it. Then you end up with the right answer that you looked up, meaning you did an experiment and you measured it.

It is cheating; it's definitely cheating. The renormalization approach works, but it doesn't predict the charge of a proton, and that's the bad news. You might have wished, as a physicist, that some day we could calculate the charge of a proton or an electron—what a lovely dream—from some fundamental theory. Well, we're not going to be able to do that in this theory because we're saying we have to cheat, and we have to go and measure the charge. We have to know what it is in order to figure out what the starting point is.

The reason that we accept this cheat, and we don't worry about it, is that there are an infinite number of things you can calculate about electrons and protons. Energies, magnetic field strengths, what happens if you put it in an electric field, what happens if you put two atoms together, and on and on and on. There are an infinite number of observable data, and you only have to cheat, it turns out, not about one thing, but about two. You have to cheat, in the sense that you have to know what the charge is going to work out to be. You can't predict that, so you have to cheat in the sense that you have to know what the mass is going to be. You also have to renormalize the mass because of Albert Einstein's $E = mc^2$. If you're borrowing energy from the vacuum, you're shifting the energy, or mass, of the system, so there are two quantities that you have to measure; you cannot predict them. But every other physical quantity can now be calculated in this theory, and so people started doing that; it was the breakthrough. People started doing these calculations, and the calculation of the Lamb shift worked, not approximately, but exactly. It was a spectacular agreement. To within the best ability of the experimentalist who measured that splitting, this quantum electrodynamics, QED, gave you the correct number as well as the same thing for every other observable that we started measuring. For the magnetic strength of the electron, and the energies of other orbits in the hydrogen atom, the theory was successful. In fact, it was spectacularly successful.

Let me give you an example of the contemporary success of quantum electrodynamics. If you try to estimate the magnetic strength of an electron, how big of a magnet it is, Mr. Dirac's theory actually tells you the answer. Mr. Dirac's theory says that it's approximately two, in a certain unit system. Now you go to quantum electrodynamics, and quantum electrodynamics says it's not exactly two; it's two point, and you start calculating. You have to add in the effect of these little vacuum fluctuations; you add up a whole bunch of them, and what you get is the following number. This is the world's best number as of today—2.00231930435. I have to stop there. That's the end, as far as we've calculated it. In order to calculate that number, somebody had to calculate the effect of vast numbers of virtual positron and electron pairs popping out of the vacuum. These had to be numerically calculated, by some poor graduate student in a research group, who spent six years of his life doing this, maybe more than six years, though that's the sort of typical span of a graduate student. The point is, that number agrees with the experiment to that many decimal places, 12 digits of accuracy. So if you're kind of wondering whether this stuff is a correct description of nature, correct meaning scientifically correct, boy, that's the most powerful evidence I can think of. It's the most successful calculation of any physical quantity of any kind that I know of in science. Quantum electrodynamics is really a great theory, and it's a shame that the mathematics is so hard. It takes a lot of training to learn how to do these calculations.

I want to tell you a little bit about Mr. Feynman's big contribution to this story, because Feynman, and Schwinger and Tomonaga essentially worked independently. Schwinger was an amazing mathematician, and I still don't understand the math that he was doing. Part of Mr. Freyman's genius was taking something that was very complicated and looking at it in an elegant and different way than anybody else, so that it looked simple. It's a very powerful and wonderful skill to have. Mr. Feynman took all of these mathematical calculations and summarized them in a little cartoon, a little picture, which is called a Feynman Diagram. So let me try to explain to you the basic idea of the Feynman Diagram, because all particle physicists use Feynman Diagrams all the time. In fact, it's not just particle physicists; just about any physicist uses these things, because what they really are is a kind of a shorthand for the mathematics that's lying underneath them. Each Feynman Diagram is really a symbolic representation of a mathematical equation, and I'll tell you what you do with that equation in a minute.

In order to draw a Feynman Diagram, you have to have some symbols. So, for particles we're just going to draw straight lines. For example, imagine an electron that's just cruising along, and let me draw a graph of its position on the vertical axis, versus time on the horizontal axis. This is a little graph of position versus time, and if the particle is just cruising along, it's a straight line and is moving forward in time and forward in position. Now suppose that this photon has a little electromagnetic interaction. The quantum view of electromagnetic forces is you spontaneously produce a photon. So Mr. Feynman's Diagram for a photon is a little wiggly line, which leave the electron and goes running off on their own. That little spot, where the electron and the photon were touching for one brief instant, that was the interaction spot, the vertex where they interacted. Now, why would this happen? It would happen if that photon were capable of being reabsorbed very quickly because you have borrowed energy from the vacuum to create that photon, and you have to give it back. Perhaps there's a proton nearby, and the proton is cruising along, and it, too, interacts with the electron; then it recoils, because it absorbs something. Even though it's a virtual particle, it recoils. Both particles recoil; that's the force that the electron and the proton feel because of each other. This is a Feynman Diagram, and each line, each vertex, each wiggle represents some mathematical operation. The beauty of this is that you don't have to be a brilliant physicist. You could even be a computer and be trained to compute that diagram, because Mr. Feynman did all the hard thinking for us, and now we have this lovely representation. When you look at that, it almost looks like a little cartoon of a collision between particles. You can visualize all the nasty mathematics in a lovely, simple way.

When you do the calculation, it involves some integrals, and some, actually, more complicated matrix manipulations, and in the end you get a number. This graph represents a number. What do you do with that number? Well, first you square it, so that it comes out positive, and that number always comes out to be between zero and one. It turns out just to work out from the mathematics, and it's a probability. It's the probability that this process occurred. This is quantum mechanics; quantum mechanics tells you about probabilities of events happening. So, if you want to know what is the likelihood that your electron and proton (which come in with some given initial properties, and go out with some measured, final properties) what's the likelihood that that's going to happen? That's what the Feynman Diagram will tell you. You can calculate the likelihood of any event, and that's the best that quantum mechanics can do for you. It's good enough to make these spectacular predictions.

Now what about the quantum field theory that I talked about? Well, that photon, traveling between the electron and proton might naively be thought of as the electric field. So there's an electric field between them, and that electric field could split and briefly create an electron and a positron pair. You could draw a Feynman Diagram for that. At a certain place in the middle, the photon turns into a positron, which is running backwards like this, and an electron which is running forwards like so. Then they re-annihilate to create a photon again, because they can only fluctuate for a brief moment in time, and then the photon continues on its way. This is now a modification story. You could calculate this diagram just as you could calculate the first one. It's a numerically harder calculation, because it turns out the more vertices you have, the tougher the calculation. Still, a computer, or anyone who is trained to interpret Feynman Diagrams could do this.

It turns out that the numerical value for this diagram is much smaller than the previous one. That's because there are more vertices, and each time you get a vertex, it introduces a numerical factor that makes the number come out smaller. So this is a low-probability event. What about the renormalization trick? It's already built into the mathematics, so you don't have to worry about that infinite sum. The game with Feynman Diagrams is that you add diagrams, and you add more and more. You can imagine drawing more and more complicated versions of this one; maybe the photon splits twice along the way, maybe there's two photons, you can imagine all sorts of complications, but those diagrams always come out smaller. The more complicated the diagram gets, the smaller its value, so you keep calculating until you get tired. Okay? That number that I told you for the magnetic strength of an electron was calculated out to some arbitrary level of complexity until people's computer programming skills sort of petered out, and the computer programming time got to be too long when they got out to 12 digits. But if we get a better, faster generation of computers, we can add one more level of complexity to the diagrams, and add another digit or two to that calculation. We can keep improving the number. The experimentalists have the really tough job here, which, actually, is to measure that thing, but that's another story. It's a spectacular job to make a measurement of a physical quantity to that level of accuracy; and, as I said, the agreement is out to 12 decimal places using this kind of Feynman Diagram. It's completely universal; if you use this diagram, you can calculate any kind of observable. Remember, what's important is what's coming in, and what's going out. You add all possibilities; that's the Feynman Rule. You add all possible intermediate diagrams, sum them up, and the answer you get is the numerical estimate that you're interested in.

So this is the modern theory of quantum electrodynamics, and it's incredibly powerful. It was developed during the post-war era, the 1950s, and really ties together everything. Now we have a theory that has electricity and magnetism; light; electrons; quantum mechanics; relativity; and fields. It's all been put together. It's a truly unified theory of electricity and magnetism. So now physicists start thinking and asking if they can use this idea to understand all the forces of nature. Electricity and magnetism isn't the only force. What about the strong nuclear force? Could we understand the interactions between protons, and neutrons and pions, using an idea such as this? Mr. Yukawa had proposed that instead of photons traveling, you had pions traveling. People tried to write down a theory like this one, like QED, and they failed. They couldn't make it work. Part of the problem was that the interaction was a strong one, so the diagrams didn't get smaller and smaller as you added more and more complexity. Another problem was that the particles themselves weren't really the true particles of nature. That is to say, the protons and the neutrons and pions were, themselves, complicated objects with internal structure. We didn't know that in the 1930s, or even in the 1950s. That was the underlying simplicity that all those accelerator experiments in the 1950s and 1960s were pointing us toward. It was impossible for us to create a relativistic quantum field theory with the ingredients that we had in the 1930s through the 1950s.

Albert Einstein had been working, in this era, on a unified theory. He was trying to unify all the forces of nature, and he failed. He had great successes in his early life, and he had some successes in his later life, but he never achieved his true goal. In part it was simply because he didn't have enough data. He didn't have enough information about what the real building blocks of nature were. Had he lived another 20 or 30 years, he might have had the tools to complete his work, who knows? In any case, we now have the ingredients; and where we're going to be heading, in future lectures, is first, to figure out what are the fundamental particles. In the case of Mr. Feynman, he was talking about electrons. Electrons interacting with electrons. Electrons are fundamental particles. There is nothing inside of an electron, as far as we know today; and so that's really why quantum electrodynamics is so successful. Because he was just lucky, I guess, that the fundamental particle he was working with really was a point-like particle. That is an implicit assumption in quantum field theory, that the ingredients you're working with are point like.

This idea of renormalization is a critical idea, because when you're hunting around for a new theory of a new force, you don't have the kind of strength and background of Mr. Faraday, and of all the people who worked to understand electricity and magnetism. You don't have such a deep theory to start with. Instead, you're struggling to find a new theory, and with the principle of renormalization, this helps you to say, "I'm not going to use that theory because it's not renormalizable." Some theories are and some theories aren't. When you work it out, sometimes you can make the infinities go away, and sometimes you can't. It's a mathematical fact about a theory. So, in fact, there are very few number of theories you can construct that are inherently renormalizable. The theory of the world that we have today to work with, the standard model of particle physics, is a renormalizable quantum field theory. It includes the strong force, the weak force and the electromagnetic forces, all put together in a theory that's like QED, but generalized to all the forces that I've been talking about.

Lecture Eleven
"Three Quarks for Muster Mark"

Scope: In this lecture, we attempt to crack the hidden code of the particle zoo and discover that the hadrons (strongly interacting particles) are fundamental but not elementary. What evidence is there to guide us to a deeper understanding? Could these particles be made of something else—mini-atoms themselves? The breakthrough idea of quarks provided the answer, along with a return to simplicity. This lecture offers a guide to the early "quark picture," the names of these new particles, their properties, and their roles in making up the world. We also look at the importance of successfully predicting the existence and properties of new fundamental particles of nature with a new type of periodic table of particles.

> Three quarks for Muster Mark!
> Sure he hasn't got much bark
> And sure any he has it's all beside the mark.
>   —James Joyce, *Finnegan's Wake*

Outline

I. We return now to the situation in experimental particle physics in the 1950s–1960s. As you recall, hundreds of particles were being discovered, and scientists were trying to pinpoint their commonalities to come up with an organizational scheme.

  A. Despite the complexity of the particle zoo, physicists had an inkling that some underlying simplicity might exist. Their clue to this simplicity stemmed from the fact that protons and neutrons didn't seem to be point-like.

   1. Let's distinguish between a fundamental particle, which we've been talking about all along, and an elementary particle. An elementary particle would be a true building block of nature—a dot, a point with properties.

   2. Electrons seemed to be elementary particles of nature, but the protons and neutrons did not. Even in the 1950s and 1960s, scientists saw clues that the proton might actually be spread out in space, that it might be made of something.

3. The atom was known to be spread out in space even when the periodic table was generated. An atom has a size of about 10^{-10} meters, which is tiny, but it's not 0. The atom is spread out over some distance scale.

B. In the late 1950s, Robert Hofstadter made a similar discovery for the proton itself and received the Nobel Prize in 1961.

1. Hofstadter used a beam of electrons, which are truly point-like particles, as a kind of magnifying glass to examine the proton. Ordinary light cannot be used to look at a proton, because the wavelength of visible light is huge compared to the tiny size of a proton. By focusing a beam of electrons on the proton, Hofstadter went back to an analogy to Rutherford's experiment to examine the atom.

2. Rutherford directed alpha particles at an atom and deduced that the atom has a nucleus inside and electrons on the outside. If the energy of the incoming beam is increased, smaller and smaller distance scales can be seen.

3. At Stanford University, Hofstadter produced a beam of electrons that was high enough in energy to probe to the size of the nucleus.

4. Remember that Rutherford observed that every now and then, a particle would bounce backward, indicating that there was something hard, like a little nugget, in the middle of an atom. Hofstadter saw the exact opposite occurrence. The electrons would go into a proton and, perhaps, bend at a small angle, but in general, they went straight through.

5. This observation is an indicator that the proton is, itself, like the plum pudding. It's just a smear of electric charge. By looking at the details of how the electrons bounced, Hofstadter was able to deduce the size and distribution of the electric charge.

6. He found that the proton was a spread-out, positive charge over a size scale of about 10^{-15} meters (a *femtometer*). That measure is 100,000 times smaller than an atom, but it's not 0.

C. Hofstadter demonstrated that the proton is not a point, but if that is true, then it is not an elementary particle. Physicists started to wonder if some other particle was inside the proton, spreading out positive electric charge.

D. Another piece of evidence, even more abstract, pointed to the fact that the proton is not an elementary particle.

   1. Recalling Dirac's equation and the development of QED, physicists wondered if the same theory that had been used to estimate the strength of magnetism of an electron could be applied to a similar calculation for the proton.

   2. But that calculation failed miserably, implying that the starting assumption, namely, that the proton is a point, is incorrect.

   3. Murray Gell-Mann and George Zweig, working independently, came up with the breakthrough idea that the proton is built up of three smaller objects. Gell-Mann called them *quarks*, because they were quirky particles that were not directly observable.

   4. Quarks are tiny objects that are bound together very tightly. According to the Heisenberg uncertainty principle, they are spread out, presumably over a distance scale of about a femtometer, and they form a cloud. That's what a proton is: nothing more than three quarks held together by some strong force of nature.

   5. There are three different types of quarks, whimsically named by Gell-Mann, *up*, *down*, and *strange*.

II. Gell-Mann wondered if all of the particles that were being discovered in bubble chambers could be explained with this idea. He proposed that every baryon, every one of the strongly interacting particles that acts like a particle, such as the proton and the neutron and all its heavy partners, are all made of just three quarks.

   A. Different combinations of these quarks are possible, such as three up quarks, two up quarks and a down quark, and so on. The zoo of particles can be understood as combinations of three building blocks.

   B. Now the question is: Will this model work? Will we be able to describe in detail all the particles we see?

   1. Let's go back to the strongly interacting particles that are the subject of experiments at the particle accelerators, the hadrons.

   2. Remember that hadrons can be either baryons or mesons. Baryons are particle-like, such as protons and neutrons. Mesons were, at least originally, thought of as force carriers.

3. Mesons and baryons are different. Mesons, for example, can be produced easily, but the total number of baryons is always conserved.

4. What is a meson, then? According to Gell-Mann's quark model, a meson should be a quark and an anti-quark tightly bound together, a binary system spinning together.

5. Given Gell-Mann's rules, any combination of quarks can be formed, and these must match up with one of the known particles.

C. Physicists first wanted to see if this model could be made quantitative; to do that, they had to come up with a list of properties of quarks.

1. For example, what is the mass of a quark? What is its charge? What is its spin? Those three properties are almost all we have to describe about quarks, because they are postulated to be point-like particles, just like electrons.

2. What is the spin of a quark? Gell-Mann proposed that all quarks have spin 1/2, just like electrons and protons.

3. How does that spin affect the particles that quarks make up, such as a meson? A meson consists of a quark of spin 1/2 and an anti-quark of spin 1/2. How can they combine? Depending on the direction of the spin, they combine to form mesons with spin 0 or spin 1 (just as $1/2 - 1/2 = 0$, and $1/2 + 1/2 = 1$).

4. What about protons? In this model, Gell-Mann proposed that a proton is two up quarks and one down quark. They can't all spin in a clockwise direction because if they did, the result would be three halves, which doesn't work. Instead, two of them are spinning clockwise and one is spinning counterclockwise. The equation $1/2 + 1/2 - 1/2$ yields the correct answer for the spin of the proton.

D. Using this model, we can look at a particle from the data and try to determine its quark makeup or do the reverse; that is, start tabulating all the logical possibilities from the model and see if every logical possibility matches with some particle that has been found. In this way, we discover that there really is a one-to-one correspondence. Every particle has a description, and every description can be found in the laboratory.

III. Even though this seems to be a good model, it does have some problems.

  A. For example, we have never seen one quark by itself in a bubble chamber or a pair of quarks. We only see them bound together in sets of three or as a quark and an anti-quark.

  B. That fact was puzzling in this early era, and in fact, for many years, the quark model was dismissed by most physicists as no more than a mathematical construct to help organize the zoo of particles.

  C. Later, we saw growing evidence for the reality of quarks, including tracks in bubble chambers (jets) that come indirectly from quarks.

IV. Let's return to our discussion about adding up the properties of quarks to get the properties of the objects themselves.

  A. Consider the property of mass. If a proton is three quarks—two ups and a down—we might ask: What is the mass of each of those three quarks?

  1. Gell-Mann suggested that perhaps the up and the down quark were very similar to each other, each of them carrying about one-third of the mass of a proton. If we put three of them together, we get a proton mass.

  2. If the up and the down quarks are similar in mass, then what would happen if we put an up and two down quarks together? The result would be a neutron. According to the quark model, given that up and down weigh about the same, the neutron and the proton should weigh about the same, and they do. This model explains, then, why certain particles come in pairs or partners.

  3. Similarly, the three pi mesons can be understood as just different combinations of up and anti-down or down and anti-up quarks, but they all should weigh about the same, and indeed, the three pions form a triplet that weighs about the same.

  B. The quark theory is really only crude with respect to masses. It does a better job with other properties, e.g. consider how well the model holds up to the data when we look at electric charge.

  1. Gell-Mann proposed that the up quark has charge 2/3 and the down quark has –1/3. Keep in mind that the proton has charge

1, and the electron has –1. The charges for the quarks seemed bizarre.

2. No one had ever seen an object in the laboratory that had electric charge of 1/3 or 2/3, which should leave a distinctive pattern in a bubble chamber. The more electric charge a particle carries, the greater the density of the bubbles it produces. The charges that Gell-Mann was proposing should leave a clear signature.

3. Gell-Mann accounted for this discrepancy by noting that quarks always come in either pairs or triplets. Remember that a proton has two up quarks and a down quark. The equation for its charge, then, is $2/3 + 2/3 – 1/3$, resulting in 1. A neutron, which has one up quark and two down quarks, would be $2/3 – 1/3 – 1/3 = 0$, which is the electric charge of the neutron.

4. As we look at other particles, we see that the charges are predicted in the same way. All the quantum numbers add up and make sense to form a coherent picture.

C. We can perform similar calculations for strangeness.
    1. Remember that some strange tracks had been found in bubble chambers in the 1950s, and Gell-Mann had proposed that they were the result of a new property of matter, a quantum number, similar to electric charge. He called this property *strangeness*, and the strangeness number was always 1, or –1 (or 2, or –2).

    2. The quark model now explains the strangeness number. We just count up how many strange quarks are present. If we have a baryon that consists of one up, one down, and one strange quark, that particle should have strangeness 1. If we have a particle that has one up quark and two strange quarks, that would be a particle with strangeness 2.

    3. Again, the calculations matched the data beautifully. The strange quark would have to be a little bit heavier than the up and down quarks because the strange particles were a little bit heavier, but this knowledge allowed physicists to make predictions of other particle masses.

    4. By looking at an up-down-strange combination, we can estimate the mass of the strange quark. If we then make a

prediction about the mass of an up-strange-strange combination, that's roughly what its mass turns out to be.

5. What if we had a particle that consisted of three strange quarks? That would be a triply strange baryon, and nobody had ever seen such a thing. Just as Mendeleev had found holes in his periodic table a hundred years earlier, this lack of a triply strange baryon constituted a hole in the organization of quarks. Later, this object was found; it's called an *omega particle*.

V. The strange quark doesn't occur very frequently in nature. It is heavier than the other two quarks but has an electric charge that is the same as the down quark. It seems almost like a heavy partner of the down quark, much as a muon is a heavy partner of an electron.

A. Why do we have strange quarks in nature? Our bodies and almost everything around us are mostly made of up and down quarks, but we still have these heavier, somewhat mysterious particles that do, occasionally, occur in nature.

B. Physicists, myself included, are still investigating the details of this story, but the fundamental idea was proposed in 1964 and it's still part of our standard model of particle physics today.

C. As we will discover, there are more particles and forces, and we have to try to fit all these ideas together. At this point in the course, we have a descriptive model, but we still need the underlying theory to explain it.

Essential Reading:

Barnett, Muhry, and Quinn, *The Charm of Strange Quarks*, chapters 2.7–2.8.

Kane, *The Particle Garden*, chapter 4, pp. 53–60.

Schwarz, *A Tour of the Subatomic Zoo*, chapter 4.

't Hooft, *In Search of the Ultimate Building Blocks*, chapter 8.

Recommended Reading:

Calle, *Superstrings and Other Things*, chapter 24, section on quarks.

Lederman, *The God Particle*, chapter 7, section on "The scream of the quark."

Questions to Consider:

1. Given that the K+ particle is a meson with strangeness +1, which one of the following choices gives the "quark makeup" of a K+? (Hint: the u quark has charge +2/3; the d and s quarks each have charge −1/3. The strangeness of an s-quark is [somewhat perversely] called −1, not +1.)

   (a) u and anti-s (b) u and s (c) d and anti-s (d) d and s (e) u and anti-d

2. The delta++ is a baryon with charge 2. It is not strange. Which one of the following choices could be the quark makeup of a delta++?

   (a) u u u (b) d d d (c) uud (d) udd (e) anti-u anti-u anti-u

3. A pi+ is a u anti-d meson. A pi− is the anti-particle of a pi+. What is its quark makeup?

Lecture Eleven—Transcript
"Three Quarks for Muster Mark"

We left our experimental particle physicists at their accelerators; it's the 1950s and the 1960s. They're collecting data, observing tracks of particles traveling through bubble chambers and cloud chambers, and just trying to collect information. What are the charges? What are the masses? What are the lifetimes? Just basic information, because people don't really understand what these things are—what they're made of, how they fit together. It looks like there's more to the world than just protons, and neutrons, and pions. There's now a whole slew, at first dozens, and then hundreds of particles; and every one of them seems to be as fundamental as any other one. They start getting exotic names, often Greek letters: the kaons, the lambda particles, the sigma particles, and you can classify them. People are beginning to make some sense, at least put some order to this chaos. Certain particles belong together as they have similar properties. So, you notice doublets, or triplets, or even eight particles that fit together, having commonalities. But still, no underlying story.

The question of the day is, "is this the way nature is?" Is nature just to have 100 fundamental particles, and that's what we're going to have to live with, and we're going to have to tabulate them, and that's the way science is going to proceed? Or, can we make sense like we did 100 years earlier, when we had 100 chemical elements: hydrogen, and helium, and lithium, and so on? At a certain point, some sense was made of that, once we realized that atoms can be understood to be built up out of a nucleus in the middle, and electrons flying around it. You count electrons, and that tells you what element you have. Is there some simple picture like that for the fundamental particles?

There was a clue that such a situation might be available to us. The clue was that protons and neutrons didn't seem to be point-like. So I've been using the word fundamental to talk about particles and I suppose one could distinguish between a fundamental particle and an elementary particle. An elementary particle would be one that is a true building block of nature, which is a dot. It's a point with properties. Hard to picture, but that's the simplest thing you can imagine that the world is made up of. An electron seems to be an elementary, fundamental particle of nature. But the protons and neutrons did not seem to be quite that way. Even in the '50s and '60s as these, the zoo of particles, were appearing, there were clues that the proton might actually be spread out in space and that it might be made of something. The atom was known already, when the periodic table was generated, to be spread out in space. An atom has a size. It's about 10^{-10} meters, which is insanely tiny, but it's not zero. It's a number, you can measure it; if you get a good enough magnifying glass, in principle, you can see that the atom is spread out over some distance scale. Something similar happened for the proton. The real measurement that indicated the size of the proton happened in the early 1960s, maybe late 1950s, at Stanford University. Robert Hofstadter was the physicist; in fact, he got his Nobel Prize for this in 1961, so it must have been in the late 1950s.

It was, in effect, a magnifying glass. But you can't use light to look at a proton, because the wavelength of light is huge compared to the tiny size of a proton. So you need something that's like a magnifying glass. What he used was a beam of electrons, truly point-like particles. It's back to Mr. Rutherford's experiment, all over again. Right? Mr. Rutherford shines tiny point-like particles at an atom and deduces that the atom has a nucleus inside and electrons on the outside. And, if you increase the energy of your incoming beam, you focus down onto smaller and smaller distance scales. So Mr. Hofstadter had available, at Stanford University, a high-energy physics lab, which could produce a beam of electrons, which was high enough energy that it could probe all the way down to the size of the nucleus. So it's really like the Rutherford experiment. Now remember, what Mr. Rutherford observed was that every now and again a particle would bounce backwards, indicating that there was something hard, a little nugget in the middle of an atom.

Well, Mr. Hofstadter saw the exact opposite. The electrons would go into a proton and then they would maybe bend a small angle, but pretty much they went straight through. So this is an indicator that the proton is itself kind of like plum pudding, only without the plums. It's just a smear of electric charge. And Mr. Hofstadter, by looking at the details of how the electrons bounced, was able to deduce the size and the distribution of the electric charge in there. It just seemed to be a fuzzball. The proton was a smeared-out, positive charge, over a size scale of about 10^{-15} meters. That's also called a femtometer, and that's 100,000 times smaller than an atom. So you're imagining a little nugget at the center of a hydrogen atom. It's 100,000 times smaller, but it's not zero.

If you think about our solar system, we've got the earth going around the sun, and the earth is some distance away from the sun, and the sun also is a smeared-out fuzzball of hydrogen gas. And the size of the hydrogen atom is about ten thousand times smaller that the distance out to planet Earth. So, other than a factor of ten, it's really quite comparable to what we've got going on. We really have a little teeny solar system, with a fat sun in the middle and then, far away, the electron acting like planet Earth. Of course, it's not like planet Earth; it's an electron. It's a point, and when you start thinking about quantum mechanics, you really can't think of it as just a little discrete object. It has a wave nature. In any case, Mr. Hofstadter has now demonstrated to us that the proton is not a point. So if it's not a point, then you're immediately thinking, "Well, it's not elementary and maybe it's made up of something. Maybe there's something buzzing around in there, spreading out positive electric charge." It's an idea that you formulate, and now people are trying to make this into something a little bit more concrete.

This is the idea of trying to figure out what something microscopic looks like, when you can't really see it directly, by scattering objects off of it. I saw a demonstration in a museum once, where they had a little object, a little geometrical object, and it was either a circle, or a square, or a triangle, but you didn't know. It was covered with some cardboard and you could figure it out by shooting little BBs at it with a little pistol. So the BB would roll, disappear behind the cardboard, and then bounce. If it bounced straight back, then you could conclude that it was hitting a straight edge. If it bounced off to the side, you could conclude that it was hitting some sort of a curved edge. So you could deduce the shape of the object, even though you couldn't see it. I think that's a nice metaphor for what was going on in these experiments of Mr. Hofstadter, to try to deduce what the proton, what this fuzzball of a proton looks like.

There's another piece of evidence, which is more abstract, that the proton is not a fundamental particle. Remember that Mr. Dirac had written down a theory of quantum mechanics and relativity, and that theory predicted the magnetic strength of an electron, and it did a pretty good job. And then, in the 1950s, Feynman and Schwinger and Tomonaga developed an even more sophisticated theory, quantum electrodynamics—fancy, relativistic quantum field theory—that took into account pretty much everything we know about basic physics, and was able to predict the magnetic strength of an electron to 12 digits. Actually, at that time it wasn't 12 digits; that's a modern calculation. But the theory was extraordinarily impressive, in its ability to predict the properties of the electron. Now, you could say, "Well gee, if the proton is a point particle like the electron, we should be able to use the same theory, do a similar calculation, and estimate the strength of magnetism of a proton." Something that was also being measured in those days. That calculation failed quite miserably. Instead of getting many, many digits of accuracy, you were off by a factor of 2 or 300 percent errors in the calculation. So that implies that your starting assumption, namely that the proton is a point, is breaking down.

The breakthrough idea was the postulation—Murray Gell-Mann was one of the people who came up with this idea, also George Zwieg, independently—that the proton is really built up out of three smaller objects. Tiny objects. Murray Gell-Mann called them quarks because they were kind of quirky, funny little things that were not directly observable. They were tiny objects very, very tightly bound together. They would buzz around and form a cloud, because they're really quantum particles, so in a certain sense they get smeared out by the Heisenberg Uncertainty Principle. They get smeared out, presumably, over a distance scale of about a femtometer, 10^{-15} meters, and that's what a proton is. It's really nothing more than three little quarks, stuck together by some strong force of nature. Mr. Gell-Mann called them quarks; Mr. Zwieg called these little things "aces". You get to make up a name and, for some reason that I can't really explain, Murray Gell-Mann's "quark" has survived and "aces" did not. Nobody calls them aces anymore, although they both published these papers in 1964.

There are three different quarks in nature. This is Mr. Gell-Mann's model, a picture of how the world works, and you've got to give them names to distinguish them. So there's three different types of quark. They're quite distinguishable objects, in principle. Being of a whimsical mind, he came up with the names "up", "down", and "strange". Okay? Now, you know, he could have called them anything. He could have called them chocolate, vanilla and strawberry. Thank God he didn't. Up, down, and strange are bad enough. There was a reason why he called one of them "strange," and we'll get to that in a minute. "Up" and "down" were just two good words that distinguished two different things. It has nothing to do with directions in space of any kind, it's just a name. So, here's a little picture of a proton: it's three quarks bound together by some sort of strong force, which we're going to have to talk about, and that is what a proton is. So Mr. Hofstadter, when he's scattering electrons off of this, is just seeing this distribution of quarks inside of a proton, spreading out their electric charge over some region of space.

Now, if this idea was just to explain the proton, I think it would have come and gone. It just would have been an idea that had a little bit of value. But Gell-Mann said, "Look, we've got 100 particles,"(well, not quite, but lots and lots of fundamental particles), "that are being discovered in bubble chambers. Can we explain them all with the same picture?" Now you might imagine a kind of an atomic model, where the proton is three quarks, and maybe one of these heavier particles, like the delta—I don't know, maybe it's four quarks, and five and six; but that's not what Mr. Gell-Mann proposed. It's not just mimicking the old idea of how atoms get built up with more and more and more electrons.

Instead, Gell-Mann's idea is actually simpler. He says there's three quarks. And, every baryon, every one of the strongly interacting particles that really acts like a particle, like the proton and the neutron and all of its heavy partners—they're all made of three quarks, just three. So, you've got different combinations. You could have an up, and an up, and an up; or an up, and an up, and a down; or—right, start making up the combinations. You could have an up, and a down, and a strange: any combination you want, that's going to make a particle. So, all the zoo of particles can be understood as some combination of three building blocks. That's a very simple, very elegant idea; people like the number three, something almost religious about it, and so now the question is, is this model going to work? Are we really going to be able to describe, in detail, all the particles we've seen?

The strongly interacting particles, (that's, sort of, the subject of all these experiments at the particle accelerators) the hadrons, are subdivided into two classes. The hadrons can either be baryons or mesons. Baryons are the ones that are kind of particle-y, like protons and neutrons; and the mesons are the ones which were originally introduced by Mr. Yukawa as kind of "force carriers". Later that idea kind of fell away, but somehow the mesons are intermediaries. There's something very different about mesons and baryons. The mesons, for example, can be produced easily. You can produce as many mesons as you want. Baryons are different. If you start with a baryon, you always end with a baryon. The total number of baryons is always conserved.

So what's a meson? Well, according to Mr. Gell-Mann's quark model, he says, he believes that the meson should be a quark and an anti-quark tightly bound together, a little binary system, just spinning around one another. Now, why he proposed that is sort of the genius of Murray Gell-Mann, and now the whole story begins to make even more sense. So, what's a pion? What's Mr. Yukawa's pion? It's an "up" quark and an "anti-down" quark; you've got one quark and one anti-quark. If you were writing this in a physics paper, you would write U and then a D with a little line over it. We would say "U D-Bar". The bar means anti-particle. So you can form any combination you want, and any combination you want should match up with one of the known particles.

So people started working on the details now. It's a model, and you want to see if the model can be made quantitative, mathematical. So, first of all, we've got to come up with a list of properties of these quarks. For example, what's their mass? What's their charge? What's their spin? And that's almost it. There's a little bit more that you have to describe when you're talking about a quark; but really, this is now postulated to be a point-like particle, just like an electron is a point-like particle. It's the real building block. It's not protons and neutrons after all; it's the things inside of them.

So, for example, what's the spin of a quark going to be? Well, Gell-Mann says, "Let's suppose for a moment that all quarks have spin one-half, just like the electron did, just like the proton did." It's a little object—even though it's a point, we're picturing it as a little ball that rotates—and it's got one-half of a unit of angular momentum. So, what would happen in a meson, where you have a quark of spin one-half, and an anti-quark of spin one-half? How could they combine? Well, you might think a half plus a half equals one; so the meson should have spin one, and many of them do. But spin has a direction. You can spin clockwise, or you can spin counterclockwise; so you could imagine a meson where you had one quark spinning clockwise and another quark spinning counterclockwise, and then it would be a half minus a half, and you'd get zero. So, according to this model, you could have mesons with spin zero or spin one. That's pretty much it, unless they actually have some orbital angular momentum to add up to more; and that's exactly what was observed. The pion, for example, is a spin-zero meson, so now you're learning about the microscopic structure of a pion by looking at its properties.

How about a proton? In this model, Murray Gell-Mann is proposing that a proton is one "up" quark, another "up" quark, and a "down" quark. Those are just the names, remember, of quarks; an up, and an up, and a down. And, how are they spinning? Well, they can't all be spinning clockwise. Because, if they were all spinning clockwise, you'd have a half, plus a half, plus a half is three halves, which doesn't work. Instead, two of them have to be spinning clockwise, and one of them is spinning counterclockwise. So you get a half, plus a half, minus a half, and you get the correct answer for a proton. So, with this model, you begin to formulate what's each particle. And, you can work it both ways. You can look at a particle from the data and try to figure out what its quark makeup is; or, you can be a theorist and just start tabulating all the logical possibilities from Mr. Gell-Mann's model, then see if every logical possibility matches with some particle that's been found. In this way you discover there really is a one-to-one correspondence. Every particle has a description; every description can be found in the laboratory. So that's very compelling evidence that this is a cool model. It's a good model.

Now, there are some issues with this model. For example, if you have these quark things, and if they're real, they're slightly unusual little particles. How come we've never seen a quark running through a bubble chamber, just one? How come we've only seen three of them bound together? Or a quark and an anti-quark? We've never seen a pair, like two quarks. We've never seen four of them. And certainly, we've never seen one all by itself. That was a big puzzle in this early era, in 1964. And, in fact, for many years this quark model was dismissed by most physicists as sort of interesting, had some useful features, but it couldn't be true. Even Murray Gell-Mann would give talks and say, "You know, I don't really believe that quarks are real. I just think they're a mathematical construct which helps us to organize this zoo of particles and understand that it's not so complicated after all. All these patterns, the octet and the triplets, and all these groupings of particles can be understood in a simple way." So, as usual, you come up with this radical hypothesis of a new particle and then you're timid about proposing that it's real. It's like Mr. Dirac, who's afraid to believe that his equation, which predicts antimatter, is predicting something real; or Mr. Pauli, who's afraid that his postulation of a neutrino is really something real. At first, people tend to think, "Well, it's just math." And then, as the evidence begins to grow, people begin to view the object in question as something real. So, we're going to be talking in this lecture, and future lectures, about the growing evidence for the reality of quarks. Quarks are actually little point-like objects that buzz around and can whack into things; and, in fact, nowadays you can see tracks in bubble chambers that aren't directly from a quark, but indirectly from a quark. It's extremely powerful evidence for the real existence of these things.

Let's talk more about adding up the properties of quarks to get the properties of the objects themselves. Let's think about mass for a second. If a proton is three quarks, two ups and a down, you have to ask yourself, okay, what's the mass of each of those three quarks? This is a subtle story which got modified over the years but, in Gell-Mann's day, he suggested that maybe the up and the down quark were very similar to each other, each of them carrying about one-third of the mass of a proton. So, you put three of them together and you get a proton mass. Okay? Now, if the up and the down quarks are similar in mass, then what would happen if you put an up, and a down, and a down together? That would be a neutron, and according to this quark model, since up and down weigh about the same, the neutron and the proton should weigh about the same—and they do. So this model explains why certain particles come in pairs, or partners. Similarly, the three pi mesons can be understood as just different combinations of up and anti-down, or down and anti-up; but, they're all going to weigh about the same. And, indeed, the three pions form a little, lovely triplet, which weighs about the same. So, again, this model has some simple premises; and, once you've written them down, you're inevitably drawn to conclusions. Then the question is, does the data match? And it does: time after time, experiment after experiment, the data matches the predictions of the simple model.

How about electric charge? These are little particles; they have to have some kind of electric charge on them. And Gell-Mann kind of works out what it should be, and says, "Okay, here's my model: the up quark has a charge of two-thirds; the proton is one; the electron is negative one; my up quark is two-thirds, and my down quark is negative one-third, so is the strange quark." So, this is weird. Kind of creeping people out. In fact, I think that reason alone was why Murray Gell-Mann didn't quite believe in the reality of his quarks; because, nobody had ever seen an object in the laboratory that had spin—sorry, electric charge—one-third or two-thirds. That would leave a very distinctive pattern in a bubble chamber. The more charge you have, the density of bubbles effectively changes. You can really tell what the electric charge of an object is. It would leave a really clear signature if you had an object of charge one-third, and nobody had ever seen anything like that.

So, Gell-Mann says, "Well, it's okay, because in my model the quarks always come either in pairs or in triplets." And so, let's think about a proton, an up and an up and a down. So that's charge two-thirds, and another two-thirds, and then minus one-third. Do the arithmetic: two-thirds, plus two-thirds, minus one-third, is one. So the charge of the proton works out. And, a neutron would be an up and a down and a down. So that's plus two-thirds, minus one-third, minus another third, gets you to zero, which is the electric charge of the neutron. So, at first, it looks like numerology, but you don't have any room to fudge. Now you're stuck. Every other particle's charge is predicted; so, you can check if it agrees, and it does. The whole zoo, hundreds of particles, all these quantum numbers are adding up and making sense. It's a coherent picture. It's a nice little model.

This model has been improved over the years, since 1964, and deepened. We're going to have to talk about the ways that this model moved from a crude picture to a real, firm, mathematical theory; but, it certainly gave people pause. They looked at the model, and everybody was intrigued by it. It was nice to think that the world really is simple, and that's there's not 100 fundamental particles of nature. So, I think, although at first people didn't necessarily believe in it, (there was a lot of skepticism about this quark model) people were at least willing to challenge it and test it. It was kind of a working framework that you could compare data with; and, it continued to succeed, every time you tested it.

This model has some predictive power. So, for example, let's talk about this "strange" quark. Why did he call it that? Well remember, people had been finding some strange tracks in bubble chambers in the 1950s. They didn't quite make sense at first. Indeed, it was Murray Gell-Mann, in the 1950s, who proposed that there is a new property of matter, a new quantum number—like electric charge, but different—which he called the strangeness number, because it told you how strange these tracks were. It was always an integer, one, or two, or negative one. And, with every particle in the zoo, when people were tabulating the properties, they'd tabulate its mass, its lifetime, its electric charge and its strangeness number. And now, in the quark model, here's the explanation for strangeness number. Just count up how many strange quarks are in there. If you have a baryon, that's up, down, strange, that should be a particle with strangeness one. If you have a particle that's up, strange, strange, that would be a particle with strangeness two. And again, it matched the data very beautifully. The strange quark would have to be a little bit heavier than the up and down, because the strange particles were a little bit heavier; but, you can make predictions. By looking at an up, down, strange, you can estimate the mass of the strange quark. And now, you can make a prediction about the mass of an up, strange, strange object, and that's exactly what its mass turns out to be. Nice, convincing evidence that this is a good model.

Here's a really interesting idea. What if you had a particle that was strange, strange, strange? Right? You can build any combination of three you want. That would be a triply strange baryon, and nobody had ever seen such a thing. So this is like 100 years earlier, when Mendeleyev first makes his periodic table of the chemical elements and says, "There's a hole here. I believe that nature should fill that hole. There's got to be some material," which later turned out to be germanium, and he was correct. It's the similar story here. There's a hole in our new little periodic table, if you like, of quarks building up baryons and mesons. Let's go look for a strange, a triply strange object. And, it was found; it's called the omega particle. Omega is the last letter of the Greek alphabet and, in a certain sense, this is the last particle that you would have expected, based on this model. Of course, there have been many more particles since then that we'll be talking about.

To be honest, or fair, the omega particle was actually discovered and predicted a little bit before the quark model, because Murray Gell-Mann had come up with what he called the "eight-fold way," which was simply an observation of the patterns that the quark model explains. So really, the quark model retro-dicted, rather than predicting, the existence of the omega. It made such complete sense, and fit in so well, that I think the discovery of the omega particle really was a pretty good (I won't call it a clinching) argument for the quark model. At least it left people with a sense that this is a viable model.

The "strange quark" is a little bit of a funny particle; it is a strange particle. For one thing, it's heavier than the other two. For another thing, its electric charge is the same as the down quark. So, it kind of looks like a heavy partner of the down quark. We've seen this before. We've discovered a muon, which is a heavy partner of an electron. In fact, oftentimes, when we discover particles, we discover that they have heavy (I don't know what to call them.) family members. It's another particle of nature that's a lot like it, but radioactive and heavier. The down quark is the stable one, and the strange quark decays weakly into a down quark, and/or into an up quark, it turns out. This is just a sort of an interesting curiosity; it's kind of another "who ordered that?" situation. Why do we have a strange quark in nature? Are there strange quarks in the ordinary world? Not very many. Your body is made mostly of protons and neutrons, which are ups and downs. In fact, everything in the world around you, to an extremely large approximation, is just up quarks and down quarks, which build protons and neutrons, which build the nuclei of atoms; and then there's electrons, and that's all you need. So protons, neutrons, and electrons are still a perfectly valid way of thinking about how the world is made of. But if you want to look at the simplest level, then you should think of ups, and downs, and electrons, and that's the light particles, they tend to be quite stable; and then we've got these heavier, somewhat more mysterious particles.

They are real, and they occur in nature. They occasionally occur in cosmic rays. There are stars that people believe have strange quarks in them. This is an active field of research in astrophysics. Some of my own nuclear physics research is looking into the amount of strange-quark content in ordinary matter. There's always going to be some small probability that, for example, a strange/anti-strange pair will pop into existence inside of a proton. So I'm investigating the strangeness of ordinary matter. I might be a strange-particle physicist if this stuff turns out to be correct. So people are still investigating the details of the story, but the fundamental idea was proposed in 1964, and is still part of our standard model of particle physics today.

It's not the end of the story. What we're going to discover is that there are more particles, and we're going to have to try to fit this stuff all together. In particular, I haven't yet said word one about the forces that bind these quarks together. Right now, we have a very descriptive model. It's kind of like Johannes Kepler, taking this mass of data about the planets and saying, "Aha! They form ellipses. Very elegant." So, I think that's what Murray Gell-Mann has done by creating this model of quarks inside of protons, and neutrons, and pions, and so on.

And what we're still lacking, and what's going to come next, is a real theory. Like Isaac Newton said, "Ah—but we can understand that elliptical orbit as arising because there's a universal law of gravity." So what we're really looking for is some underlying, theoretical, mathematical framework. The quarks will be the players, just like the electrons were the players for QED. We're going to look for the analog of the theory of quantum electrodynamics. For this new strong force of nature that creates the baryons, the mesons, the hadrons that build up nuclei.

Lecture Twelve
From Quarks to QCD

Scope: Fully understanding the world requires describing not just the *players* but the *forces* they feel. If quarks are the particles, how do they interact with one another? The answer to that puzzle has a curious (and, at first, misleading) name: They carry a new kind of charge, a *strong charge*, described by *color*. A fledgling theory arose in the 1970s, called *QCD*. We'll unpack the etymology of that acronym to make sense of it, just as we did for QED. QCD is a formidable mathematical theory, but the results are elegant and comprehensible. Quarks are *confined* together, they stick together fiercely, and if we try to pull them apart, we inevitably make more (ordinary) particles. On the other hand, when they get very close to each other, the strong force fades away to nothing, and we can make quantitative predictions with relative ease. We will look at further experimental evidence for the existence of quarks and discuss why we might believe in these particles that can never be seen, even in principle.

> Our job in physics is to see things simply, to understand a great many complicated phenomena, in terms of a few simple principles.
>
> —Steven Weinberg, Nobel Prize Lecture, 1979

Outline

I. Gell-Mann and Zweig introduced the idea of quarks into the physics literature in 1964. This idea helps to describe and explain a good deal of data, but it was quite controversial for many years.

    A. Already by the 1960s, there was a sense in the physics community that a crude model was not good enough. Physicists wanted to be able to make concrete and rigorous statements about how the world works. This goal went hand-in-hand with the belief that the world really is simple, and that simplicity can be described and explained.

    B. The goal of particle physics, starting from the introduction of quarks, was to come up with a theory, a mathematical framework, in which one could calculate and predict any observable involving subatomic particles.

C. Physics already had a framework that was successful for describing electrical forces, quantum electrodynamics. Physicists looked to QED as a kind of template, wondering if a quantum field theory could be developed for the strong force.

    1. The particle of interest in quantum field theory is the quark, because the theory is put forth to examine strong forces; it focuses on the hadrons and their constituents.

    2. In QED, every particle possesses electric charge. Charge is essential in QED because it tells us the strength of the interaction.

    3. Unfortunately, we didn't know the analog of charge for the strong nuclear force. We knew that quarks carry something, but what? They have electric charge, which means that they have electrical interactions, but electrical interactions don't explain why the quarks bind together.

    4. In fact, electricity would make quarks tend to fly apart. Two up quarks, for example, have the same charge. Electricity makes them want to fly apart, but they can be bound together in a proton with a down quark. We need to identify some much stronger force to overwhelm the electricity and bind quarks together.

D. Some evidence and theoretical arguments in the 1960s pointed to a new quality of nature that would play the role of electrical charge for quarks. Scientists eventually pulled this evidence together into a formal theory, which was even more complicated than QED.

II. What is this extra quality that scientists were looking for?

A. Think about the delta particle, a kind of heavy version of a proton. It decays strongly and rapidly and turns into a proton. In the quark model, the delta is three up quarks.

B. The delta particle has spin 3/2. It is a combination of three up quarks, all spinning clockwise. A quantum physicist would object to this statement because an up quark has spin 1/2, and we can't put two up quarks (or three) into the same quantum state.

C. The only way out of this dilemma is found in quantum mechanics, which allows particles to be put together if they are distinguishable, that is, if there's something different about them.

D. We know that we can put an up and a down and a strange quark together. We also know that we can put two ups and a down together as long as some of them spin clockwise and some, counterclockwise. But how can we put three up quarks together? The answer is that there must be some other property of quarks, a new quantum number.

E. Apparently, this new number must have three possible values. The new variable was called *color*, even though it is not related to visible color. (It's *just* a name.) Physicists now spoke of blue, green, and red quarks.

F. If we start from the idea that quarks possess color, then we can explain how three seemingly identical quarks can be put together to form a delta particle. Various other pieces of evidence also validated the idea that color was, in fact, a new degree of freedom.

G. Some scientists thought that this new property might also be the charge that is responsible for binding quarks together. This charge is called *color charge*, and the force that arises when color charges are near each other is *color force*.

H. Another puzzle was why quarks combine in threes to make baryons.

    1. If we think of a color wheel, the metaphor is a good one. A color wheel has three primary colors, and if the three primary colors are added together, the result is white, which is a sort of neutral color.

    2. If we combine three quarks and they form white, then we have no net color left. Thus, a white object and another white object, such as two protons, do not attract each other particularly strongly. A red quark and a blue quark, on the other hand, feel an enormously strong force, a hundred times stronger than the electrical force, which is what binds them.

    3. We must note that protons do have a little distribution of color charge inside them that makes them begin to attract as they get closer. They feel no force until they are about a femtometer apart, then they stick very strongly.

I. How does this theory work for mesons? These particles are built not of three quarks but of a quark and an anti-quark.

1. This metaphor works for that situation, too. A color and its anti-color are straight across from each other on the color wheel, and they also add to white.

2. A meson, which is a quark and an anti-quark, would be red and anti-red, for example. If these are combined, they will attract very strongly because they're different colors, but they also form white. As with a baryon (such as a proton), a meson will feel nothing most of the time until it gets close, then it feels the strong force of nature.

J. At this point, the idea of color was still not a theory. It lacked a mathematical framework that would be similar to the one for QED. In this framework, however, instead of having two electrical charges, we would have three color charges. The name for this quantum field theory of color forces is *quantum chromodynamics*, or *QCD*.

III. QCD took a long time to develop, in part because the mathematics of three is more complicated than the mathematics of two.

A. In QED, when we have a charged particle, it has an electric field around it. That's a classical way of thinking about the force created by electricity.

1. How do you visualize photons in this classical picture? If you jiggle the charge, you're jiggling the electric field, which propagates like moving water on a pond and makes the wave that travels out. When you jiggle an electric charge, classically, you get an electromagnetic wave. In the quantum mechanical view, the "jiggle" is the photon.

2. In the same way, if we have a colored object and there's a color field around it, if we were to jiggle it, that would be the classical way to propagate the strong force.

3. What is the particle of the strong force? If we have to give a name to the "jiggle" in the color field, we call it a *gluon*.

B. The gluon is essentially the force carrier of this strong color force, and it really is like glue. It makes things stick together incredibly strongly. We might picture a proton with three colored quarks surrounded by this color field.

C. One of the complexities of this color theory is the following:

1. If we have an electric charge and it emits a photon, that particle, the jiggle of the field, is itself electrically neutral. An electron has charge, but the photon does not.

2. When we work out the math with the color forces, however, we discover that the jiggles in the color fields are themselves colored; a gluon has color charge. This means that as a gluon is traveling along, it can interact with another gluon. Gluons can make gluons, can make gluons!

3. This is part of the reason that the strong force is so strong, so much stronger than the electric force. We start with two color charges, and they produce a color field between them, which enhances itself spontaneously.

4. If we think in terms of Feynman diagrams, we might draw one diagram with some gluons on it, then we would have to draw another diagram with more gluons and gluons meeting gluons and gluons forming all sorts of complicated structures. Further, because the force is so strong, the diagrams don't get smaller as we add more complexity. The numerical calculations in QCD are extremely difficult.

D. We can calculate some of the consequences of QCD. For example, imagine an experiment in which you are trying to pull two quarks apart.

1. As you pull, the gluonic field is stretched. Gluons are bosons. They are attracted to other gluons. You will start to get more and more gluons, stretching in a line between those two quarks.

2. The more you stretch the field of gluons, the more glue you have and the tighter the quarks will pull. As you pull these quarks apart, a force of nature similar to a rubber band is pulling them back together.

3. If you do the calculations for this experiment using QCD, you discover that it is impossible to get the quarks out of a proton. The more you try to pull them apart, the stronger the strong force becomes. It actually grows with distance, unlike all the other forces we know about, such as gravity or electricity, which become weaker as objects get farther apart.

4. What had seemed like a dilemma for Gell-Mann's quark model, namely, "Why haven't we ever seen a quark if they are

real?" is now shown to be a mathematical consequence of the theory.

IV. Let's turn to another question about QCD: Why should we believe in a theory that's so hard to calculate with?

    A. Let's approach this question from another direction: What if we try to squeeze two quarks together?

        1. We can do this if we work with a proton that is struck with very high-energy particles. High energy means that the distance scales are very short.

        2. If we're looking at two quarks close together, the situation is again much like a rubber band. When we stretch the rubber band, it tends to snap back, because the system has a lot of energy and the forces are strong. If we try to squeeze a rubber band, however, there's no resistance at all.

        3. The same thing happens in QCD. When the quarks are at distances much smaller than about a femtometer (10^{-15} meters), the forces become very weak.

        4. Inside of a proton, the three quarks are almost free particles. If they get a little bit too far away, they are snapped back, but while they're in the proton, they're sort of loosely bound.

    B. From the early 1970s, when this theory was proposed, until today, physicists built higher energy machines and finer microscopes and discovered that the quarks are less and less strongly bound. When the bonds are looser, the calculations are easier to perform.

    C. This is a wonderful twist. Usually, as we build more sensitive machines, we get more complicated results. In this theory, however, the opposite is true. If we think in the language of Feynman diagrams, the coupling gets weaker; therefore, those complicated diagrams become numerically less important and the calculations become easier and more reliable.

    D. Again, this is not proof that QCD is correct, but it is certainly a compelling reason to take it seriously. The more high energy the experiment is, the more accurately we're able to test and verify it.

V. There are other reasons that scientists give credence to quarks and QCD, relating to the concept of confinement.

    A. What would we see in the laboratory if we hit a quark very hard?

1. If we hit one quark with an electron beam and it goes flying off, the other two sit behind for a moment because of their inertia, but one of them is running away, stretching that rubber band. A stretched rubber band, or in this case, a stretching gluonic field, has a great deal of energy in it. All the energy that the experiment gave to the particle is now going into the gluonic field.

2. What happens in quantum mechanics if a lot of energy is in a small space? A quark/anti-quark pair can be produced. Matter and antimatter can be produced spontaneously out of that energy.

3. As the rubber band stretches, it might snap back, which would leave behind a rattling proton, but it might also essentially break. The rubber band breaks and, at the point of breaking, a particle/anti-particle pair is produced.

4. Remember, we have one quark running away and two quarks left behind. We produce a quark/anti-quark pair. How will they line up? The quark that is produced will stay behind and the anti-quark will go along with the quark that is running away. We're left, then, with a baryon. We started with a baryon of three quarks, and we end with a baryon of three quarks.

5. However, we've also produced a quark/anti-quark pair, a pion or a meson, a heavier version of a pion. If we have enough energy, the rubber band might snap a number of times while the quark is running away, resulting in a stream of pions running away from a baryon.

6. Again, that's exactly what we see in laboratories. These are called *quark jets*. When an electron or a particle is smashed into a baryon, we can actually see a straight line of particles flying off, constituting almost direct evidence of the existence of quarks.

B. We identify other fairly direct arguments for the existence of quarks.

1. Think about Rutherford's experiment in the 1900s, the prototype particle physics experiment: striking a target with a beam of particles to see how they scatter. The idea is to determine what's inside the target by looking at how particles bounce off it.

2. This experiment has been repeated countless times in different guises. Hofstadter, for example, did a similar experiment with electrons hitting protons and determined that the proton was like a smear of electric charge. The energies he was using, however, weren't high enough to see down to the level of a quark.

3. What if we use higher energy still? Remember, the higher the energy, we use, the smaller the particles we can see. We would think that at a high enough energy, we would begin to notice that there really are nuggets, little hard objects, inside the proton.

4. In the 1960s, a fabulous physics facility was built, called the Stanford Linear Accelerator Center (SLAC). SLAC is essentially a series of straight-line accelerators, resulting in an enormously high-energy beam of electrons.

5. In the early 1970s, this beam was aimed at protons, similar to the Hofstadter experiment but at higher energies and with greater magnification. As in Rutherford's experiment, every now and then, an electron would bounce backward. This offers direct physical evidence that the electron was hitting something solid instead of running into a smear.

6. At first, scientists were skeptical that this phenomenon was the result of quarks; instead, the particles that were being hit were called *partons*, for "parts of protons."

7. As time went by, however, and the experiment was refined, physicists began to make quantitative conclusions about, for example, the charge of this object. In fact, the charges turned out to be 2/3 and –1/3, which is just what the quark model said they should be. The spin also turned out to be 1/2, just as the quark model predicted.

C. Over a period of five to 10 years, more data were collected from this experimental procedure, which is called *deep inelastic scattering*. The result of these experiments was that we could see the consequences of quarks on the whole system, even though we couldn't see the quarks themselves.

Essential Reading:

Barnett, Muhry, and Quinn, *The Charm of Strange Quarks*, chapters 3.3 and 4.1.2

't Hooft, *In Search of the Ultimate Building Blocks*, chapter 13.

Recommended Reading:

Lederman, *The God Particle*, chapter 7, the section on "Rutherford returns" and from "The strong force revisited" to the end.

Riordan, *The Hunting of the Quark*, chapters 5–7 and 10.

Wilczek and Devine, *Longing for the Harmonies*, Seventh Theme.

Questions to Consider:

1. Considering the history of growing evidence for quarks, at what point (if ever) do you think you would have joined the "bandwagon" of physicists who believed in the existence of quarks?

2. Do you think any physicist will ever detect a "free quark" in the lab? Why or why not? What consequences would such a discovery have on the theory of QCD?

Lecture Twelve—Transcript
From Quarks to QCD

Murray Gell-Mann and George Zweig introduced the idea of quarks into physics literature in 1964. It was an elegant idea. It certainly helped describe and explain a lot of data. But, it was quite controversial, and subject to a lot of debate and analysis for many years, in part, because it was just a model. It was just a clever but crude collection of ideas, which made some sense of a lot of data without deeply explaining it. And, I think, a crude, qualitatively successful, sometimes quantitatively successful, model is good enough in a lot of branches of science. But particle physics is beginning to expect a little bit more, I think. Already, by the 1960s, there's a sense that we should really be able to make concrete and deeply rigorous statements about how the world works. It's kind of a philosophical attitude. A belief that the world really is simple, and that simplicity can be described and explained mathematically. And so, this is now the goal, I think, of particle physics, starting from the introduction of quarks: to come up with, not a model, but a theory, a mathematical framework, in which one could calculate and predict any observable that you could ask for involving subatomic particles.

Now we already had a framework, a mathematical theory, which was very successful for describing electrical forces, and that's quantum electrodynamics. Quantum electrodynamics is a theory, you can write it down on a piece of paper. It's got particle constituents, that is to say, electrons and any charged particle; and it's got formulas which tell you how to make quantitative predictions about the behavior of those electrons in the presence of other electrical charges. It includes the effects of vacuum fluctuations, virtual particles appearing out of nowhere and then re-annihilating very quickly, which make tiny, but sometimes important modifications. So that's kind of a template that physicists are looking at, and wondering, "Can we make a quantum field theory for the strong force?" Now the particles that are going to go into your quantum field theory will be the quarks. That's the idea, because we're looking for strong forces. The electron has nothing to do with strong forces. The electron isn't made of quarks. It's got nothing to do with quarks. It's a whole separate story by itself. We're focusing on the hadrons and their constituents now.

In the theory of quantum electrodynamics, there is a very, very important quantity that every particle possesses, and that's electric charge. It's a number. The electron has negative one in a certain unit system, the proton, positive one, and so on. Every particle has a certain amount of charge. Charge is absolutely essential in quantum electrodynamics because it tells you the strength of the interaction. The electrical interaction depends on your electric charge; the more charge you have, the stronger the interaction. It's really the key ingredient. It's not the electron. It's the charge on the electron that's really essential. Unfortunately, we don't know the analog for the strong or nuclear force. We know that quarks carry something, but what do they carry? They do have electric charge, so they're going to have electrical interactions; but, electrical interactions don't explain why the quarks bind together. In fact, electricity would make them want to fly apart. Two up quarks, for example, have the same charge. Electricity makes them want to fly apart, but they can be bound together in a proton, along with a down quark. You need some stronger, much, much stronger force to overwhelm the electricity and bind them together.

There was some evidence, and more theoretical arguments which had developed in the '60s, which pointed to a new quantity of nature which would play the role of electrical charge. It would be like a strong charge. And, it took a long time for this story to develop. It took about ten years for people to really put it all together into a formal theory; and the formal theory that developed was even more complicated that quantum electrodynamics, in a lot of ways. We're not going to go into the mathematical details, but again, I think the essence is something that is very much graspable. So, what is this extra thing that people were looking for? Well, let me tell you about a puzzle that led to a, sort of an idea, which led to the idea of what the electric charge is. It was a chain of logic. Think about the delta particle. I've never really said much about it, except that the delta particle is a kind of a heavy version of a proton. It decays strongly and rapidly and turns into a proton. In the quark model, the delta is an up, and an up, and an up. And, the delta particle has spin three-halves, so it must be an up quark which is spinning clockwise, and another which is clockwise, and another which is clockwise. That's what a delta particle is; it's just this little combination of three identical up quarks spinning all the same way.

Now if you're a quantum physicist, you say, "Hold on a second. No way!" Because an up quark has spin one-half, you can't put two up quarks into the same quantum state. You can't put two up quarks together. You certainly can't put three together if you can't even get two of them together. So there's a dilemma here, and there's only one way out of this dilemma in quantum mechanics. And that is, you are allowed to put particles together, in quantum mechanics, if they're distinguishable, if there's something different about them. So it's no problem to put an up, and a down, and a strange together. And it's even possible to put an up, and an up and a down together, as long as some of them are spinning clockwise and some of them are spinning counterclockwise. But how on earth can you put three up quarks together?

The answer would have to be that there is some other property of the quarks. They contain some other information that we haven't thought about yet. It would be a new quantum number. And apparently there have to be three different possible values, not just two. It can't be like spin, where you can be clockwise or counterclockwise. You need a new variable that has three degrees of freedom, instead of two. What shall we call this new degree of freedom? People had a clever idea. Let's call it color. So we could have a blue quark, and a green quark, and a red quark, and now they're distinguishable and they can go in the same place. Now please bear in mind, it's not really blue; it's got nothing to do with the color that you see with your eyeballs. That color, blue, that you see with your eyes represents very long-wavelength electromagnetic radiation. It's just a metaphor. Red, green, and blue are three colors on the color wheel, and just a nice way of representing different properties of these quarks.

So if the quarks possess what we now call color, which again is this physics word that doesn't mean visible color, then we can explain how it is that you can put three, what look like identical quarks, together and form a delta particle. And, there were other various pieces of evidence that color was, in fact, a degree of freedom. Sometimes you can do experiments, bouncing off quarks, where you can count how many different quantum-mechanical degrees of freedom they have, and the answers seem to be three, rather than one. So people got the clever idea that this new property of quarks might also be the charge of the quarks that's responsible for them binding together, so they called it color charge. And, the color force is the force that arises when you've got color charges near each other. And this turned out to be a brilliant idea, and it is the basis for the theory that we now have, which explains, in great detail, the interaction of the quarks.

Now one of the big puzzles of the day is, why do the quarks combine in threes to make baryons? If you think of a color wheel, the metaphor gets better and better. On a color wheel, there are three primary colors. Only three. Exactly three. And, if you add the three primary colors together, you get white, which in a certain sense, you could say, is colorless. Or you could say it's all colors. But in any case, it's a sort of a neutral color. And so, supposing you took three quarks, red, green and blue; if you combine three quarks together, and they form white, then you have no net color left, and so a white object and another white object will not attract one another particularly strongly. And that's why a proton over here and a proton over there don't feel each other at all. Because they're white objects, they don't feel this strong color force. A red quark and a blue quark, those are different colors. They feel an enormously strong force, 100 times stronger than the electrical force, which is what binds them.

Now, it's not true that they feel absolutely nothing, because the red's in one spot, and the blue is slightly displaced, because these are quarks buzzing around one another. So, as two protons get closer and closer together, even though they're overall neutral, the fact that they have a little distribution of color charge inside of them makes them begin to attract as they get closer, which is just the way protons behave. They feel no force until they get about a femtometer apart, and then they stick very strongly, because now all of a sudden these colored quarks are right next to each other, and they notice that there's colors nearby, and everybody binds together. So this metaphor of colors is a nice one to help explain the properties that had been observed already, starting in the 1930s, about how the strong interaction works.

How about the mesons? The particles that are built not of three quarks, but of quark/anti-quark? Well that works too with this metaphor. If you have a color and its anti-color—that would mean straight across it on the color wheel—those also add to white. So a meson, which is a quark and an anti-quark, would be red and anti-red, for example. And those will combine together as they will attract very strongly, because they're different colors. But, when you look at the two of them together, they form white; and so a meson, like a proton, will most of the time feel nothing until it gets really close, and then it feels the strong force of nature. This is still not a theory. What one need to do is write down a mathematical framework that's just like quantum electrodynamics, only instead of having two electrical charges, you're going to have three color charges; and now you're going to try to kind of mimic the mathematics of quantum electrodynamics. And people came up with the clever name of quantum chromodynamics. Chromo, meaning color. So, this is the quantum field theory of color forces. Quantum chromodynamics took a long time to develop, in part, because the mathematics of three is a lot more complicated. It's not just 50 percent more complicated. It's a lot more complicated than the mathematics of two different charges.

Let me say a word about QED. When you have a charged particle, it has an electric field around it. That's one of the ways, a classical way, of thinking about the force created by electricity. And what is a photon? How do you think visually about a photon in this classical picture? Well, jiggle that charge. If you jiggle a charge, you're jiggling the electric field, and it's a jiggling electric field, which propagates out like jiggling water on a pond and makes a wave that travels out. So when you jiggle an electric charge, classically, you get an electromagnetic wave. Quantum mechanically, you emit photons. So, I think of photons, the particle of light, as the jiggles in the field. And so, if we have a colored object, and there's a color field around it, if you were to jiggle it, that would be the way that you would propagate the strong force. And what would the particle of the strong force be? Mr. Yukawa thought it was the pions, but we don't think that anymore. The pions are, themselves, just quark and anti-quark pairs. We're looking a level deeper now, and we're saying it's the color field. And, you have to give a name to this jiggle; instead of a photon, we call it a gluon. It's another one of those silly names that physicists came up with, but it's nice. It's the name for the force carrier of this strong color force of nature, and it really is like glue. It makes things stick together incredibly strongly.

If you're trying to picture a proton, you've got three colored quarks, and the colored quarks have these color fields around them. There's a little cartoon that you can draw. This is not a physicist's diagram, it's just a sketch of a proton, with a red quark, and a blue quark, and a green quark in it; and then these little lines going between them are supposed to represent the color field, in some artistic sense. It's the gluons that are binding the quarks together.

One of the complexities—and, in fact, very important aspects of this color theory—is the following; if you have an electric charge and it emits a photon, the particle, the wiggle of the field, is itself electrically neutral. An electron has charge, but the photon does not. The photon interacts with charges, but it doesn't have its own electric charge. But when you work out the math with the color forces, with quantum chromodynamics, you discover that the wiggles in the color fields are themselves colored. So a gluon has color charge on it, and now you think, "Oh my goodness! That means that a gluon, as it's traveling along, can interact with another gluon. Gluons can make gluons, can make gluons." This is part of the reason why the strong force is so strong, so much stronger than the electric force. You start off with two color charges, and they produce a color field between them, which enhances itself, spontaneously.

If you're thinking in terms of Feynman diagrams, which would be the way that you would try to calculate these forces, an analogy with Feynman diagrams in QED, life becomes complicated. You draw one diagram with some gluons in it, and then you have to draw another diagram with more gluons, and gluons meeting gluons, and gluons forming all sorts of complicated structures. And, because the force is so strong, the diagrams don't get smaller and smaller as you add more and more complexity. So it becomes very, very difficult to do numerical calculations in quantum chromodynamics. I could take a graduate student in a field-theory course— so that would be a, maybe a second-year physics graduate student—give them a homework assignment to evaluate a simple quantum-electrodynamic process, like an electron scattering off of another electron, and they could do that in a week or two. But I can't do that in quantum chromodynamics; it would be a whole thesis project. And even now, people are still, today, doing research, trying to figure out how to mathematically compute these complex diagrams.

There are many consequences. Some of them can be computed. It's not that we can't do anything, it's just that you can't do *everything* in quantum chromodynamics like you can in the theory of electricity and magnetism. For example, supposing you take two quarks, or three quarks. You have some object, and there's all these color, glue fields; and you start to pull apart the quarks. Now, that's a sort of a metaphor. How would you really do that in the lab? You might send an electron beam in, which whacks into one of the quarks and sends it flying by the electrical interaction. In any case, you could imagine an experiment where you try to pull two quarks apart. Well, what's going to happen? As you pull, the gluonic field gets stretched. Gluons are bosons; they love to be near other gluons. So you'll start to get more and more and more gluons, kind of stretching in a line in between those two quarks. You get a very intense blob of gluons in there. And the more you stretch it, the more glue you're going to have, and the tighter it's going to pull. So it's like a rubber band. As you pull these quarks apart, there's a rubber-band force of nature which is pulling them back together. So that makes you think that it's going to be really hard to get a quark out of a proton.

So you sit down. You start calculating with this fledgling theory of quantum chromodynamics, and you discover, in fact, it's impossible. The more you try to pull them apart, the stronger the strong force gets. It actually grows with distance, unlike all the other forces we know about, like gravity or electricity that get weaker as you get farther apart. And so there's no limit to how strong this can get; the quark is confined inside of the proton. There's no way to get it out. You could stretch it, you could add energy to the system, but you'll never get a free quark. So the theory of quantum chromodynamics is actually making a quantitative—maybe a qualitative prediction, that you won't observe a free quark in the laboratory.

So what had seemed like a dilemma for Murray Gell-Mann's quark model; namely, why haven't we ever seen a quark, if these things are real, is now actually a mathematical consequence of the theory that's developing. So that's nice. It's not proof that quarks exist; but it's at least preventing you from throwing away the idea of quarks, because the model itself is consistent with the fact that you will never see a quark by itself. In fact, this idea explains exactly why you either get a quark/anti-quark, which is overall colorless, or three quarks, which is overall colorless, but nothing else. You don't get a quark and a quark, because you can't add two primary colors and get white. If you don't have white, you've got incredibly strong forces, and that's going to suck another quark in there to make a baryon. It might even create a quark out of the vacuum if it has to, because you can't just have a colorful object all by itself in the laboratory. So this theory is beginning to explain the details of what Mr. Gell-Mann had just proposed as an almost ad-hoc model. That's what you want. You want your theory to explain the features of the model, and make them quantitative.

There's another aspect of quantum chromodynamics that makes it extremely appealing. What I want to head towards now is, why should we believe in a theory that's so hard to calculate with? Why should we believe in QCD? Why do we believe in the existence of quarks? We're going to be seeing more and more evidence, in future lectures, for why we believe in these things as being real; but, let me tell you some more reasons right now. What about if you go the other direction? What if you take two quarks and you try to squeeze them together? You can do that by looking at a proton with very high-energy particles. High energy means short distances, so you're looking with a microscope at smaller and smaller distance scales. If you're looking for two quarks close together, it's again very much like a rubber band. When you stretch a rubber band, there's lots of energy in that system and it wants to snap back, and the forces are strong. But if you try to squish a rubber band, there's no resistance at all. It's floppy. And the same thing works out in the theory of quantum chromodynamics. As the quarks get at distances much smaller than about a femtometer, than about 10^{-15} meters, the forces actually get very weak. So inside of a proton, the three quarks are almost free particles. They're just roaming around. If they get a little bit too far away, they get snapped back. But while they're in the proton, they're quite happy, and sort of loosely bound.

What this means is that over the years, from the early '70s, when this theory is proposed, until today, as you build higher and higher-energy machines, and you're looking basically with a finer and finer microscope, what you find is that the particles you're observing, the quarks, are getting less and less strongly bound. They're getting more loose, and the looser-bound they are, the easier it is to calculate. This is a kind of a wonderful twist. Usually, as you build higher-energy machines, new accelerators, all hell breaks loose. You get new particles, you don't know what's going on; it gets more complicated. But now we've got a theory that goes the other way around. The higher-energy, the higher-tech your detectors and accelerators get, the better we can do the calculations, the more accurately we can predict what's going to happen. So we can actually make predictions, numerically quantitative predictions, about behaviors at high energy much, much better than we can do at the old, 1960s, low energies. So this is, again, not proof that quantum chromodynamics is correct, but certainly a nice, compelling reason to take it seriously. The more high-energy the experiment gets, the more accurately we're able to test and verify it. If you think in the language of Feynman diagrams, the coupling gets weaker, and therefore those complicated diagrams become numerically less important, and your calculations get easier and more reliable.

Let's talk about other reasons why people believe in quarks and QCD: What about this story of confinement? What would happen if you really whacked a quark hard? What are you going to see in the laboratory? You send in an electron beam, and one of the quarks is going to go flying. The other two are kind of stunned and sit behind for a moment, because of their inertia, and one of them is running away, stretching that rubber band. A stretched rubber band, or in this case, a stretching gluonic field, has lots of energy in it. All the energy that you gave to the particle is now going into the gluonic field, and in quantum mechanics, if you've got a lot of energy in a small space, what can happen? You could produce a quark/anti-quark pair. You can produce matter and antimatter spontaneously out of that energy. Right? It's $E = mc^2$, put together with quantum mechanics. So, as the rubber band stretches, it might snap back, which would just be leaving you behind with a rattling proton, but it might also essentially snap. The rubber band snaps, and at the point of snapping, you get a particle/anti-particle pair. Let's think in detail how this would work.

You left two quarks behind, and one quark is running away. You produce a quark/anti-quark pair. How are they going to line up? The quark that's produced is going to stay behind, and the anti-quark that's produced is going to go along with the quark. So what are you going to have? You're going to have a baryon left behind. You started with a baryon of three quarks and you end with a baryon of three quarks. But, now you've also produced a quark/anti-quark pair, a pion, or something like a pion, some meson, a heavier version of a pion. And if you have enough energy, the rubber band might snap a bunch of times while that quark is running away, and so you should get a little stream, in a straight line, of pions running away from a baryon. And that's exactly what was seen in laboratories. These were called quark jets, and when you smash an electron or a particle into a baryon, what you see is bubble-chamber pictures—or more sophisticated versions of bubble chambers, nowadays, electronic versions of bubble chambers. And you can actually see a straight line, a jet of particles flying off. And you're almost directly seeing a quark, but not quite, because the quarks always combine together to form mesons or baryons in the laboratory. But it's extremely close to direct evidence, and in fact, you can see multiple jets. As you go to higher and higher energies, you might share that energy between two quarks. We've see two jets and three-jet events, and they all agree quantitatively. You can make predictions, from the theory of quantum chromodynamics, like how many particles in the jet, as a function of energy. Which way do they go? How do they recoil? And it all works. It all ties together very nicely.

Here's another argument, also a fairly direct argument, for the existence of quarks. Think about Mr. Rutherford's experiment, going way back to the early 1900s. This is the prototype particle physics experiment. You have a beam of particles, you strike a target, and you look and see how they scatter. What do they do? Do they go through? Do they bounce backwards? You're trying to figure out what's inside your target by looking at how particles bounce off of it. Now this experiment has been repeated a million times, in a million different guises. For example, Mr. Hofstadter, in the '50s, did that experiment with electrons hitting protons and he discovered that they were acting like the proton was a kind of a smear of electric charge. That's because the electron's energy was not too high, and not too low. It was high enough to see down inside the proton, but not high enough to see down all the way to the size of a quark. So, it was just seeing the average of these quarks that were buzzing around, and so it's a smear.

But what would happen if you go to higher energy still? The higher your energy, the smaller you're looking. You would think that at a high enough energy, you would begin to notice that there really are little nuggets, little hard objects inside the proton. That's our model. That's what Murray Gell-Mann, and ultimately QCD, propose: that the world is made of little, real, hard objects called quarks, and they bind together by the strong force. That's what a proton is, for example. If they're real, you should be able to bounce off of them. So, there was a laboratory built in the 1960s called the Stanford Linear Accelerator Center. It's a fabulous physics facility; it's still going strong, and there've been many Nobel Prizes that have come out of what is now called SLAC. It was built in 1966.

It's not a cyclotron. People had evolved beyond always building cyclotrons. It's called a linear accelerator because it's accelerating electrons. And, if you make an electron go around in a circle, even a big circle, it's going to radiate away some of its energy, and that's wasteful. But, if you make it accelerate in a straight line, it doesn't radiate away so much of its energy. So what Stanford has done is, for two miles long you've got a little particle accelerator, and then another one, and then another one and another one, all in a straight line. So, by the end, you've got this enormously high-energy beam of electrons. And in the early '60s—sorry, late '60s, early '70s—this beam was aimed at protons; so it's like a repeat of the Hofstadter experiment, but at higher energies, at bigger magnification. Now what did they see? Well, they saw, every now and again, an electron would bounce backwards. So this is like Rutherford all over again. We're seeing a little nugget inside of the proton. So that's direct physical evidence that you're whacking into something solid, instead of just hitting some sort of a smear.

Now at first, when people saw this, they said, "Yeah, could be the quarks," but people didn't quite believe in quarks yet. So they said, "Look, let's be skeptical as we can be. Maybe it's quarks. Maybe the quark picture isn't right, and maybe there is something hard and small and charged inside of a proton, but it's not a quark. It's just a part of a proton. Maybe there's not threes, and all this stuff that we've been deducing." So they were a little bit skeptical at first. Instead of calling them quarks, they called them partons, for the parts of the proton. I believe it was Mr. Feynman who actually came up with that name. And as the years went by, and the experiment refined, you could begin to make quantitative conclusions about, for example, what's the charge of the object that you're bouncing off of? The more charge it has, the stronger an electron's going to bounce; and so, you can begin to collect data, which tells you that. And in fact, the charges turned out to be two-thirds and negative one-third, which is just what the quark model says. And you can look at angular momentum of the process and deduce what's the spin of those little quarks. And again—the partons, whatever it is that you're hitting—and again, it came out to be one-half, just like the quark model.

So, I would say over a period of five to ten years after those original experiments, which were obviously indications that there's something in there, the idea that they might be quarks got stronger and stronger. And now, we have this enormous plethora of data from a bunch of facilities, not just Stanford, that indeed all the properties that we attributed to the quarks are being demonstrated. These experiments are called deep inelastic scattering—that's the buzzword in the physics community. Deep, because it's ultra-high-energy and you're plowing right down into the core of a proton, looking for the quarks. And inelastic is the physicist's technical word for "smashing into smithereens". Right? You're taking an electron, and you're whacking it into a proton and just blowing that thing apart. And you're looking for, not all the spray that comes out, although people are looking at that now, but you're just trying to follow that electron and see what it does. It's like throwing a rock through a plate-glass window, and you're not watching the shards of glass, you're just looking to see what the rock does. You can learn something about glass by seeing how the rock behaves as it goes smashing through.

Let me give you a little metaphor that I like to think of when I'm trying to ask myself whether I believe in the physical existence of quarks. Once upon a time, people knew about planets in the solar system, going from Mercury in the middle all the way out to Neptune, near the outside edge. People didn't know about Pluto, at a certain point in time. And if you look carefully at the orbit of Neptune, you discover that it doesn't quite obey Newton's Laws. It doesn't quite go in a perfect ellipse. And you ask yourself, "Why not?" I can think of two explanations. Maybe Newton's Laws are wrong, our model of the solar system is wrong; or maybe there's a simpler explanation maybe there's another planet out there that we haven't seen yet—Pluto. And, in fact, we could deduce its position, and its mass, and its orbit, just by looking at the other planets. And to me, this is the situation with quarks. We can't see them directly—we can't even see atoms directly. But we can see their consequences on other systems very directly, quantitatively, and there's no doubt in any physicist's mind, at this point—at least, many physicists' minds—that quarks are as real as Pluto.

Lecture Thirteen
Symmetry and Conservation Laws

Scope: What does the word *symmetry* mean to a physicist? Almost what it means to you: an aesthetic property of a system, a pattern that appears the same when viewed from a different perspective. The role of symmetry in our understanding of the universe has evolved considerably with time to become a guiding theme of physics. Emmy Noether's theorem played some role in this development; we'll learn how she connected symmetry to *conservation laws*, which are of enormous practical value. Spatial symmetries are easy to picture, but the more abstract, "internal" symmetries are not. How do we work these into a theory? What do they tell us about the world? Why are physicists so enthralled with symmetry?

> Symmetry seems to be absolutely fascinating to the human mind. We like to look at symmetrical things in nature, such as perfectly symmetrical spheres like planets and the sun, or symmetrical crystals like snowflakes, or flowers which are nearly symmetrical. However, it is not the symmetry of the objects in nature that I want to discuss here; it is rather the symmetry of the physics laws themselves.
> —Richard P. Feynman (*The Character of Physical Law*, chapter 4)

Outline

I. In the next two lectures, we turn from the constituents to the laws of physics and the tools that physicists use to understand how the world works. One of the most powerful of these is the principle of symmetry.

   A. The meaning of the word *symmetry* in physics is fairly close to the meaning that you might think of but a bit more general. To a physicist, *symmetry* relates to looking at an object or a system from a different perspective. If an object looks the same from two different perspectives, then we say that it is symmetrical.

   B. A snowflake has six-fold symmetry, which means that if it is rotated by one-sixth of a turn, it looks the same as it did at the start.

C. We can also talk about more abstract symmetries. If we say that a mathematical theory has symmetry, we mean that even if we make some change in perspective, the answers remain the same.

D. The mathematics of symmetry is called *group theory* and was once obscure. We now think of it as simply the mathematical formalism that describes the different kinds of symmetries that exist in the world.

II. Let's begin by looking at some perfect symmetries.

    A. Imagine that you see a group of children on a field playing some sort of game. Some of them are wearing green shirts and some are wearing red shirts. What is the nature of their game? You might try to answer that question in a number of ways.

        1. If all the children change into green shirts and continue to play in the same way, you learn that the game has symmetry, because a change in color is completely irrelevant to the rules.

        2. This fact teaches you something about the game; that is, it probably doesn't involve teams. You don't know everything about the game, but symmetry is one tool that helped you learn one aspect.

        3. If the children change shirts, then don't know what to do, you learn that the game probably does involve teams, and that color matters. Lack of symmetry has taught you something about the game.

    B. That analogy was a bit of a stretch, but it is a good one for a discussion of the strong nuclear force, which involves protons and neutrons combining to form a nucleus.

        1. Remember that in the 1930s, scientists were trying to understand the strong nuclear force. As in the game analogy, they discovered that the label "proton or neutron" was irrelevant. The forces of attraction held true whether the particles in question were protons or neutrons.

        2. In other words, protons attract protons, protons attract neutrons, and neutrons attract neutrons. There was no change in the rules when the nature of a particle was changed from proton to neutron.

        3. The strong force has a proton/neutron symmetry, which taught us something about the nature of protons and the nature of the strong force. Later, we came to understand that this force was further explained by the existence and nature of quarks.

C. What does symmetry teach us? In this case, we learned something about the nature of the nuclear force. Symmetry simplified our worldview. It reduced the number of degrees of freedom that we had to account for in describing a nucleus mathematically.

D. Think back to the snowflake example, which is geometrical. If we rotate a snowflake by 60 degrees, it looks the same. That fact teaches us something about the crystals that build the snowflake; in this case, we learn something about what goes on inside the structure.

III. We can also look for symmetry in theories, such as in Emmy Noether's mathematical theorem of symmetry and conservation laws from 1915.

    A. Emmy Noether was raised and lived most of her life in Germany in the late 1800s through the early 1900s—a difficult time for a woman to be a mathematician. She met many challenges in her life to get her Ph.D. in mathematics and secure a position at a university.

    B. Noether developed a theorem to connect symmetry to conservation laws. Generally, it argues that if we have a specific symmetry, then we are guaranteed to have a conservation law associated with that system.

        1. Remember that a conservation law says that if we start with some quantity and it undergoes any number of changes, in the end, we will still have the same original quantity.

        2. For example, if we start with some energy, we might change its form or create matter/antimatter, but we will always have the same total energy in the end. The same is true of electric charge.

    C. Noether's theorem connected an abstract principle of a theory, its symmetry, to a practical benefit. Let's look at some specific examples.

        1. Imagine that I am doing a physics experiment in which I knock billiard balls against one another to tabulate the laws of physics.

            a. If I do the same experiments in a different location, the laws of physics are the same. The laws are unchanged when the system is viewed from a different perspective, which is the definition of symmetry, in this instance, *translational symmetry*.

    b.   Noether's theorem states that every symmetry of nature has an associated conservation law. We can't guess what will be conserved, but we can use the theorem to derive it.

    c.   We discover that translational symmetry leads to conservation of momentum. If an object is moving to the right and it crashes into something, whatever is left of the object after the crash will still have some motion toward the right.

    d.   Knowing about translational symmetry and conservation of momentum, we can test these ideas in the laboratory. How far can I take my equipment and still have the laws of physics be the same? As far as we know, the laws of physics operate the same as they do here even in the Andromeda Galaxy.

  2. If I do my experiments today, then I do them tomorrow, I also find that the laws of physics are the same. This symmetry is *time translation*. At different times—which would be from a different perspective—the laws of physics are the same. The conservation law associated with time translation is conservation of energy.

IV. Let's discuss one more symmetry of nature that is very abstract and ties in with Noether's theorem—*gauge symmetry*.

  A. Gauge symmetry arose in the theory of electricity and magnetism. In the 1800s, people observed that the laws of electricity and magnetism were symmetrical. They were invariant under a certain change.

    1. Think of a simple electrical circuit. If we have a car battery and some wires going to a light bulb in the headlight of the car, when we hook them up, the light bulb glows.

    2. The laws of electricity and magnetism allow us to make predictions about how bright the bulb will be, how long it will burn, how hot the wires will become, and so on. These are consequences of the laws of electricity and magnetism.

    3. To do the calculations to make these predictions, we need to know that one side of the battery is at 0 volts and one side is at 12 volts. Physicist A does his calculations from the premise that the left side of the battery is at 0 volts and the right side is at 12 volts. Physicist B does her calculations from the premise that the left side of the battery is at 12 volts and the right side is at 24 volts. The numbers are different, but the difference is 12.

    4. Does Physicist B come to the same conclusions about the observables as Physicist A? The answer is yes. It's completely irrelevant what numerical value is assigned to the left pole and the right pole; all that matters is the difference.

    5. That's the essence of gauge symmetry: No matter what point is picked to be 0, the results are the same.

B. The conservation law associated with gauge symmetry is conservation of electric charge, which is quite useful in a wide variety of experiments, not just those in particle physics.

V. We'll conclude this lecture by returning to the development of QED.

A. Remember that Feynman, Schwinger, Tomonaga, and others were trying to come up with a quantum mechanical theory that would take into account the 19th-century understanding of electricity and magnetism and make it consistent with quantum mechanics.

B. These scientists already knew that their quantum theory must obey the law of gauge symmetry.

C. To ensure that the quantum theory was gauge symmetric, the mathematics forced scientists to introduce a new particle of nature, the photon.

    1. The photon comes out of the *principle of gauge invariance*, which is another term for *gauge symmetry*.

    2. All the properties of the photon also follow from the requirement that the theory be gauge symmetric.

D. Once we have a gauge symmetry that we believe is present in nature, we can start to predict not only the existence of particles, but their properties.

E. In later years, when people began thinking about the strong force between the quarks, the color force in QCD, they didn't have an earlier understanding to work from. They had to create the theory from scratch, and they used gauge symmetry as a guiding principle.

F. These scientists started from the idea that if gauge symmetry is true for electricity and magnetism, it might also be true for the color force. Further, if gauge symmetry is true for the color force, then there must be a particle of nature that carries the strong force. We now call this particle the *gluon*, and we know that it has certain properties, which all come out of the theory.

G. Even later, when scientists were trying to understand the weak force of nature, which was even more complicated, they again asked if it was related to gauge symmetry and again predicted the existence of a particle. This particle, the carrier of the weak force, is the Z particle.

Essential Reading:

Barnett, Muhry, and Quinn, *The Charm of Strange Quarks*, chapter 5.

't Hooft, *In Search of the Ultimate Building Blocks*, chapter 10.

Weinberg, *Dreams of a Final Theory*, chapter VI.

Recommended Reading:

Calle, *Superstrings and Other Things*, chapter 25, first half, up to "The color force."

Greene, *The Elegant Universe*, pp. 124–126.

Krauss, *Fear of Physics*, chapter 5, first half, up to "Symmetry breaking."

Lederman, *The God Particle*, chapter 7, section on "Conservation laws."

Wilczek and Devine, *Longing for the Harmonies*, Eighth Theme (through chapter 25).

Questions to Consider:

1. The laws of physics have many invariances; for example, they are the same no matter what your position is in the lab or what time it is in the lab. Are there more?

    (a) Do the laws of physics change depending on your orientation in the lab, that is, which way you are facing?

    (b) Do the laws of physics change depending on your speed as you walk across the lab, assuming that you walk in a straight line with a steady speed?

    (c) Do the laws of physics change depending on whether you sit still or accelerate across the lab? (This is a subtle one!)

2. Do you agree with the U.S. Patent Office policy that perpetual motion machines (which explicitly violate energy conservation) should be dismissed out of hand, without any further investigation? Why or why not?

3. Can you think of a symmetry in the world of particle physics that is exact? How about one that is approximate but not exact?

Lecture Thirteen—Transcript
Symmetry and Conservation Laws

I want to shift gears here a little bit for the next couple of lectures, and talk—not about the constituents, the particles and quarks and electrons that make up the world—but to talk a little bit more abstractly about the rules, the laws of physics. The tools that physicists use to try to understand how the world works. Both are important for particle physicists—to know what we're made of, and also how those things fit together in a theoretical framework—to understand the workings of the microscopic world. One of the most powerful tools that particle physicists use nowadays is the principle of symmetry. It's very useful. It has both practical consequences and aesthetic consequences. And, I want to spend this entire lecture just talking about what physicists mean by symmetry, and why and how it's useful if you want to understand the world of particle physics. In fact, it's much more general than just particle physics. Any kind of physicist, or even a biologist, or a chemist, any kind of scientist nowadays, is going to know about symmetry and care about symmetry. As you'll see, it's a very powerful guiding principle.

Physicists often use words in ways that are unfamiliar. When I talk about a force, I mean something very definite. It's a push or a pull. In the English language, the word force can mean, I don't know, forces of nature, the great force of an army, or "may the Force be with you". In English, these words have lots of meanings. And, in physics, we always make a definition and it's usually mathematically based. The word symmetry is pretty much what you think it is; it's just a little bit more general. To a physicist, symmetry means when you take an object, or a system, or anything, and you look at it from a different perspective. If it looks the same from the two different perspectives, then we say it's symmetrical. So think of a snowflake: A snowflake has a six-fold symmetry, and that means that if you rotate it by one-sixth of a turn, it looks exactly the same as when you started, or you could tilt your head by a sixth of a turn; it would look the same. So, that's a symmetrical object, because when viewed from a different perspective it's unchanged. Now, you can talk about symmetries of objects, but you can also talk about more abstract symmetries. You can talk about the symmetry of a mathematical theory; and you don't mean that if you tip your head, the symbols look the same. What you mean is that if you make some change in perspective, some change in how you look at that theory, the answers come out the same. We'll talk about some concrete examples of symmetry in theories coming up very shortly.

The mathematics of symmetry is called group theory. And, once upon a time, this mathematics was very obscure. It became important for physicists to learn about group theory in the early 1900s, when quantum mechanics was developing. And, group theory is really just the mathematical formalism that describes what different kinds of symmetries there are in the world; turns out there's different sorts, there's geometrical symmetries and more abstract kinds. And nowadays, it's an essential ingredient in the toolbox of pretty much any working scientist. We're not going to learn group theory in this course, but I want to talk about the basic ideas, the underlying principles behind the story.

The snowflake is a nice concrete example, because you can visualize it. Geometric symmetry is pretty easy to visualize. Another symmetry is left/right symmetry on my body. This side, and that side, look pretty much the same. We'll talk a little bit more about imperfections, or breaking of symmetries in the next lecture, when we start looking in fine details. But for today, I just want to focus on perfect symmetries. So I want to make a, a kind of, first a metaphor, or—I don't know—some sort of a concrete example of a symmetry which is not geometrical in character. And I'm stretching a little bit here, but I want you to imagine that you're in a foreign country, on a holiday, and you see a bunch of kids out on a field, and they're playing. It's obvious that there's some sort of game going on; they're in a confined area, and you're watching and you notice that some of the kids have green shirts and some of the kids have red shirts. And otherwise, you don't know what's going on down there. There's just a whole bunch of screaming and fun activities. There are various ways you could try to figure out what's going on—what is the nature of this game? Well, if you can't speak the language, you can't ask somebody what the rules are. Now, supposing you notice there's a pile of extra shirts off on the side, and so, you know, you sort of do the foreign-language pantomime thing, and you get all the kids with red shirts to put on green. So now everybody's wearing green shirts. Okay?

This is an experiment that you just did to learn about the symmetry of the situation. So, now you let them go. Watch what happens. There's various possibilities. One is that they continue to play just the way they were before. They're still running around and screaming, and now you've learned that there is a very deep symmetry in this game; that color is completely irrelevant. So, my definition of symmetry was, you change something, but then the result looks the same. Here we changed the color of half of the players, and the end result is the same. This is a complete symmetry of color in this game. And what have you learned? Well, it can't be a competitive game, unless the kids know each other so well that it doesn't matter what the color of the shirt is; but one possible conclusion is it's one of those new-fangled cooperative games, where it doesn't really matter, there's not teams. The kids just wear whatever shirt they feel like; and it was just coincidence, when you first looked, that it was fifty-fifty, and they're playing some game that's not competitive. So you learn something; and of course, it's not going to teach you everything about the game. Symmetry is just one tool among many. There's another possibility, which is that everybody's green, and then they just all kind of stand around. They don't know what to do. Then it's more likely that the game was competitive, and you need to have two teams, and now the kids are confused because they don't know who's on what teams. So again, by looking either at symmetry or a lack of symmetry, you can learn something about the rules of the game.

Now, you might continue to do further symmetry experiments. For example, instead of just making everybody green, you could swap, so all the reds become green and all the greens become red. And again, there's various possible outcomes. It's possible that the game will proceed just as before. So now you would say there is "a symmetry." It's not the deepest possible symmetry; but there's a nice symmetry, which is under color interchange. The rules of the game are unaffected. So, apparently color doesn't matter. Most games are like that; in a soccer team, it doesn't matter if your team is red or green, all that matters is that you have two different colors. And so again, you've learned something about the nature of the game.

Now, you could imagine a situation where you flipped the colors and the game changes in some way. It might be that blue-eyed kids have to wear red shirts, and if they don't they're all confused, because that's the way it's always been. That's what, maybe even the rules say. It would be a slightly unusual situation, but you would learn something. This is a little bit of a stretched example of symmetry, but I wanted to come up with something that wasn't just geometrical, where you rotate an object, or flip it, and it looks the same. And in fact, this analogy is a good one for the strong nuclear force, where you have protons and neutrons that combine together to form a nucleus. So we'll take one step back up from the quarks for a moment.

In the 1930s, people were busily trying to understand the nature of that force. How does a proton attract another proton, or another neutron? How do neutrons attract one another? That was nuclear physics, and people were very interested; it's the essence of the strong force. And it was, in a certain sense, like my little game analogy. What you discovered was, color was completely irrelevant; or in this case, whether you're a proton or a neutron, which team you're on, didn't matter. Protons attract protons, protons attract neutrons, neutrons attract neutrons; there's no change in the rules of the game when you change the nature of a particle from proton to neutron. So there's a proton/neutron symmetry of the strong force, which we were observing, and that taught us something very important about the nature of protons and the nature of the strong force. Later, we come to understand that more deeply, as a connection that says up quarks and down quarks are almost the same object, and the strong force doesn't care about whether you're an up quark or a down quark; it really only cares about the strong color charges inside, which is something totally different than whether you're a proton or whether you're a neutron. Now, it's also true that what I said is only approximate. There's a slight difference between proton-proton and proton-neutron, and that slight breaking of symmetry is also going to teach us something. That will be the subject of the next lecture, where we look at what violations, or breaking of symmetry, teaches us.

But for right now, I just want to continue to focus on what symmetry teaches us. So in this case, we've learned something about the nature of the nuclear force. It has simplified our worldview. If you're a nuclear physicist, and you're trying to describe a nucleus mathematically, you don't have to label each particle proton or neutron—at least at some crude level you don't have to—because it doesn't matter. So you've simplified your job, you've simplified the mathematics; you have reduced the number of degrees of freedom that you have to keep track of. And that's a very useful and powerful goal, in any science; to simplify the explanation of phenomena. So that's one value of symmetry. Think back to that snowflake example, which is geometrical. If you rotate a snowflake by 60 degrees, it looks the same as it did before, and that teaches you something about the individual crystals that are building up the snowflake. I'm not saying that all snowflakes look the same, but they have this general property of symmetry, which is always common. And, the fact that essentially all snowflakes have that symmetry teaches us something about the shape of the crystals inside. Even before we were capable of looking with electron microscopes, or the laws of quantum mechanics, to figure out exactly what an ice crystal looks like, we already knew something about it, just by looking at this macroscopic object that had symmetry. So sometimes, rather than saying it simplifies, you can say that symmetry teaches you about what's going on inside; another powerful use of symmetry.

There's more. Symmetry has many, many functions and values in the world. Of course, it's also nice to look at something that's symmetrical. Something about our nature as human beings; when I look at a snowflake, it just looks beautiful to me. And in part, it's because of that symmetry. It seems like it makes the object feel simpler to me. And so, when we're looking for theories, I think there's a just kind of a natural human desire to look for symmetrical theories in some way.

There was a mathematical theorem, which was published in 1915 by a mathematician named Noether, and it's quite a famous mathematical theorem. I usually don't want to teach math in this course, but in this one, without going into the details, again we can talk about the essence, and it's had a lot of implications in particle physics. Emmy Noether was raised, and lived most of her life, in Germany in the late 1800s—early 1900s; and it was a very difficult time for a woman to be a mathematician. The German institutions were incredibly sexist. She was, in principle, not allowed to take college courses, or get her Ph.D.; and she kind of did it by going to professors who were a little bit more liberated than the institutions themselves, and they allowed her to sit in on the classes. So she basically worked her way through to a Ph.D. in mathematics against the, sort of desires of some of the higher-ups. And even after that, when she had her Ph.D., she took a position at a university, and it was essentially an assistant professor position; she was teaching, she was doing research, and she didn't get paid. It took her many, many years before she could finally convince people to pay her for this job. But despite all of this adversity, she was a fantastic mathematician. She was a mentor to many, many young students—unfortunately mostly men—who later became great mathematicians as well.

And this theorem in 1915 was a mathematical statement that connected symmetry to something a little bit more concrete, which are conservation laws. Now, the details are a little fancy, and in fact, the theorem is probably a little bit less general than what I'm going to imply here. You really need some fairly special assumptions and special cases for this theorem to hold. But the general statement is this, Noether's Theorem argues: If you have a very specific symmetry, then you are guaranteed to have a conservation law associated with your system. Let me remind you what a conservation law means. Conservation law means that you start with some quantity, and then all sorts of complicated stuff can happen; and then, in the end, you've still got the same amount of that original quantity. For example, energy, you start out with some energy. It might change forms, $E = mc^2$; you might create matter and anti-matter, it might re-annihilate; but in the end, you always will have the same total energy as what you started with. Or electric charge, that's also conserved. If you start off with an interaction of particles, it doesn't matter what the nature of that interaction is: electrical, magnetic, strong, weak. It's always true that whatever charge you begin with—it could be a big blob, and a big mess, and lots of confusion—and then when the dust settles, whatever comes out has the same charge as what you started with.

Conservation laws are wonderful for physicists. We teach them to freshman students, and we keep reminding them, over the years as they go on, there are more and more sophisticated and complex kinds of conservation laws. They make understanding complicated systems very easy, because it eliminates from consideration an enormous amount of physically impossible outcomes. When you've got particles smashing together, you might throw up your hands in dismay and say, "Anything could happen. It's quantum mechanics probabilities. Right? Anything could happen." And the answer is, well, anything that can happen will happen; but, many things can't happen. You can't create charge out of nowhere. So physicists love conservation laws, and it's practical. You know, police go to an accident scene, and they use conservation of momentum—looking at the tire tracks—to conclude whether the original driver was speeding; and, so, they can decide whether to give somebody a ticket by using a conservation law. This is, you know, real-world stuff.

What Noether's Theorem has done is, it's connected an abstract principle of a theory, its symmetry, to this practical benefit. And let me give you some specific examples of Noether's Theorem, because it really is pretty deep mathematics, and a little bit hard to visualize. Example number one: I sit at this table and I start doing physics experiments. Okay, I knock balls against one another, billiard balls; and I start tabulating laws of physics, Newton's Laws, and the laws for electricity and magnetism, and the laws for the strong force and the weak force, just by doing experiments. Okay? I'm not talking about the details of the experiment. I'm talking about the laws of nature that I'm concluding from them. Now I pack up my gear, and I move next door, and I do the experiments again. People have done this, and the laws of physics next door are the same as the laws of physics here. Newton's Laws are true in Australia, and Japan, and the United States; they're the exact same laws of nature. So this is "a symmetry." Okay?

When you look at a physical system, the laws are unchanged when you view it from a different perspective. In this case, your perspective is looking at it from a different place. We call that translation symmetry. Translation not having to do with language, but just meaning translating yourself along some line. This is "a symmetry" of nature; and Noether's Theorem says, when you have "a symmetry" of nature, there must be an associated conservation law. Now what you can't do is to just guess what is going to be conserved. But Noether's Theorem is really a mathematical theorem, and you can derive what is conserved. And when you do the math, you discover that this particular symmetry, translational symmetry, leads to conservation of momentum. When objects are moving along, and you have a crash, if they started off with some motion to the right, to my right, then in the end, they—the mess, the blob of stuff that's all crashed together—will still be moving to the right, unless there's an external force. But that's a complication to the story.

Conservation of momentum, very useful; connected to symmetry of where you where looking at the laws of physics, very abstract. And, now you can test conservation of momentum in the laboratory; but you can also test for translational symmetry, and it's indirectly doing an experiment for you. So, for instance, how far can I go and still have the laws of physics be the same? It's an interesting question. Astronomers have looked at galaxies far away. The Andromeda galaxy is the nearest one, but it's extraordinarily far away, millions of light-years; and yet, I'm going to argue that the laws of physics in the Andromeda galaxy are apparently the same as the laws of physics here. Now, how do I know that? There are no aliens who've told me the laws of physics there. But, I can look at a star in the Andromeda galaxy, and it's got the same colors, and the same light, and the same temperature, and the same distributions. All of the laws of physics, and pretty much all laws of physics, have to be put together to explain something as complicated as a star. That ultimate object is the same in the Andromeda galaxy as it is here. So this is very powerful evidence that the laws of physics are the same all the way over there; so conservation of momentum seems to be an incredibly, well-proven fact of nature, and it's well proven through this mathematics. It's kind of a wild concept.

The connection between position and momentum is definitely not obvious. I'm not implying that you should understand Noether's Theorem intuitively. Why momentum? And, as I said, the answer comes out of the math. There's a hint from quantum mechanics. The Heisenberg Uncertainty Principle says, an uncertainty in position is related to an uncertainty in momentum. So already, you can sort of imagine that momentum and position are, in some interesting way, hooked together. Noether's Theorem has really provided that hook in a rigorous way.

There are other symmetries of nature where Noether's Theorem applies. For example, I do my experiments today, and then I come back tomorrow and I do them—and I already did them yesterday—and the laws of physics aren't changing. There is "symmetry". The symmetry is time translation. At different times, that's a different perspective. The laws of physics are the same. So, what is the corresponding symmetry—sorry, corresponding conservation law? Noether's Theorem tells us the answer. You work through the mathematics; the conservation law associated with time-translation symmetry is conservation of energy. So you can check conservation of energy in the laboratory, or you can verify that the laws of physics are not changing with time; they're coupled together. And once again, if you look at the Andromeda galaxy, think about that for a second. The light that was emitted from those stars was emitted millions of years ago, traveling through space all this time, and now we see it. So not only are the laws of physics the same far away, they were also the same long ago. So this is a very, very powerful argument, and we've seen now out to billions of light-years; and the laws of physics appear to be completely unchanged, even from then to now. This is a powerful confirmation of this law of conservation of energy. So, it's not just physics experiments that make us believe that, it's also this mathematical theorem.

And I think this is why the U.S. Patent Office refuses to consider patents where you plug—well you don't plug something in, and energy comes out. A perpetual-motion machine has been dismissed by the U.S. Patent Office as clearly unscientific. It violates the principle of conservation of energy. And you might say, "Well, but what if somebody's really clever? Can't you get around the conservation of energy?" And my answer is, well, I don't think so; because (a) we've tested it directly, and (b) we know that the laws of physics aren't changing with time. If you build such a machine, then it also means that the laws of physics are changing with time, so maybe your machine isn't going to work tomorrow, and I'm certainly not going to buy it. So this is Noether's Theorem. Albert Einstein loved this idea, he was very impressed with Emmy Noether's theorem; and it became a part of his general theory of relativity, where he was looking at deep symmetries of space-time, and this theorem allowed him to connect these to very powerful conservation laws, which are useful in the mathematical statement and application of relativity.

I want to tell you about one more symmetry of nature, which is very abstract, and it's going to hook in with Noether's Theorem, but it also has a, sort of a story to tell all of its own. This is the toughest part of this lecture, for sure, so if you don't get it, just roll with it. Okay? Try to absorb what you can because I can't describe the symmetry to you in, sort of, obvious terms. This is a very abstract symmetry of nature. It's got the name "gauge symmetry;" and even the name, it's some historical name, has something to do with railroad tracks and the gauge of the railroad tracks, but I don't have any idea why that word is associated with this symmetry. "Gauge symmetry." Let me try to give an example that kind of hints at what gauge symmetry is. It arose in the theory of electricity and magnetism. In the 1800s people observed that the laws of electricity and magnetism were symmetrical, they were invariant under a certain change. So let's think of a simple electrical circuit. You've got a car battery, and some wires going to a bulb, like the front light bulb of your car. And you hook it up and the light bulb glows. The laws of electricity and magnetism allow you to make predictions: how bright will the bulb be, how long will it burn, how hot do the wires get, and so on. These are consequences of the laws of electricity and magnetism. Now, when you have a battery, one side's at zero volts and one side's at 12 volts; and you need to know that in order to do the calculations. So Physicist "A" does their calculations, saying, "This side of the battery, the left side, is at zero volts, and that side of the battery, the right side, is at 12 volts." Now Physicist "B" comes along and says—they don't know what Physicist "A" has done—Physicist "B" says, "The left side of the battery is at 12 volts, and the right side of the battery is at 24 volts." Okay, the difference is still 12, but they've picked different numbers. And now here's the big question: Does Physicist B come to the same conclusions about the observable, the light bulb and the wires? And the answer is yes. It's completely irrelevant what numerical value you assign to the left pole, and the right pole. All that matters is the difference. So the number doesn't matter; that's "symmetry" of nature. You can pick this point to be zero, or a different point to be zero; that doesn't matter. You can change that and the end results are the same.

So as I said, it's a rather abstract symmetry of nature. That's the essence of gauge symmetry. Gauge symmetry is a little bit deeper than that, but that's the idea. You could also think about analogs with gravity. Okay, I put an object like my fist at the top of the table and I say, "How high is that fist?" Well, I could say it's zero; it's zero above the table. Or, I could say it's two feet high, it's two feet above the ground. Or, I could say it's 150 feet above sea level. It really doesn't matter what number I associate with how high this object is; because, in the world of physics, what matters is always changes in height. As I lift it up, how much work did I have to do? I either go from zero to one, or I go from three to four, or I go from 150 to 151, or whatever numbers I'm using. It really doesn't matter. That's kind of like gauge symmetry. It's "a symmetry" of nature, so Noether's Theorem says there must be an associated conservation law. And, as usual, you'll never guess it; you've got to work through the math. And the conservation law that springs out of the mathematics is quite spectacular. Gauge symmetry in electricity corresponds, according to the theorem, with conservation of electric charge. So, that's one of those eye-opening surprises in a junior-level physics class, when you first teach Noether's Theorem, that makes people kind of get excited about this theorem. Because conservation of electric charge is so powerful, and so useful, in a wide variety of experiments, not just particle physics. And, it comes from this abstract mathematical symmetry of the starting theory that doesn't really seem to have necessarily anything to do with counting up charges.

Gauge symmetry goes much beyond Noether's Theorem, however; and that's where I want to finish up today's lecture. We constructed a theory— we, Feynman, and Schwinger, and Tomonaga, and the various other people like Dyson who were many people involved in the discovery and elaboration of quantum electrodynamics. Remember what they're doing. They are trying to come up with a quantum-mechanical theory that takes all of the 19th-century understanding of electricity and magnetism, including the concept of fields and electric charge, and make it consistent with quantum mechanics. So, when they tried to do that, they already knew that gauge symmetry was a good symmetry of electricity and magnetism. They knew that, whatever quantum theory they were going to write down, it must obey the law of gauge symmetry. That is to say, it must be symmetrical. Whether you call this point zero volts or ten volts doesn't matter, only differences matter. That's got to get built into their theory. So as they're constructing the theory of quantum electrodynamics, they have to be very careful—they're modifying it to be consistent with quantum mechanics— they have to be careful not to modify it in such a way that you lose gauge symmetry. And what they discovered—this is very difficult to grasp the first time through—is that, in order to make sure that the quantum theory is going to be gauge symmetric, they have to introduce a new particle of nature, the photon. It's not that you stick it in the theory because you know they're there. You have to put it in, in order to preserve gauge symmetry. It comes out from the Principle of Gauge Invariance. That's the fancy word for gauge symmetry.

So think about it. Not only does the existence of the photon come out, but its properties: it has no mass, it has no charge, but it does have spin. The photon has spin one; it's a boson. All of those facts about the photon are coming out, simply from the requirement that the theory be "gauge symmetric". Think of the power here. Once you have a symmetry that you believe is present in nature, you are starting to predict not only the existence of particles, but also their properties. And, so in later years, when people began thinking about the strong force between the quarks, the color force in quantum chromodynamics, people are struggling at first to construct a theory. They don't have electricity and magnetism—classical electricity and magnetism—to work from; now you've got the harder job of creating the whole theory from scratch, and gauge symmetry is a guiding principle now. You say, "Gosh. Gauge symmetry was true for electricity and magnetism; what if it's true for the color force as well?" It's an idea, and now you work out the consequences. If gauge symmetry is true for the color force as well, then there must be a particle of nature that carries the strong force—we now call it the gluon—it must have certain properties. It has no mass. It has spin one. It carries color. All that comes out of this theory. So, not only the existence, but also the properties of the gluon are forced upon you.

So when people discovered this, and it was matching with the theory and the data, people got really excited and they said, "Hey, gauge invariance is a great thing!" So then, later, when they started trying to understand the weak force of nature—which was even more complicated and even more esoteric and even more bizarre—people said, "Well, what if gauge symmetry is a principle of nature?" And this time around, they predicted the existence of a particle—the carrier of the weak force, not the W but the Z particle—which nobody had ever seen, nobody had ever thought of. It was just because of this principle of symmetry that they were able to predict the existence of a new particle; and a new particle means a new force of nature. It was a new aspect of the weak interaction that nobody had ever thought about before. And, it was real. It was there. They went out in the laboratory, took about ten years, and they found it.

So I think at this point, symmetry, as a general principle, is deeply required and deeply admired. It's useful. It's powerful. It's part of the way that physicists look at the world. As we reach out and try to understand new physics, physics beyond the standard model of particle physics—because we're still working on our understanding of the world—symmetry is still, and probably always will be, now, a guiding principle. You look for symmetries, and when you find them, you use them as a tool to understand the world.

Lecture Fourteen
Broken Symmetry, Shattered Mirrors

Scope: Imperfect symmetry is called *broken symmetry*, which can be either slight or extreme. In either case, we inevitably learn something useful about the world. What is *mirror symmetry* (also called *parity*), and what would it mean if it were broken? We look at the surprise of parity violation and the role of broken symmetry in modern particle physics. This topic is difficult but significant to particle physics; we will accept some confusion without ending our exploration.

> I heave the basketball; I know it sails in a parabola, exhibiting perfect symmetry, which is interrupted by the basket. It's funny, but it is always interrupted by the basket.
> —Michael Jordan (former Chicago Bull)

> A broken symmetry breaks your heart.
> —Abdus Salam (*The World Treasury of Physics, Astronomy, and Mathematics*, p. 669)

Outline

I. In the last lecture, we noted that symmetry can serve as a simplifying principle. If we find that a complicated system is symmetrical, our description and understanding of the system can be made easier. We can also make predictions using symmetry, as we did with Noether's theorem and the laws of conservation.

   A. In the real world, many times, symmetry is broken. Consider your own body. Are your hands the same? In some respects, yes, but in other ways, no. You can't put a right glove on a left hand, for example.

   B. Think again about watching children play a game, this time, soccer. The team of boys is wearing blue uniforms, and the team of girls is wearing pink. If we have the teams switch uniforms, we should find that an exact symmetry exists between uniform colors.

   1. Suppose that after watching many games, we observe that the pink team seems to win more often. That's a breaking of symmetry.

2. We might conclude that in America, young girls mature faster than boys and tend to be more coordinated and stronger. Thus, the pink team may beat the blue team more often if they're second graders, but not when they're tenth graders.

3. This subtle breaking of symmetry teaches us something about the players and the culture they live in.

4. If the players are high school students and the boys are forced to wear pink, the girls might laugh at them and the game might fall apart. This is a more radical breaking of symmetry. We didn't expect to swap the two colors and stop the whole game.

C. Physicists often look at systems first for symmetry, which is easy to find when the symmetry is exact or almost exact. Often, if we find a slight break in symmetry, we notice it and it teaches us something.

1. For example, as we've noted, the laws of the strong interaction are independent of whether a particle in a nucleus is a proton or a neutron. This is a deep symmetry of nature, but it's not exact.

2. A nucleus with a proton and a neutron is stable, but one with two protons is not. Swapping a neutron for a proton changes electrical properties. Because electricity is very weak compared to the strong force, it doesn't have a big influence, but it can play a role.

3. We can point to another symmetry between protons and neutrons that is also broken, very subtly: If they were completely symmetrical, they would weigh exactly the same, but a neutron weighs about 0.1 percent more than a proton.

4. This breaking of symmetry teaches us that the down quark might have a slightly different weight than the up quark. After all, a neutron is made of two downs and an up, and a proton is made of two ups and a down. A tiny difference between the quarks, then, could lead to a tiny difference between the two objects.

5. We can also use this breaking of symmetry to learn something about electricity. At first, we might say that the role of electricity inside a nucleus is insignificant. The strong force binds everything together tightly.

6. However, electricity exists inside the nucleus. Positive charges repel, but the neutrons are not repelled, because they're neutral. Again, we learn about the connection between electricity and the strong force in the nucleus by looking at the differences between nuclei that have different numbers of protons and neutrons.

II. We also need to look at some situations in which symmetry is so badly broken that we hardly recognize it ever existed.

    A. For example, imagine that you're at a fancy dinner and you sit down at a large, round table. In front of each seat is a plate and in between the plates is a glass of water.

        1. When you sit down, there's a glass of water on your left and a glass of water on your right. You might think that because the table is perfectly symmetrical, you are free to choose either glass.

        2. Sooner or later, however, somebody around the table will reach for a glass. If the first person picks up a glass with his or her left hand, the symmetry is broken. You no longer have a choice.

        3. A latecomer might arrive and see a full glass on the left and an empty glass on the right. This latecomer believes that the table has no symmetry.

    B. Let's look at another example that is more related to physics.

        1. Suppose you look inside a kitchen magnet. Physicists understand magnetism as arising from the magnetism of individual atoms, but let's just take for granted that an individual atom acts like a tiny magnet with a north pole and a south pole.

        2. The atom can be oriented in any direction, and if it's warm, it moves around. First, it points north, then west, then east.

        3. Now suppose that you have a crystal, which is a large number of atoms. If all the atoms are pointing in different directions, the result is just a crystal, not a magnet.

        4. What happens if all the atoms in the crystal are magnetized in the same direction? If the atoms are pointing in opposite directions, they cancel, but if they're pointing in the same direction, they add up. Thus, a kitchen magnet is a material in which all the atoms are pointing in the same direction.

5. Let's think a little bit more about why this happens. If you warm up your kitchen magnet, it will demagnetize, because the thermal energy results in the atoms pointing in different directions. If you cool it down, the atoms will tend to begin lining up again in a sort of chain reaction, and the magnet will remagnetize itself.

6. A tiny physicist living inside the magnet might deduce that north and all the other directions of the compass are radically different, because all his experiments tend to point toward north. He might conclude that the laws of nature are not symmetrical. If another tiny physicist comes along and heats up the world around them, she could prove that their world has broken symmetry, but that the laws of nature do not.

C. Let's go back to mirror symmetry, in which you notice that your left hand and your right hand are nearly the same, but they're opposite. Physicists call mirror symmetry *parity*, or *p symmetry*.

1. Remember that a human body is almost, but not quite, left/right symmetric, but that's not a law of nature. A human body is an object.

2. The breaking of symmetry of a human doesn't break any symmetry of laws of nature.

D. To further understand this idea, imagine that you are using a pool table to do physics experiments in a room that has a big mirror on the wall. Instead of watching the table, however, you watch the mirror and take all your data from the mirror-image balls.

1. When you watch the laws of physics in a mirror world, you find that they are exactly the same as they are in our world. You could check electricity, magnetism, and the strong nuclear force, and you would find that the outcomes are identical but reversed in the mirror and on the pool table.

2. Any physicist would agree that the laws of physics are unchanging with time or space. In fact, in 1956, two physicists, Lee and Yang, wrote a paper in which they observed that this fact of mirror symmetry had been verified countless times for gravity, electricity, magnetism, optics, and the strong nuclear force.

3. But nobody had ever checked the weak interaction. What if we look at a beta decay? That's a weak nuclear reaction. Is it true that the reaction and its mirror reaction both occur with

the same probability in the real world? Lee and Yang posed this question.

4. Of course, most physicists believed that parity symmetry would be unbroken. A year later a physicist named C. S. Wu set up an experiment to test mirror symmetry.

   a. Let me try to summarize the idea for you: If we have a nucleus that is spinning in the clockwise direction, we might say that it has a certain "handedness." If a particle is moving away from me and spinning clockwise, I call it a right-handed particle. Similarly, a left-handed particle would be spinning in a counterclockwise direction from my point of view.

   b. Wu examined some nuclei that had a certain handedness and were in beta decay. After the decay, she was left with a nucleus that had been transformed, and beta rays (electrons) go flying out.

   c. Imagine watching this experiment in a mirror. If the nucleus in the real world is spinning clockwise, the mirror nucleus is spinning counterclockwise. If the real nucleus is labeled north and south, the mirror nucleus is the opposite, south and north.

   d. If the world is mirror symmetrical, we would expect that the outgoing electrons would be equally likely to go north as to go south, or east or west. If all the electrons went north in the real world, that's what we would call south in the mirror world and would be a drastic breaking of mirror symmetry.

   e. Wu's experiment found that *almost* every electron goes to the north and none of them goes south. It was almost the maximal breaking of mirror symmetry imaginable.

5. As we know, the weak interaction is already rather odd; it changes the strangeness of particles; it transforms neutrons into protons; and apparently, it knows the difference between something spinning right-handed and something spinning left-handed.

6. Many, many experiments were done after Wu's, and they all verified that in any weak interaction, left-handed spinning particles and right-handed spinning particles behave differently.

7. In fact, when we start doing experiments with neutrinos, we would expect that right-handed and left-handed neutrinos would be equally likely to exist in the universe, but there are no right-handed neutrinos. To an excellent approximation, the neutrino massively breaks mirror symmetry.

8. We have also found that matter/antimatter symmetry is broken, although more subtly than parity symmetry in this instance.

E. In physics, we look at these broken symmetries and ask: Is that an accident, or is there a reason that this symmetry is broken? Does it have some fundamental origins? Can we learn something from it?

Essential Reading:

't Hooft, *In Search of the Ultimate Building Blocks*, chapter 7 (slightly hard going).

Weinberg, *Dreams of a Final Theory*, chapter VIII.

Recommended Reading:

Krauss, *Fear of Physics*, chapter 5, second half, after the introduction of "Symmetry breaking."

Lederman, *The God Particle*, "Interlude C."

Wilczek and Devine, *Longing for the Harmonies*, chapter 26.

Questions to Consider:

1. When you look at a mirror image of a map of the United States, it is hardly recognizable. Is this an example of parity violation; that is, does it demonstrate that the laws of physics are not mirror symmetric? Why or why not?

2. Consider an idea from Richard Feynman: Suppose you could have a radio conversation with a distant alien race. We have no visual connection, audio only. The members of this race are so distant that we can't even see the same stars. After much effort, you might learn to communicate with them. (Perhaps you could tap 1, then 2 so that they could understand how we count. Then you might say "hydrogen" and tap 1, then "helium" and tap 4, and so on, tapping the atomic weights of the elements. They might then figure out the names of elements.) Imagine, ultimately, that you could converse freely with them. Then one day, they ask you to describe yourself, and you say, "My heart is on my left side." The aliens reply, "We don't know the word *left*. Which side is left?" Think carefully about this: How could you possibly explain this to them?

(Hint: If parity was not violated, I don't believe that there is any conceivable way that you could!)

Lecture Fourteen—Transcript
Broken Symmetry, Shattered Mirrors

Our last lecture was about symmetry. It was a little bit rough going in spots. There were some heavy-duty mathematics, Noether's Theorem and the principle of gauge symmetry. And today's lecture is going to also be about symmetry. About the breaking of symmetry and what that can teach us about how the world is made, and as a guide for learning about laws of physics and our understanding. Sometimes, when the going gets really abstract in a colloquium, I get a little frustrated. I go to colloquia all the time. And, if it's about nuclear and particle physics, oftentimes I can kind of nod my head and follow the whole time. And if it's—especially if it's another branch of physics, or some other science—I can get completely lost in parts of colloquia. And, I want to remind you that it's okay to get lost from time to time, as long as I can bring you back at some point, either in this lecture or an upcoming lecture. Oftentimes, when you get lost, it's because you're learning about some brand-new topic that's a little bit abstract, and as you continue along you will learn some principles and some ideas, so that when you go back and look at what you had heard before, and it was confusing, it all begins to tie in together. There's not always a starting place when you're learning about some complicated thing. It's just a big old mess, and you just have to jump in somewhere, and begin to learn more and more of it, and then the big picture begins to appear. So, having said that, I want to talk about the breaking of symmetry today; and we're going to talk about how that fits in with the world of particle physics.

Let me start off just by reminding you of why symmetry is such a lovely principle. Not just its aesthetic appeal to us as human beings, it can serve as a simplifier. If you look at a complicated system, but it's symmetrical, then your description of the system, and your understanding of it, can be made easier, simplified. For example, a neutron-proton system, like a nucleus, is not quite as complicated as you thought. You could just call them all nucleons, and work out a theory that really only has one object in there. It can also teach you about microscopic details. Think of a crystal, where the six-fold symmetry of ice teaches you something about the shape and character of the individual molecules, which might otherwise be invisible to you. And, it can lead to predictions, and understanding in a very deep way. For example, Noether's Theorem allows you to take "a symmetry" and predict a conservation law, which is practical, useful. Or even better yet, gauge symmetry, an extraordinarily abstract symmetry of nature, can lead you to predict the existence of a new particle, of new forces in the world. Symmetry is a wonderful thing.

And alas, in the real world, many times, symmetry is broken. Look at your body. Okay, it's got a left side and a right side, and it seems to be quite symmetrical. Look at your hands, in particular. Hold them in front of you. Are they the same? In many respects, they are the same—except, they're totally different! You can't put a "right" glove on a "left" hand. They're completely different, but they're also completely symmetrical. And then you look more carefully, and you notice, "You know, I've got a scar on one hand, and not on the other; my body has a heart on one side and not on the other." So there are small, subtle "breakings" of this symmetry, and many of them are accidental. Sometimes, symmetry begins and then some accident of nature breaks it, and that's a common situation, and that might teach you something. By looking at a scar on my hand, you learn something about the history of my life. It can also be that you learn much deeper principles, and that's where we're headed today.

Let's think again about watching games, and thinking about the colors of the uniforms as a slightly stretched metaphor for thinking about symmetries. So, let's go now to an American game, soccer, and we've got two teams; and imagine that one of the teams is wearing pink, and one of the teams is wearing blue, and they're a bunch of little kids and they're playing soccer. Now, the question is, is there an exact symmetry between swapping the color pink and blue? In principle, there should be. The rules of the game are exactly symmetrical; if you look in the rulebook, it's going to tell you it doesn't matter what color you are, and if everybody on one team swapped with everybody on the other team, nothing should change. So now you watch this game, and let's suppose that after watching many games you observe that the pink team seems to win more often. Okay? That's a breaking of symmetry. It might just be a subtle breaking of symmetry, and what you might learn then is, in America, that little girls in the second grade mature a little bit faster than little boys, and they tend to be more coordinated and stronger. And so, the pink team may, in fact, beat the blue team more often, if they're second graders; and when they're tenth graders, maybe it's the other way around. So this is a subtle breaking of symmetry that teaches you something about the players, and maybe the culture, that they live in.

And, you could also imagine playing that game I described last lecture, where you swapped the uniforms. If you take a bunch of high school students, and you swap the pink for the blue, the game might just fall completely apart. Because all the little girls are laughing at all the little boys, and they can't play soccer anymore. Because in our culture, it's a little bit bizarre, boys aren't supposed to wear pink. And so again, by looking, this would be a more radical breaking of symmetry. You swap the two colors and now the whole game falls apart. It's not at all the way you expect it.

So physicists are often looking at systems, and looking first for symmetry. It's easiest to find symmetry when it's exact, and it's pretty easy to find when it's almost exact. In fact, sometimes it sticks out more when it's almost exact. I think artists have really mastered this. Many, many paintings will have some symmetrical aspects; but, they are almost never perfect. Geometrical. Exact. And, it's that slight "breaking of symmetry" that kind of draws you in, and gets your attention. So, I think, many times in physics when we're looking at situations where there's a slight breaking of symmetry, we notice it. Not only does it point us towards the symmetry, but it also teaches us something. For example, I've talked about neutrons and protons in a nucleus, and I said, the laws of the strong interaction are independent of whether you're a proton or a neutron. It's a deep symmetry of nature. That it doesn't matter which you are. But it's not exact. In fact, a nucleus with two protons is not a stable nucleus; a proton and a neutron, is. It's a very subtle difference in most cases. Swapping a neutron for a proton changes electrical properties. Electricity is very weak, compared to the strong force, so normally it doesn't have a big influence; however, it can play a role. There's another symmetry that's broken between protons and neutrons, very subtly: if they were completely symmetrical, they would weigh exactly the same. And in fact, a neutron weighs about 0.1 percent more than a proton. It's a tiny breaking of symmetry, and it's very interesting. Where does it come from? Well, it comes from a couple of places. It teaches us, for example, that the down quark might weigh just a little bit differently than the up quark. Because, after all, a neutron is made of two downs and an up, and a proton is made of two ups and a down; so a tiny difference between the quarks would lead to a tiny difference between the two objects. You can use this small breaking of symmetry to learn something about the quarks that build up the objects, even if you can't see them directly.

You can also use this breaking of symmetry to learn something about electricity. What is the role of electricity inside of a nucleus? At first, you might say, "Never mind: the strong force is so strong, everything is tightly bound who cares about electricity?" But, it's there. It's real. Positive charges repel, but the neutrons are not repelled, because they're neutral.

And again, you can learn about the connection between electricity and the strong force in the nucleus by looking at the differences, between nuclei that have different numbers in protons and neutrons. So, when symmetry is slightly broken, it's easy to find, it teaches us lots of things; and, we can use both the symmetry and the breaking to learn stuff.

Life gets tough when symmetry is badly broken. When the symmetry is so broken that you'll hardly even recognize that there was any symmetry there to begin with. Let me try to make this concrete with a couple of examples: the first one, a little bit more metaphorical, and the second one, a little bit more physical. Imagine that you're at a fancy dinner, and you sit down at a very large, round table, and there's a bunch of seats; and in front of each seat there's a plate, and in between the plates is a full glass of water. Now, when you sit down, there's a glass of water on your left side, and there's a glass of water on your right side; they're between the plates. And you say, "Ah, this is a perfectly symmetrical table. I am completely free to choose left, or right, to grab the glass of water." Now, there might be a social convention, I never know, so what I do is, I just wait. And, sooner or later, somebody around the table is going to grab their glass—and I'm paying attention, everybody else is doing the same thing. If the first person reaches out with their left hand and takes a sip, everybody's obligated. The symmetry is broken. You no longer have a choice. Because if one person goes left, and another person goes right, somewhere around the table—think about it—somebody is not going to have a glass of water. And so, this is a case where symmetry is broken. Imagine a latecomer who sits down, and now everybody at the table, except for them, has taken some sips. So, the newcomer sits down, and they see a full glass on their left side and an empty glass on their right side. They say, "This table has no symmetry whatsoever; completely broken symmetry of water glasses." And they might, if they're not so bright, they might not even be able to fathom the idea that the table began symmetrically. And they might think, "Well, gee. You have no choice in life. You have to drink from the left glass, because that's where the water is." This analogy is not mine; it was actually invented by Abdus Salam, who is one of the co-winners of the Nobel Prize for his understanding and explanation of the weak force of nature, which manifests some very serious broken symmetries. And we'll come back later, and talk more about Salam's understanding of the symmetries of forces, which are analogous to this left/right choice that you have to make at the dinner table.

Let me give you another example that's a little bit more physics-y. Supposing that you look inside of a kitchen magnet. So, a kitchen magnet is a small chunk of metal, and when you stick it to the refrigerator, it sticks. Why? What's going on? Well, physicists understand magnetism as arising from the magnetism of individual atoms. And I don't want to, you know, go on this detour to talk about why an individual atom is itself magnetic. We need a course on electricity and magnetism to talk about that. But let's just take for granted that an individual atom is acting like a little teeny magnet, and it's got a north pole and a south pole, and it's an atom; so it can be oriented any which way. So, it's like a little, randomly oriented atom—a little randomly oriented magnet—and in fact, if it's warm, it's jiggling around. First it's pointing north, and then it's pointing west, and then it's pointing east. Now, supposing that you have a crystal, which is a whole bunch of atoms. There are various possibilities. One is that all of the atoms are pointing every which way. Then you just have a crystal. It's not a magnet at all. But what happens if all of the atoms in the crystal are magnetized in the exact same direction? Well, magnetism adds up. If they're pointing in opposite directions they cancel; but, if they're pointing in the same direction, they add up. And so a kitchen magnet is a material where all the little atoms, all the little mini-magnets, are all pointing in the same direction.

Okay, now let's think a little bit more about why this happens. If you have a magnet, and it's warm—if you warm up your kitchen magnet, it will demagnetize because there's so much thermal energy in there, all the atoms are pointing every which way. Now imagine cooling it down. Maybe one of the atoms will be pointing in some direction, and its neighbor will say, "Hey, you know, magnets love to be pointing in the same direction. Compass needles love to point up with the Earth's magnetic field. That's what magnets do, they like to line up." So if one atom is pointing in some direction, its nearby neighbors—if they're close enough and if they are strongly enough interacting—may momentarily line up. Now imagine a neighbor of that pair. So you've got two that lined up, and now there's a nearby third one who says, "Hey, two of my neighbors are lining up. I definitely want to line up," because now there's an even stronger local field. So, it's kind of a chain reaction. Once a couple starts to line up, then it spreads. Everybody lines up. It's like lemmings. Once somebody does it, everybody wants to do it. And this is, then, the spontaneous magnetization of the magnet; and magnetic materials are materials where the atoms are close enough, and strongly enough coupled, that they can do that.

Now, imagine that you're a little teeny-weeny physicist, and you live inside the magnet, and it's magnetized. Right? You're living on the refrigerator, and you've spent your whole life in there, and you're trying to deduce fundamental laws of physics inside the magnet. At first you're not too bright, and you're making observations, and trying to come up with rules, and it seems like many, many experiments that you do point north, and it's special. For example, if you build a little compass needle, it always points in one direction. Always. And it doesn't matter where you go in your universe. Right? You're really small. Your whole universe is inside this magnet, even if you get far away from your home. Right? It's like a human being leaving Earth. If you left Earth, your compass needle would point in a random direction, but not if you live inside Magnet World. North is special to you. And furthermore, if you take little charged particles, and you make them travel, you make beams when they're going north. They go in straight lines. But, when they're going west, they go in circles. So north and west are radically different directions to you. So, maybe you become religious and you say, "God is to the north." And maybe you're a physicist, so you say, "The universe is not symmetrical. The north direction is the special direction. And the laws of physics reflect the rule that, if you're going north, life is simple; but, if you're going west, life is complicated." And, then comes along a little genius mini-physicist, a little micro-Salam who says, "No, no, no. I believe that the world is truly symmetrical, that atoms can point any which way. It was an accident of history that they happened to line up northwards. And really, when we describe the laws of nature, they should not care about which way is north and which way is west. It's just that our world is a broken-symmetry world."

So this metaphor implies that you live in a world that's complicated, and some direction is special. It requires some insight to realize that the fundamental laws are truly symmetrical. And this little mini-physicist could prove that he or she is correct by heating up the world around them; and, if you heat up the world around you, then all of a sudden north is no longer special. And now the symmetry becomes manifest. It becomes obvious. This kind of completely broken symmetry, shattered symmetry, is one that exists in the real world; and let me give you an example. This one is, again, slightly difficult to picture. But, it's the way the world is. So let's go back to mirror symmetry, where you look at your left hand and your right hand and you say, "Pretty much the same thing, only opposite. Left and right have been swapped." Physicists have a fancy word for mirror symmetry. It's called "parity symmetry." And, I actually don't have any idea why we call it parity; the etymological origins of that word don't seem to connect with mirrors. But anyway, that's what we call it, parity or "p" symmetry. And I'm not talking about human bodies now. A human body is an object. My body is not left/right symmetric, but that's not a law of nature. You could imagine, in some genetically engineered future, that human beings have a heart on the left and a mirror heart on the right. It would be nice. Then you could have a heart attack, and it wouldn't matter, because you've got a spare. So, you could imagine a world that's the same as ours, the same laws of physics, where human beings were exactly symmetrical. So, the breaking of symmetry of a human doesn't break any symmetry of laws of nature. I'm talking about something much deeper.

I do some experiment on the table, like I bang billiard balls. Playing pool is a good physics experiment. Physicists like to play pool because, in principle, if you know the laws of conservation of energy, and conservation of momentum, and angular momentum, you can predict exactly what's going to happen. Now it's just the physical skill of getting the balls to do what you know they're supposed to do. It's a metaphor for doing an experiment testing fundamental laws of nature. Now, imagine that you're in the pool hall doing your physics experiments, and there's a big mirror on the wall. And, instead of watching the table to write down your data—to verify that energy is conserved, and F equals ma, and all the laws of physics you're trying to derive—suppose that instead, you watch the mirror and you're taking your data from the mirror-image balls. Now there's some weird aspects, like the three ball has this funny symbol on it that's a backwards three; but that's just a human convention, you can get used to that. There's not a law of physics that says threes have to be one way. And indeed, when you watch the laws of physics in Mirror World, the laws of physics over there are exactly the same as they are in our world. You check mechanics. You check electricity, and magnetism, and the strong nuclear force. And you discover, in experiment after experiment: you do the experiment, you do its mirror image, the outcomes are identical, but mirror reversed. There's no difference in the laws.

So you say, "Ah; this seems pretty reasonable. Just like the laws of physics are the same in different places, the laws of physics are invariant under translation. And, just like the laws of physics are the same at different times, so to the laws of physics are also invariant under mirror reversal." So this looks like a good symmetry of nature. And, if you could go back in time and chat with Albert Einstein, and you said to him, "Albert, do you believe that mirror symmetry is an exact symmetry of nature?" He would say, almost certainly, "Of course it is! It's obvious! It's as obvious as the fact that the laws of physics are unchanging with time, or with space." To almost any physicist—you can't even imagine—I mean, how could it be that mirror billiards spontaneously fly up into the air? No. I don't think so. It just doesn't make any sense.

In 1956, two physicists, by the names of Lee and Yang, wrote a paper in which they observed that this law of—this fact of—mirror symmetry had been verified countless times for all the forces that I've mentioned. In fact, gravity, electricity, magnetism, optics, and strong nuclear force were all mirror symmetric. But, apparently, nobody had every actually checked the weak interaction. What if you have beta decay? That's a nuclear reaction. Is it true that the reaction, and its mirror reaction, both occur with the same probability in the real world? Lee and Yang just sort of posed this. They suggested that, maybe not; because there were some puzzles in bubble chambers that were a little bit confusing, and this was a possible way out. Now at first, I don't think they were suggesting that it had to be. They were just asking, could it be? And most physicists said, "Oh, come on! For goodness sakes, it's obvious, parity is a good symmetry." I have a quote from Wolfgang Pauli: "I am prepared to bet that the experiment will be decided in favor of mirror invariance, in spite of Yang and Lee. I don't believe that God is a weak left-hander." Okay, what does he mean by that? God is the metaphor for laws of physics. And, calling him a weak left-hander means that, in the weak force, your left hand and your right hand would be fundamentally different beasts. They wouldn't just be mirror images of one another.

Well, Mr. Pauli was wrong. The experiment was done just a year later. It was done by a physicist at Columbia by the name of C.S. Wu; and, Madame Wu was a brilliant experimental physicist, expert in beta decays. She set up an experiment to test mirror symmetry. Let me try to summarize the idea for you; the details are a little bit complicated. Take a nucleus that spins. Okay, maybe it's spinning in the clockwise direction. So, it's got a certain "handedness". Let me say why left hand and right hand is a nice metaphor. If I take my right hand and I point my thumb forwards, my fingers naturally curl in a particular direction. From behind, they seem to be curling clockwise. So, when a particle runs away from me, and it's spinning clockwise, I call it a right-handed particle. Similarly, my left hand, they go the other way; if I point my thumb away from me, like the direction of travel, a left-handed particle would be a particle that is spinning in the counter-clockwise direction, as viewed by me. So we can talk about left-handed particles or right-handed particles. So, C.S. Wu has taken some nuclei which are spinning in a certain direction, they have a certain handedness, and they beta decay. That's the weak force in action that makes them spontaneously, after a certain amount of time, fall apart. And, what comes out? Well, you're left with a nucleus that's been transformed; and beta rays, electrons, go flying out.

Now imagine watching this experiment in a mirror. If the nucleus in the real world is spinning clockwise, the mirror nucleus is spinning counter-clockwise. So, if the real nucleus is labeled north and south, your mirror nucleus is exactly the opposite, south and north. So now, what would you expect if the world were truly "mirror symmetrical?" I would expect that the outgoing electrons would be equally likely to go north as to go south, or east, or west. They should be equally likely to go in different directions. Because, if all the electrons went north in the real world, that's what we would call south in the mirror world. That would be a drastic breaking of mirror symmetry.

Now you could imagine that mirror symmetry is just slightly broken. Maybe, out of every hundred, fifty-one go north and 49 go south. That would be a very small breaking of mirror symmetry. And, what Wu's experiment found, was radical. Okay? Every electron goes out to the north, and none of them goes out to the south—almost. It was almost the maximal breaking of mirror symmetry that you could imagine.

Mr. Pauli wrote afterwards, "After the first shock is over, I begin to collect myself. It was very dramatic." In fact, Mr. Feynman also made a bet—he made a fifty-dollar bet. This is in the 1950s. I'm guessing that was a lot of money back then. He bet that parity would be a good symmetry of nature, even in these weak decays. Everybody expected that. So, it was really a big surprise. And, our understanding of the world shifted. Now comes this weak interaction, which is already kind of an oddball: it changes the strangeness of particles; it transforms neutrons into protons; and apparently, it knows the difference between something going, spinning, right-handed, and something spinning left-handed. God is a weak left-hander, if you want to make that metaphor.

Many, many experiments were done after this, and they all verified that, in any weak interaction, left-handed spinning particles and right-handed spinning particles behave differently. In fact, when you start looking at experiments with neutrinos, it's an intrinsically weak interaction now, because the neutrino can only interact via the weak force. So, if any particle of nature is going to be the kind of most obvious place to look for parity violation, or breaking of mirror symmetry, the neutrino is where you want to go. And even today, with neutrino physicists, one of the things that they're trying to understand is mirror symmetry.

It turns out that every neutrino in the world is left-handed. Remember, neutrinos are particles. They don't have electric charge, and they don't have mass, but they do have spin. If a neutrino runs away from me, it's spinning in the left-handed sense; it's spinning counter-clockwise as it's running away from me. And, if mirror symmetry was a good symmetry of nature, well, shoot; you would expect half the neutrinos in the universe would be left-handed, and half would be right-handed. They should be equally likely. And, in fact, it's as unlikely as can be. There are no right-handed neutrinos. I make a very strong statement about that; and actually, there's some current research that is hinting that maybe there's a tiny, tiny possibility of right-handed neutrinos. That's a branch of physics we'll come to in one of the last lectures, to talk about this sort of exotic idea. But to an excellent approximation, the neutrino massively breaks mirror symmetry. They're all left-handed.

Now, what about the anti-neutrino? The anti-neutrino is right-handed. Every anti-neutrino that we've ever found, or measured, has always been right-handed. So, the mirror image of the particle, in this case, is the anti-particle. That's not normally true. The mirror image of a billiard ball is not an anti-billiard ball. It's just another billiard ball. The mirror image doesn't change electrical charge. So, there's something very unusual, and very special, about neutrinos and the weak force.

And now, I told you at the beginning of this lecture that when you find a broken symmetry—and this one is as massively broken as can be—you might ask yourself, "Is it really true that nature has some beautiful left/right symmetry, and that, as the universe cooled down from the original big bang, somehow, left froze out by accident?" Kind of like in my little magnet world, it started off completely symmetrical, and then, as it cools down, one direction, north, kind of freezes out as being special. The answer is, we don't know. This is a branch of theoretical physics research that people are interested in today. We would love to be able to construct a deeper theory of nature in which left and right are completely symmetrical because it just seems so nice and elegant. It's an accident of something, history for example, that the left hand became special. I mean, otherwise, why is left, in a certain sense, better than right? It's crazy.

So, we're working on it. We don't have a deep explanation at the present time; but, what we do have, is a description. So, we're kind of like Johannes Kepler. We have observed behavior in the subatomic world. We have observed that all neutrinos are left-handed. We have constructed a model. I keep referring to the standard model of particle physics. That's the name for everything that I'm trying to build together here. A quantum-mechanical, relativistic theory that describes the quarks, and the leptons, and all the forces of nature, at least all the subatomic forces of nature: strong, weak and electromagnetic. And this standard model exists; you can write it down on a piece of paper. You can do calculations with it. And, built into that theory, is the fact that the weak interaction violates parity symmetry. And, it's not an explanation; it's just a description. So we're hunting for that.

There are other examples of broken symmetry in the world, which we're going to want to come back, in future lectures, to talk about. One of them is the symmetry between matter and anti-matter. Forget about handedness. Forget about parity. Just think about a proton and an anti-proton. Okay, at first glance, you would say those are completely symmetrical particles. They have the same properties, except for the sign of their electric charge. That's the difference. But, they have the same mass. They obey the same laws of physics. Well, we thought that was true. We thought that matter and anti-matter was a perfect symmetry, until some experiments, about 20 or 30 years ago, that showed that, in fact, just like parity is broken, so is matter/anti-matter symmetry, in a very subtle way. We have to come back and really devote more time to talk about the ways in which matter/anti-matter symmetry are broken. This is not an egregious breaking. This is a very small breaking. And, it turns out, to have many important consequences in our understanding of where the world came from.

When you have a spontaneous breaking of symmetry—like one person picking the left-hand glass and then everybody does—or a magnet cooling down, and all of sudden there's one direction, north, that's special—it's very interesting to ask, "Was it really an accident, or was there a reason?" Maybe there is a social convention, I don't know what it is, but maybe you're supposed to reach for the right-hand glass, and that's what people do, if they're savvy. Maybe, if you've got a magnet in the real world, where we live next to planet Earth, that is a magnet, that points north; well then, when you cool this hot magnet down, it's probably going to point in the same direction as the Earth's magnetic field, because there's an external reason for everybody to line up in that particular direction. It's not exactly an accident.

Now why is the Earth's magnetic field pointing the way it does? We have to look and see if there are some deeper reasons. Sometimes broken symmetry has some deep reasons, sometimes it really appears to be a true accident; and it's one of the goals of theoretical researchers, when you see a broken symmetry, to try to ask, "Where did it come from? Does it have some fundamental origins? Is it an accident? Can we learn something from it?"

Lecture Fifteen
The November Revolution of 1974

Scope: In November 1974, two simultaneous experimental discoveries rocked the world of particle physics. A new quark had been found. It came from two radically different places—a cutting-edge detector at Stanford and a challenging experiment at a powerful facility in Brookhaven. The new *charmed quark* had been anticipated (at least, by a few) and fit so well into the quark model and the new theory of QCD that it changed the scientific paradigm for many physicists nearly overnight. In some sense, this event was the dividing line between early particle physics and modern high-energy physics.

> The man was tired, for he had diligently worked the area for weeks. He stooped low over the pan at the creek and saw two small glittering yellow lumps. "Eureka!" he cried, and stood up to examine the pan's content more carefully. Others rushed to see, and in the confusion, the pan and its contents fell into the creek. Were those lumps gold or pyrite? He began to sift through the silt once again.
>
> —Harvey Lynch (SPEAR Logbook, Nov. 9, 1974)

> BJ, I think you'd better go down to the lab now.
>
> —James Bjorken's wife (on the evening of the discovery of the psi particle at SLAC)

Outline

I. By the early 1970s, the quark model was definitely established, but it was by no means dogma in the world of physics. As we've said, particle physicists were aware of the model, but many believed that it might be just a mathematical framework, not a reflection of a real object.

    A. At the same time, the theory of quantum chromodynamics, the mathematical framework for understanding the strong force, was in development. This theory showed why quarks were not being observed in the laboratory, but it had not been widely disseminated.

B. The experiment that revealed that quarks were physically real occurred in the mid-1970s, at a time now called the *November Revolution*.

II. The November Revolution is a play with two acts. Experiments were going on in parallel at two large experimental facilities in the United States.

A. We'll begin with the facility on the West Coast, the Stanford Linear Accelerator Center (SLAC).

1. Using this accelerator, electrons are accelerated up to an energy of a couple of billion eV, then smashed into targets.

2. The physics community had debated whether smashing electrons was productive. Why not accelerate protons? When a proton hits a target, it interacts strongly, but the electron does not.

B. The competing philosophy was manifested on the East Coast by an accelerator at Brookhaven National Lab on Long Island.

1. This facility accelerated protons up to a much higher energy, closer to 30 billion eV. These experiments produced many more events, because the interactions were strong.

2. This interaction is also more complicated, because the object being smashed is more complicated and results in much more debris. The interpretation was more difficult on the East Coast.

C. At Stanford, the director of one of the research groups, Burton Richter, had an interesting idea.

1. Richter knew that when energy is pumped into a small region of space to smash electrons into a target, e+/e− pairs can be produced.

2. Richter wondered what would happen if the positrons were channeled by some sort of magnetic field and trapped. He proposed accelerating the positrons further and running them in the same circle with electrons. Because they have opposite charges, they would run opposite each other, and at a certain point, they would collide.

3. The result of this experiment would be, for the first time, matter/antimatter collisions right in front of a detector. The electron and the anti-electron would annihilate. Pure energy would exist for a brief moment, then the energy would be converted into creating new particles of nature.

D. Brookhaven was using its high-energy proton beam to produce large numbers of new particles, including new matter/antimatter pairs.

    1. A physicist there named Samuel Ting was looking at all the debris coming out of the smashed protons to find a muon and an anti-muon coming out back-to-back.

    2. What would make a muon and an anti-muon appear out of this debris? Presumably, they would be created from some small bundle of energy, maybe a particle, for example. Imagine that a particle is created; it lives for a short time, then decays. One of the things that it might decay into would be matter/antimatter.

    3. Muons leave a distinctive signature in a detector. Remember that the muon is like a heavy electron; it does not interact strongly.

    4. In the summer of 1974, Ting's research assistants noticed something interesting in the data. According to conservation of energy, total energy going in must equal total energy going out. As they graphed the data, the researchers noticed a number of events at one particular energy. Many muon pairs were being produced with the exact same total energy. Such an event is now called a *bump*.

    5. This bump indicates, most likely, that there is a fundamental particle of nature with that energy. In this case, it was a very unusual signal for a variety of reasons. Ting understood that he had made a radical discovery of a particle that was super narrow and super strong, but he did not immediately publish his data.

E. On the West Coast, Richter's group now had an e+/e− collider and was analyzing data from a reverse of Ting's experiment. Instead of looking for a particle/anti-particle pair coming out, they were starting with a particle/anti-particle pair, then looking for what was produced.

    1. The procedure is to tune the energy of the beams to know exactly how much energy is coming in to the system, then to look at all the particles that come out; most of the time, not much happens.

2. If you hit a resonance—the energy where a particle could be created—then you'll create that particle. It will exist for a moment, then it will fall apart and other particles will come out.

3. Experimenters at Stanford had been steadily ramping up the energies in the accelerator and collecting data. Earlier, some researchers had noticed an unusually high count rate of events in the detector at around 3 billion eV. The decision was made to back down to that energy to see what was happening.

4. On November 10, 1974, SLAC was running at 3.12 GeV. When they changed the beam energy a little bit, the count rate exploded. Particles were appearing everywhere. When they adjusted the beam energy again, they saw hundreds and hundreds more events than the regular event rate.

F. At this point, the two stories converge; the particle that was being discovered at Stanford was the same particle that Ting had discovered earlier at Brookhaven. Ultimately, the particle came to be called the *J/psi*, a combination of the names given to it at both laboratories.

III. Why was the discovery of this particle so exciting?

A. First of all, its narrowness was a big surprise. In this context, *narrowness* means that it has an extremely definite energy or, if we're thinking quantum mechanically, it lives for a very long time.

1. If we think about the Stanford experiment, we note that e+/e− pairs annihilate, and a new particle is formed. We can determine its quantum numbers, including its mass, which was 3 GeV.

2. We would expect that the J/psi would decay immediately because all the other particles in the zoo decay very rapidly, and the more massive they are, the more rapidly they decay. This particle lived for a very long time. Of course, its lifespan is still fractions of a second, but in the particle physics world, that's a long time.

3. No one had ever seen a meson like this one, but almost immediately, physicists figured out what it was. In fact, some scientists had predicted what it was before it was even seen.

B. In the years preceding this discovery, two predictions had been made.

    1. Remember the quarks that we've talked about. We have up quarks and down quarks, which are similar. They're both quite light and they make up most of the ordinary world. One of them, the up quark, has charge 2/3. The down quark has charge –1/3.

    2. We also have the strange quark, which has a charge of –1/3. It's just like a down quark, only heavier and unstable.

    3. The two light quarks and one heavy quark make a sort of funny pattern. If we want to have symmetry, we ask why we don't have a heavier version of the up quark with a charge of 2/3.

    4. Even in 1964, physicists had thought about that question as soon as the three quarks had been proposed. Sheldon Glashow and James Bjorken published a paper in 1964 suggesting that this particle existed. In fact, they thought that the existence of such a particle would be charming because it would make the picture so elegant. Thus, they called it the *charm quark.*

C. As the years went by, stronger arguments were made for the existence of a charm quark. Glashow, along with two other scientists named Iliopoulos and Maiani, made a subtle theoretical argument for what came to be known as the *GIM mechanism.*

    1. Physicists had been looking for some events in bubble chambers that were not occurring, but everybody thought they should. A kaon particle, for example, can decay in many different pathways. It might turn into a pion, an electron, or something else. But one of those pathways did not occur, and it should have.

    2. Glashow, Iliopoulos, and Maiani came up with an explanation that involved the use of Feynman diagrams. Remember that we have to draw every possible Feynman diagram and add them up before we know the probability for some event.

    3. If we include virtual charm quarks—if we imagine creating a charm quark, then having it disappear again very quickly— mathematically, that event would cancel all the other Feynman diagrams and the final answer would be nearly 0. That would explain why there were no occurrences of a certain kaon decay in the bubble chamber.

4. This theory pins down the math of the charm quark. In order for the math to work out, a certain mass must be predicted for the charm quark. In retrospect, after the J/psi particle was found, its mass was exactly twice the predicted mass of the charm quark.

D. What is the J/psi particle, then? It's a charm/anti-charm meson. It's a new meson that is built out of the fourth quark, the charm quark, and its anti-quark.

   1. In a certain sense, this object doesn't have net charm. We call it *hidden charm*. What's called *naked charm* would be if we produced a meson that was charm and anti-up or charm and anti-down. We might have naked charm in a baryon if we produced, for example, a charm up, up; or a charm down, up; or any of the quark combinations that we could imagine involving charm.

   2. Thus, a new industry began in particle accelerators. Physicists began to look for all these new possibilities that could be seen in the laboratory, the new mesons and the new baryons.

   3. The fact that new particles appeared just as expected from the quark model clinched the idea of quarks. In the November Revolution, the idea of quarks was transformed from an abstract and slightly radical proposition to a mainstream, well-established philosophy. The discovery of the J/psi was the critical event in confirming the physical reality of quarks.

Essential Reading:

Barnett, Muhry, and Quinn, *The Charm of Strange Quarks*, chapters 1, 6.1–6.4.

't Hooft, *In Search of the Ultimate Building Blocks*, chapter 15.

Recommended Reading:

Lederman, *The God Particle*, chapter 7, from the section on "The November Revolution" up to "The third generation."

Riordan, *The Hunting of the Quark*, chapters 11, 13, and especially 12.

Questions to Consider:

1. In the quark model, how would you describe the J/psi particle? Is it a meson or baryon? A fermion or boson? A hadron or lepton?

2. Why do you suppose Sam Ting waited so long to publish his discovery of the J/psi?

3. These days, we might sometimes feel that small scientific discoveries receive a lot of "hype." Could the November Revolution have been one of the first public relations hypes of physicists—an attempt to excite the public and ensure continued funding? What arguments can you make against such a suggestion (or for it?)

Lecture Fifteen—Transcript
The November Revolution of 1974

The quark model, in the early 1970s, was definitely established; however, it was by no means dogma in the world of physics. I think all particle physicists were aware of the model. They were thinking about it; however, there was a sense that these things might just be mathematical. It might just be a framework for understanding the complexity of the world; but, not necessarily, a reflection of some real object. And you have to appreciate, on the one hand, there's this zoo of particles, a hundred particles with all sorts of complicated properties, that all make sense. It all fits in, at least to a crude quantitative level, if you believe in quarks; but nobody's finding a quark in the laboratory. You never see the trail of a particle in a detector that you can identify as a quark.

Now, at this same period of time, the theory of quantum chromodynamics, the rigorous mathematical framework for understanding the strong force, is in a process of development. And, the people who are working on it are not completely disseminating those ideas, so most of the physics community has a kind of a crude model with which to work, and they're not seeing the detailed mathematical formalism. The formalism is beginning to show that there's a reason why you wouldn't see a quark all by itself; but that idea hadn't really spread so much, and I think the physics community, in order to believe in quarks as something physically real, was in need of some concrete experiment. And, that experiment finally occurred in the mid-1970s. So in today's lecture, I want to talk about what physicists now refer to as the November Revolution of 1974.

To talk about this, it's really a play with two acts. There are two big experimental facilities in the 1970s in the United States where experiments were going on in parallel with one another. They looked quite different, and we need to talk about both of them, because they both contributed to this big discovery.

Let me start on the West Coast, at Stanford. We've talked a little bit about the Stanford Linear Accelerator Center, SLAC: this is a long, two-mile long, linear accelerator that takes electrons, accelerates them up to an energy of a couple of billion electron-volts. It's a couple of GeVs. So that's a pretty darn high energy, and what they had been doing with those electrons is smashing them into targets, primarily, and then looking to see what comes out. There had been some debate in the community about whether that was really such a great idea. Why not accelerate protons? Because when a proton hits a target, it interacts strongly, and you'll get a whole bunch of stuff happening. It interacts, almost, with a hundred-percent probability—not quite true—but it interacts very strongly. The electron does not interact strongly. It's a lepton, so it only will interact electrically; and that's a hundred times weaker, at least; and so, many of your electrons just go flying right through your detector and just get wasted, they get absorbed by some beam dump. So I think when SLAC was constructed, there was some debate in the community about whether there was going to be very many good experimental discoveries made here. And I think SLAC proved, by the number of Nobel prizes that were awarded for what they did over the years, that indeed it was a great idea. The nice thing about SLAC is that an electron is a very, very simple particle. It's just a little point-like electric charge. So, although it doesn't have such a great probability of interacting, when it does, you understand in great detail what's coming in, and you can focus your attention only on what comes out.

The competing philosophy, which was manifested physically in the East Coast by a big accelerator at Brookhaven National Lab—that's on Long Island—they accelerated protons; and they accelerated the protons up to a much higher energy, more like 30 billion electron-volts, almost ten times as much energy; and now you smash the proton into a target—and lots more energy, many more events happening, because it's a strong interaction—but now you've got a complicated object coming in. If you believe in the quark model, you've got three quarks coming in, and then you have to figure out, well, which one interacted? And, right. There are all sorts of debris coming out, some of which is from the event, some of which is just coming from what you started with. So the interpretation is a little bit more difficult on the East Coast; however, you've got more energy, and more events to work with. So it's two complementary experimental approaches.

On the West Coast, at Stanford, they had this machine. They'd been running it for several years. And the director of one of the research groups, Burton Richter, had a great idea. He knew that when you smashed electrons into the targets at the end, you produce a lot of new particles. You're dumping a lot of energy into a small region of space, and some of the particles that come out are anti-electrons, positrons. Because, when you put energy into a small region of space, you can produce e-plus/e-minus: electron and positron pairs. So, what would happen if you could take some sort of magnetic field and channel those positrons, and trap them, or store them? This was the idea. It was technologically feasible in that era. And so, Burton Richter wrote a proposal to the U.S. government to upgrade the Stanford facility, to get some of these positrons and make them run around in a circle. At first, he proposed accelerating them some more, and you would have some electrons running around in that same circle—because they have opposite charges, they run around opposite one another—and at a certain point, you could run them head-on into one another. Okay, think of this: for the first time, you could have matter/anti-matter collisions right in front of your detector. Very, very clean, very pure. The electron and the anti-electron will annihilate; you will have pure energy for a brief moment, and then, particles will come flying out. $E = mc^2$. You can convert that energy into creating new particles of nature. What a great way to learn if there's new particles out there.

So, he submits this proposal to the governmental funding agency. You know, it takes a lot of money to build a new facility like this; and it's quite normal, at national research laboratories, after they've been running for a while, to propose upgrades. It's like, you know, I've got a mountain bike, and after four or five years, I go out and I buy some front shocks for the mountain bike. It's some money, and it's a little bit extra weight; but, it really improves the character of the bike, and I can do all sorts of new things with it that I couldn't do before. It's the same idea with the facilities. The U.S. government had to approve this construction project, and they liked the project. It was a great idea. But, there were various technical problems, and there were some other projects that they needed to fund, so they put it off. And Burton Richter kept, year after year, reapplying, improving his design. At first, it was going to be a twenty-million-dollar upgrade, and he was cutting corners and figuring out ways to make it cheaper. But, you know, this was the late '60s, early '70s: there was a new accelerator being built in Cambridge, and another one later in Chicago, and there was the Apollo mission going on, and there was the Vietnam War; and, it never got funded. Burton Richter kept on trying.

And the amazing thing is, the hero of this story is a bureaucrat, a comptroller in the U.S. government. His name was John Abbadessa, and at a certain point, he said to Burton Richter, "You know, technically speaking, you have to go through this review process if you want new construction. But, if you just want to take existing operating funds and re-channel them to upgrade a detector or something, we don't need to give you official approval for that. That can be an internal SLAC decision." So, Burt Richter went to the laboratory director, Panofsky was his name—they called him

Pif—and Panofsky agreed. So, basically, instead of getting official approval, and lots of new external money, they just channeled some of their existing operating funds, and they built this thing. You know I don't want to imply that it was illegal, it was definitely legal; however, it was a little bit on the sly. And, if you look at this machine that they built, it's like cinderblocks on the parking lot. They ended up spending about 5 million dollars, instead of the original 20, and they got their particle/anti-particle rings that would collide.

Now, it had been very frustrating for Burt Richter, because the Europeans had taken this idea and run with it several years earlier. So by the time they had their machine, the Europeans, who—is it competition or cooperation between the United States, and Europe, and Japan, and everybody who's got particle detectors? I don't know. It's both. It's all of the above. Everybody wants to make the discoveries themselves, and of course, we also work all together. But it's a little frustrating when you have this great idea, and then other people are building it, and getting to run their detectors. So we were a little bit behind.

Now let me leave the Stanford story and tell you about the East Coast story. We've got Brookhaven Lab, which is this thirty-GeV proton beam; and so the protons smash into a target, and they produce a whole bunch of particles. You knock chunks of nuclei out, you create new matter/anti-matter pairs; so there's a whole spray of particles. You've got these gigantic detectors trying to collect all the data, in order to analyze what's going on. And the physicist there, in the experiment that I want to talk about, his name was Samuel Ting. He's a professor from MIT working at Brookhaven for his research; and his idea, his experimental program, was to look amidst all this complicated debris, all the stuff that's flying out, for a muon and an anti-muon that come out, in a certain sense, back-to-back. Okay, what would make a muon and an anti-muon appear out of all this debris? Well, it's matter/anti-matter; and if they're coming out back-to-back in some reference frame, then presumably they were created from some little bundle of energy, maybe a particle, for example.

So imagine, you create a particle, it lives for a short time and then it decays; and one of the things that it might decay into would be matter/anti-matter. So, the muons leave a very distinctive signature in your detectors, because they're so different. Remember the muon, that's that "who ordered that?" particle; it's like a heavy electron. It's the particle that does not interact strongly. It's just like an electron. So the tracks that it leaves are very, very different from the tracks of almost all the other particles in the zoo, which very quickly decay.

So Mr. Ting is analyzing these muon pairs, and in the summer of 1974, his graduate students notice something very interesting in the data. So, here's how you analyze this data, one of the ways. You measure the total energy of the muon pair, the total energy of your system. Okay, conservation of energy says that total energy must equal to the energy that you started with; and what you do is, you sort of draw a little graph, a histogram. And on the one axis, the horizontal axis, is the total energy that you observed; and the vertical axis is just, how many times did you see an event like that. So, you see a couple of events at this energy and a couple of events at that energy, and all of a sudden, at one particular energy, they saw a whole bunch of events. Many, many muon pairs, over and over again, were being produced with the exact same total energy. Nowadays, we call this a bump. It's a bump in the graph, and it indicates, most likely, that there is a particle of nature, some fundamental particle of nature, with that energy. So it's created, it's a resonance, so it's easily created; and then when it falls apart, it always produces a muon pair with that energy. It was a very unusual signal, for a variety of reasons. One of the reasons was, it was super-narrow. People talk about "binning" their data, which means how much energy—there's always some uncertainty in the energy that you measure—and so you're going to have some range of energies that you're going to put into your plot, and they could make that range very, very narrow; and still they were seeing a bump. It was a very definite energy that they were seeing.

Now, this is very exciting. You have to appreciate that Sam Ting understands that this is a pretty radical discovery. Even though there have been hundreds of particles discovered by 1974, there's something really special about this one, because it's super-narrow and super-strong. Many, many orders of magnitude different than anything anybody had ever seen before. Okay. This is Nobel Prize material that you're looking at. So the big question is, why didn't Sam Ting jump up and down and publish his data? Because, you know, that's what physicists do when they discover some spectacular data, they let everybody know, and they publish.

Sam Ting was an unusual fellow. He's a brilliant physicist, very well respected, very serious, and very critical. He had, for many years, been very good at being able to point out what was wrong with everybody else's experiments. So, I think, he maybe had some physicists who weren't exactly friends of his; and so, perhaps part of the story was that he really wanted to be super-cautious about this. It would be humiliating if you published and it turned out to be an accident; maybe a glitch, maybe a statistical fluke, maybe just a mistake in your analysis. So, you might argue that he was being extremely cautious, and he just wanted to be darn sure.

So, they began running again, and they changed their detectors. They fixed, they changed, and his graduate students are working very hard. He was known as a tough advisor. People were working long hours. And, I think, there was a lot of pressure on him to publish. Of course there's another idea, another possibility—you can read stories about this whole November Revolution business, and it's fascinating to think about the characters and personalities involved here—it may be that Sam Ting realized that he was probably the only person in the world with this data. Okay? There were other accelerators where, in principle, if this particle existed, you could find it. But, it was so narrow, finding it would require an incredible stroke of luck. It was a stroke of luck for him, and he figured, "I've got time. It's like finding a needle in a haystack. I've got months before anybody else even has the slightest chance of finding this thing." So, given that, he could collect all sorts of data, and really figure everything out about this particle. So, that when he finally did publish, it would be a sort of a complete explanation of what he had, which would really guarantee fame and glory; or, at least, some scientific success. In any case, he continues analyzing this data, and September goes by, and October goes by, and they're continuing to collect data—no publications.

Let me go back to the West Coast. Burt Richter's group now has this machine with the e-plus/e-minus collider, and they're now analyzing data. So what do you do? It's kind of the reverse of Sam Ting's experiment. In Sam Ting's experiment, you put a whole bunch of junk together, and you look for a particle/anti-particle pair coming out. At Stanford, you're putting a particle/anti-particle pair in; they're annihilating and forming energy, and then you're looking for all the junk that comes out. So it's kind of like the same experiment, only backwards. And what you do is, you tune the energy of your beam—so you know exactly how much energy is coming in—and then you just look at all the particles that come out; and most of the time, not much happens. They swing by each other. Maybe they bounce. Maybe they produce another particle/anti-particle pair. At most energies, there's not much interesting going on. But, if you happen to hit a resonance, if you happen to hit the energy where a particle could be created, then you'll create that particle. It'll sit there for a moment, and then it'll fall apart, and all sorts of stuff will come out.

So that's what they're looking for. You ramp up the energy, and you collect data for a little while. Ramp up the energy some more, collect data for a while. They'd been actually doing this for a year, by the fall of 1974. And, there was a strong temptation in the laboratory to upgrade to higher energies, because the Europeans—in particular in Germany, there was a facility called DESY, it's a German acronym; Deutsche Electron something or other—the DESY accelerator was planning on having a higher-energy electron/positron collider. And so, SLAC really wanted to get there first. So there's some political pressure in the laboratory to fix up your magnets, make them a little stronger so you could have higher energies, and ramp up to higher energies.

There was another conflicting opinion in his group, because the summer before, they had been collecting data, and at about three billion electron-volts at around three GeV, there had been a slightly, unusually high count rate of events in their detector. So some people were saying, "You know, we should really go back and check out what was going on at three GeV, just to see if there was something that we missed. And it might also help us to calibrate our machine." So, Burt Richter weighed the possibilities and finally compromised and said, "We'll spend a few days going back down in energy to three GeV"—going down in energy, sort of a scary political decision. But, he says, "We'll do that, but just for a couple of days; and if we don't see anything, then we'll ramp up to the higher energies." So they went down to three GeV.

Let me get the numbers and dates straight. On November the 10th, 1974, they're running at 3.12 GeV; and what they saw, on their computer screen, was a sort of a real-time image that showed the events that they were collecting. So, the screen would light up every now and again, you know, indicating that something interesting had happened; and it was lighting up a little bit more frequently than their usual baseline. So it did look, indeed, like there was something there. When you're a particle physicist, you have to be careful. You're kind of flipping coins, right? It's quantum mechanics. There's some randomness associated with the production of particles. You've got energy. It will produce anything that it can produce, but only with a, sort of a statistical distribution.

So, if you're tossing coins for a living, and you toss six heads in a row, you know, do you get excited and think that you've got an unusual coin? Well, not if you've been tossing coins thousands and thousands of times; six in a row is not all that unusual. So they really had to check their data, and run for a longer amount of time. They changed the beam energy a little bit.

They went from 3.12 to 3.11, and now the count rate just started exploding. The display was like a fireworks display—flash, flash, flash! Particles were just appearing like crazy. So, they had found something. They changed the beam energy again, from 3.11 to 3.104, right? They're honing in on something. And now it's just crazy—hundreds and hundreds of times the regular event rate.

And here's where the two stories converge, because this particle that was being discovered at Stanford was exactly the same particle that Sam Ting had discovered months earlier at Brookhaven. Of course, there's no communication between the two labs—yet—because Sam Ting had been keeping a lid on his data in the hopes of putting it all together and getting—making absolutely sure that it was right before he announced it. The climate at Stanford was very different. Stanford is—was really one of the archetypes of the kind of modern sociology of particle physics. Extremely open laboratory; physicists there were highly collaborative, the teams were international. Everybody's on the phone. It was the most expensive telephone bill in SLAC's history up to that date. They're calling all their physics buddies in Europe, and on the East Coast, and throughout the whole world, telling them about what they had, because they were very excited. It was obviously a big discovery. And I think this is more in keeping with the usual climate of scientific discovery in particle physics today. This collaborative nature is, in fact, what created the World Wide Web—not in 1974, but at a later time. CERN, in Europe—that was the European center for nuclear research, which was kind of the European version of our big laboratories—they, along with the big American laboratories, really wanted to share data as rapidly and easily as possible. I remember, as a particle physics post-doc, using the World Wide Web before there were any commercial websites; all that you had on the web were physics laboratories, where you could look up data and collect information, pretty much on the fly.

It just so happened, by sheer coincidence, Sam Ting was in an airplane on that day, on his way to Stanford. Because every year, maybe even a couple of times a year, physicists get together—it's a kind of a working group—to discuss what's going to happen at the National Laboratory. It's kind of a business meeting, and Sam Ting was part of that collaboration. And, so he was on his way for this regularly scheduled meeting when this discovery was happening at Stanford. Now, the graduate students and post-docs at Brookhaven start hearing telephone calls, because everybody's talking about this thing. And, they're just freaking out: I mean, imagine how you would feel if you've just spent the last several months of your life, spending, you know, eighteen-hour days collecting this data, thinking this is your big break, and now somebody else is going to scoop you! So, they're all frantic. And, how do you reach Sam Ting? He's on an airplane. There are no telephones on airplanes in those days. So, they left a message for him at the TWA gate, so that when he got off the plane he would get this message.

He gets the message late in the evening, and realizes—I think probably to his horror—that he's about to be scooped. So he goes to his hotel room. He hasn't even gone to SLAC yet. He doesn't really know any of the details. And, he starts making telephone calls himself, which he probably should have done much earlier. You know, I don't want to criticize Sam Ting, he's a great physicist, and he had, in fact, made the discovery first; and, it was well analyzed and ready for publication by this point. But in any case, he made telephone calls to his colleagues at Brookhaven, and at MIT; and really wanted to make it clear that this discovery was his.

And the story goes that the next day, Sam Ting walks into Burton Richter's office, and Sam Ting says, "Burt, I have some interesting physics to tell you about." And Burt says, "Sam, I have some interesting physics to tell you about!" And so they're, of course, each sharing the same physics, and they're both incredibly excited about it. Both of these papers ended up getting submitted to the journal at the same time, and they were published back-to-back; and a few years later they shared the Nobel Prize, Sam Ting and Burt Richter. It's sort of interesting to hear all this kind of human aspects, which is always present; and the physics story often, kind of misses out on this interesting character stuff.

There's one more element to the human nature of this story. If you discover a particle, you get to name it; and it's kind of a big deal for physicists. So Sam Ting wanted to name the particle the "J". Now, that's not the usual convention—people usually use Greek letters for new particles—but J, it turns out, is very close to the Chinese character for his name. And, you know, he's a physicist—like many physicists, kind of a big ego, maybe—and so that's what he wanted to call it. That's what his team had been calling it. At Stanford, they had been calling it the psi, the Greek letter; and it just so happens that P-S are two of the letters in the name of their ring. It was called SPEAR. The Stanford Positron-Electron Accelerator Ring. So they liked the name psi particle. And so there was, for many years, some sort of an East Coast/West Coast thing, where if you went to talk about this particle on the West Coast, it was the psi, and on the East Coast it was the J. Nowadays. The official name of this particle is the J/psi; so, that was the compromise.

Why was this so exciting? Why is this a revolution in physics? Well, you might just say, "Come on—it's another particle. You've just been telling me about, you know, hundreds of particles that are produced." There was something extraordinarily special about the J/psi. First of all, its narrowness was a big surprise. Narrowness, meaning it has an extremely definite energy; or, if you're thinking quantum-mechanically, it lives for a really long time. If you're thinking about the Stanford experiment, e-plus/e-minus annihilate, you form a new little particle; you can figure out its quantum numbers, it has no electric charge, and you can figure out its spin, it's a boson. So it's a meson; it's a strongly interacting particle, of very high mass. 3 GeV. You expect that it should decay immediately, because all of the other particles in the zoo decay very, very rapidly; and, the more massive they are, the more rapidly they decay. But this thing lived for a really, really, really long time. Of course, it's still fractions of a second; but, in the particle-physics world, that's forever. It's practically stable. So what is going on? This is a meson like no one has ever seen before. And, there was sort of a buzz in the physics community; but almost immediately, people figured out what this thing was. In fact, some people had predicted what it was before it was even seen. Let me tell you about two predictions that had been made in years before.

First of all, I want you to think about the quarks that we've talked about. There's an up quark and a down quark. They're both quite similar, they're both quite light, and they make up most of the ordinary world. One of them, the up quark, has charge two-thirds; the down quark has charge minus one-third. So if you're imagining a little table, you've got an up quark and a down quark next to each other; and then there's this heavier version of the down quark. It's also got charge minus one-third, just like the down quark. That's the strange quark. It's just like a down quark, only heavier, and unstable.

So, there's this sort of funny pattern. You've got two light quarks, and only one heavy quark. So if you want to believe in symmetry, you ask yourself, "Well, how come there's not a heavy partner to the up quark? How come there's not a heavier version of a quark with charge two-thirds?" And right away, in 1964, people had thought about that, as soon as the three quarks had been proposed. Shelly Glashow and James Bjorken were the physicists who published a paper in 1964 suggesting that maybe this would exist. And in fact, they thought it would be very charming if such a particle existed, because it would sort of make the picture so elegant; so they called it the charm quark. You've got the strange quark; now you've got the charm quark. It's the '60s. People are being a little bit silly in their naming conventions. It was a nice idea; but shoot—people hardly even believe in quarks in those years. It's so speculative. And now you're speculating about a quark that nobody's even seen or has any evidence for whatsoever, based on theoretical symmetry grounds. So, I think the idea was out there, but not really very well accepted.

As the years went by, there were stronger arguments made for the existence of a charm quark. Glashow, again the same Glashow who had proposed the charm quark in the first place, along with a fellow named Iliopoulos, and a fellow named Maiani—and these names are hard to pronounce, so people referred to the three of them as G-I-M or GIM—the GIM paper, or the GIM mechanism, was a kind of a subtle, theoretical argument. They were looking at some events in bubble chambers that were not occurring, but everybody thought they should. You have a kaon particle, and it can decay, and it decays in many different pathways. It might turn into a pion, it might turn into an electron; there's all sorts of possibilities for how the kaon might decay, and one of those pathways didn't occur. It should as there was no violation of any obvious conservation law. And Glashow, Iliopoulos and Maiani came up with an explanation. It involved drawing Feynman diagrams. You have to draw every possible Feynman diagram, and add them up, before you know the probability for some event. And, if you include virtual charm quarks—if you imagine creating a charm quark and then having it disappear again very quickly—it turns out, mathematically, that they would cancel all the other Feynman diagrams, and you would get zero as your final answer. So, that would explain why you were getting no such events.

Very esoteric. At the time, it actually pins down the math of the charm quark. In order for the math to work out you would have to predict a certain particular mass for this charm quark. And sure enough—in retrospect— after the J/psi particle was found, the mass of the J/psi was exactly twice of the predicted mass of the charm quark. So what's the J/psi particle? It's a charm/anti-charm meson. It's a new meson, which is built out of this brand-new quark, the fourth quark, the charm quark, and its anti-quark. Now, you've got charm and anti-charm; so in a certain sense, this object doesn't have net charm. We call it naked, sorry; we call it "hidden charm". "Naked charm" would be if you produced a meson that was charm and anti-up, or charm and anti-down. Or, you could have naked charm in a baryon, if you produced, for example, a charm/up/up, or a charm/down/up, or any of the quark combinations that you could imagine involving charm. So now, a whole industry begins in particle accelerators. Let's go look for all these new, possible things that you can see in the laboratory: the new mesons and the new baryons—and they were all found.

So you have to appreciate the significance of the J/psi. It's a new quark, so that's exciting. And, I think in a certain sense, the reality of these quarks— the fact that all the new particles appeared just as expected from the quark model—really clinched the idea of quarks. So, in this November Revolution, it wasn't that it was a completely new idea. The ideas had been all around. But what was really exciting, and radical, and transformational, in the physics community was that, all of a sudden, the idea of quarks went from a sort of an abstract, and slightly radical, proposition to pretty much a mainstream, well-established, almost orthodox philosophy. I think the discovery of the J/psi was the big critical event in making us really believe in the physical reality of quarks.

Lecture Sixteen
A New Generation

Scope: The last of the great surprises in particle physics was a new layer of particles with the discovery of the tau lepton. As we know, once we find one new particle, symmetry arguments compel us to look for others, which we found—the tau neutrino and the bottom and top quarks. What do we learn when particles are predicted before they are discovered? We conclude by revisiting the periodic table of particle physics, including all the new particles.

> Inventions have long since reached their limit, and I see no hope for further development.
> —Julius Sextus Frontinus (highly regarded engineer in Rome, 1ˢᵗ century A.D.)

> *Pastore*: Is there anything connected in the hopes of this accelerator that in any way involves the security of the country?
> *Wilson*: No, sir; I do not believe so.
>   …
> *Pastore*: Is there anything here that projects us in a position of being competitive with the Russians, with regard to this race?
> *Wilson*: Only from a long-range point of view, of a developing technology. Otherwise, it has to do with: Are we good painters, good sculptors, great poets? I mean all the things that we really venerate and honor in our country and are patriotic about. In that sense, this new knowledge has all to do with honor and country but it has nothing to do directly with defending our country, except to make it worth defending.
> —Excerpt from testimony before the Congressional Joint Committee on Atomic Energy, April 16, 1969

Outline

I. Some older physicists might look back on the 1970s as a golden age of particle physics. The discovery of the J/psi during the November Revolution was the highlight, but many other discoveries were also made, some of which we will discuss in this lecture.

A. The quark model had at first been tentative but was later fleshed out. This was followed by the theory of QCD, which transformed the idea of quarks into a mathematical framework that could be used to make quantitative predictions about quarks.

B. The significance of the J/psi particle was that its behavior and properties made sense in relation to the theory of QCD.

    1. Think, for example, about the most distinctive characteristic of the J/psi particle: It is "narrow," which means it lives for a long time. Why? It is a charm and an anti-charm. Why don't the two particles annihilate, like matter and antimatter? The answer comes from the mathematics of QCD.

    2. QCD says that when two quarks are far apart, they attract each other very strongly, but if they're close together, they don't feel a strong attraction. The closer they get, the freer they become. Further, in quantum mechanics, the more massive the quarks in the meson are, the smaller the effective distance between them becomes.

    3. The charm and anti-charm quark are, in a real sense, very close to each other in the J/psi particle. Because they are so close, they are almost completely free particles.

C. After the discovery of the J/psi, the theory of QCD began to spread rapidly and become well established in the physics community.

II. Experimentalists began working to prove or disprove the theory of QCD, which led to some "big surprises" in the field. Ultimately, these surprises fit beautifully with the theory of quarks and QCD.

A. A couple of years after the discovery of the J/psi, a physicist named Marty Perl at Stanford discovered a new lepton.

    1. Obviously, we knew about the electron and the muon, which was a heavy version of the electron. Perl had now found a third lepton. It had electric charge, just like the electron. It had spin 1/2. It did not interact strongly. It was truly a lepton, just like the electron and the muon. It was like a third generation of these other lighter leptons. Perl called this particle the *tau lepton*, or *tau particle*.

    2. What does the discovery of this particle tell us? Imagine trying to organize the fundamental particles into a kind of periodic table.

    a. Because the quarks and leptons are so different, we'll keep them separate. The lightest quarks, the up and the down quarks, are in a row by themselves. They're distinct from each other. They have different charges. Next, we have the two lightest leptons: the electron and the electron-flavored neutrino. Those four particles compose the world we live in.

    b. Next, we've got a kind of photocopy, or second generation. The strange quark is a heavy version of the down quark. It has the same charge and the same spin, but it's heavier and radioactive. In 1974, we also discovered a heavy version of the up quark, which is the charm quark. Our table now has up and down quarks and charm and strange quarks.

    c. What is analogous in the lepton sector? There, we have the muon and the muon-flavored neutrino. We have a complete second generation of quarks and leptons.

3. This second generation of particles is still kind of mysterious. They come raining down on us in cosmic rays—strange quarks, muons, and muon neutrinos but not so many charm quarks. They also exist fleetingly as virtual particles. They may exist in certain unusual astronomical phenomena, such as massive neutron stars.

4. Perl now discovered one more lepton. Imagine that beneath the electron and the muon in our table is a new particle, the tau particle. This discovery led us to expect the discovery of a row of new particles corresponding to the tau neutrino. Further, if we have a new generation of leptons, we should be able to find a new generation of quarks, heavier still and even more radioactive.

B. The discovery of the third generation of quarks came quickly by Leon Lederman and a team at Fermilab.

    1. Fermilab is a giant ring, similar to the Brookhaven facility, but it is no longer a cyclotron. It is made up of a number of magnets that guide protons around in a circle. There are also electric fields so the protons gain more and more energy as they go around the ring.

2. In 1977, this facility could produce protons with an energy of 200 GeV. Aiming that amount of energy into a target should enable production of a new particle of nature.

3. Think about our periodic table. We have an up and a down quark in the lightest generation, and the down quark, in a certain sense, is a little bit lighter than the up quark. In the next generation, the strange quark is quite a bit lighter than the charm quark. If the pattern continues, the new quarks should match the known ones in charm and spin, but be heavier still.

4. These new heavy quarks in our chart are now called top and bottom, or occasionally, truth and beauty. Lederman found the lighter of these, in the form of a bottom and an anti-bottom quark bound together in a meson, now called the upsilon. This is analogous to the discovery of the J/psi, also a meson (a charm and an anti-charm quark bound together).

C. These new bottom quarks are so massive that the theory of QCD could start to make some quantitative predictions about the lifetime and the decay products and the quantum numbers—almost everything that could be measured—about this bottom/anti-bottom meson.

D. Thus, we had discovered a third generation of electron-like particles, or lepton, the tau. We had also discovered a third generation of quark. What would be the next? We should find a tau-flavored neutrino, which would be difficult to detect but was found a couple of years ago. There should also be a final quark that comes under the up quark and the charm quark, with charge 2/3; this would be the top quark.

1. The top quark was expected to be heavier than the bottom quark, but nobody knew exactly what its mass should be.

2. Remember that the theory of QCD is a renormalizable quantum field theory, meaning that in making calculations from it, some infinities arise mathematically, and some properties, including the mass of fundamental particles, cannot be computed.

3. QCD can't predict the mass of a top quark, but we can make some educated guesses about it by looking at data.

   a. Physicists already had an idea for the mass of the charm quark by considering events in which a charm quark might appear virtually, very briefly, then disappear again.

      b. We can create a charm quark out of nothing if we give it back again, even if we don't have enough energy to actually make it come free. The act of pulling the charm quark, or any quark, out of the vacuum affects observables.

      c. If we calculate with Feynman diagrams, we must include every possible splitting of the vacuum into particle/anti-particle, which will change the probability a little bit. This kind of calculation is called *precision particle physics*.

      d. If we make a highly precise measurement, then make a careful calculation, we can compare the measurement with the calculation, and we may be able to deduce the mass of this one last unknown quark, the top quark.

E. The top quark was finally found in 1995 at Fermilab, which is now called the Tevatron, because its energy has been upgraded from billions of eV to trillions. That's the energy required to produce these ultra-massive top quarks and anti-top quarks.

III. Let me tell you a little bit about the sociology of modern particle physics and these gigantic physics facilities.

A. The accelerators at Fermilab and SLAC are gigantic. The area where the collisions take place has detectors the size of buildings and teams of as many as 500 physicists working on these experiments.

B. At Fermilab, two very narrow beams counter-rotate in the accelerator. The protons go one way; the anti-protons go the other way. The goal is to make the protons and anti-protons run into each other.

      1. These particles have a probability of interacting, but the probability is not 100 percent. Some of them will interact, and some of them will pass on by and continue to circle.

      2. The rings of the accelerator are designed to be slightly off kilter so that most of the time, the protons and anti-protons pass by each other. In a couple of spots, the rings cross paths; those are the interaction regions where the detectors are set up.

      3. At the time of the search for the top quark, two teams were working at Fermilab completely independently. Both teams ultimately came up with data, and their two publications agreed, within experimental uncertainties, on the mass of the top quark.

C. We are now in a period of consolidation.

    1. The tau neutrino was really the last discovery to be made, and it clarified and consolidated the standard model of quarks and leptons. We now believe that we have a complete list of particles, with no evidence that there is a fourth generation.

    2. In fact, we have now seen hints that a fourth generation does not exist. A fourth-generation neutrino could be very light and, if it existed, we would already have found evidence of it.

D. We now seem to have a well-established framework for understanding the world, including six fundamental quarks, up and down, charmed and strange, top and bottom, and six fundamental leptons, electron and electron neutrino, muon and its neutrino, tau and its neutrino.

E. However some people are wondering how many fundamental particles we can account for before we start to ask if the story goes even deeper. Someday, we might be able to go beyond the current standard model.

Essential Reading:

Barnett, Muhry, and Quinn, *The Charm of Strange Quarks*, chapter 7.

Kane, *The Particle Garden*, chapter 5, to p. 8.

Schwarz, *A Tour of the Subatomic Zoo*, chapter 7.

Recommended Reading:

Lederman, *The God Particle*, chapter 7, section called "The third generation"; chapter 8, section called "Search for the top."

Riordan, *The Hunting of the Quark*, chapter 14.

Questions to Consider:

1. Is it possible that we might someday find a totally new *fourth* generation of particles? What arguments can you make for and against this possibility?

2a. Why did it take so long to find the top quark and the tau neutrino? Both were discovered at Fermilab within the last several years. Do you think that the difficulties associated with discovering these particles were fairly similar or totally different?

2b. The top quark mass is about 170 GeV, compared to a proton mass of about 1 GeV. What element in the periodic table has a mass that large? (Bear in mind, the top quark is still considered a fundamental, elementary, pointlike object.) How many constituents does the "mass equivalent" element you found have?

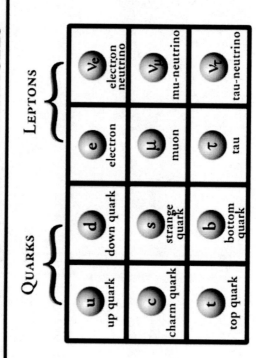

PERIODIC TABLE OF PARTICLES

QUARKS

LEPTONS

u up quark	d down quark	e electron	ν_e electron neutrino
c charm quark	s strange quark	μ muon	ν_μ mu-neutrino
t top quark	b bottom quark	τ tau	ν_τ tau-neutrino

Lecture Sixteen—Transcript
A New Generation

I think some physicists, especially older physicists, might look back at the 1970s kind of wistfully, as a golden age of particle physics. There were a lot of big discoveries being made, the J/psi November Revolution kind of being the highlight; but there were many others, some of which we're going to be talking about today. It was not just surprises; it was also a sense that it was all beginning to come together. The quark model, which at first had been tentative, had become fleshed out. And then a theory, quantum chromodynamics, had been developed, which really took what was effectively a model, an idea about quarks, and turned it into a rigorous, mathematical framework, where you can do calculations, and make quantitative predictions; and they were starting to come true. One of the most impressive things about the J/psi particle was not just that we had discovered a new quark, a charm quark and its anti-quark; it was, the behavior and property of the J/psi really made sense, if you look at the theory of quantum chromodynamics.

Think, for example, about the most distinctive characteristic of this new J/psi particle: it lives for a long time, it's very stable. Why is that? It's a charm quark and an anti-charm quark. Why don't they just annihilate like matter and anti-matter, and the thing disappears right away? And the answer is, it comes from the mathematics of quantum chromodynamics. QCD says, when you have quarks that are far apart, they attract each other very strongly; but if they're close together—it's kind of like a rubber band force of nature—they don't really attract one another very strongly at all. They become free the closer they get. And in quantum mechanics, the more massive a particle is, in a certain sense, that means distance is very small. The charm and anti-charm quark are, in a real sense, very, very close to one another in the J/psi particle; and because they're very, very close, they feel each other almost not at all. They're essentially completely free particles. They don't suck each other, one into the other, and annihilate, for a long time, because of the properties of quantum chromodynamics.

QCD was just a fledgling theory in 1974. There was a small community of theoretical physicists who knew about it, some of whom had even, really, predicted the existence of a particle like this with its properties. In fact, there was one theorist who had bet his hat, at a conference a year or two before the discovery of the J/psi, that such a particle, something like it would be discovered. I think that as soon as the particle, the J/psi, was discovered, the word of quantum chromodynamics began to spread very rapidly, because this was now a great success, a quantitative success for this new theory. And so, not only is the quark model well established by the November Revolution; but also, the theory of quantum chromodynamics becomes well established. Now, to physicists, when a theory is well established, it doesn't mean you quit thinking about it; in fact, physicists are the most skeptical, cynical, and aggressively challenging conservatives that you can imagine. If you say, a theory's out there and it's correct, all the experimentalists want to do is prove you wrong; and they're working really hard to try to find some data that proves you wrong. Of course, in the process, if they keep agreeing with the theory, it's just stronger and stronger and stronger evidence that, in fact, the theory really is correct.

So today I want to talk some more about the final, what we might call "big surprises." Okay, we just talked about the big surprise of finding a charm quark, which wasn't really such a surprise after all. In fact, it fit in, in a nice way, with the scheme that we're developing. And the same goes for the new big surprises that I want to talk about today. They were surprises, they weren't expected, but they fit in beautifully, into the theory of quarks and the theory of quantum chromodynamics. So, instead of getting everybody all upset, because they were unexpected; instead, it made everybody feel more confident and satisfied that their theory was indeed an accurate description of nature.

The first surprise came at Stanford, again. A couple of years after the discovery of the J/psi, a fellow named Marty Perl, at Stanford, was doing an experiment, and he discovered a new particle. And, it took a while for him to convince himself, and then convince the world, that this new particle that he was seeing was real. But it's very real, and since those days it has been seen many, many, countless times in lots of accelerators. It was a new lepton. Okay, we had the electron, and we had the muon, which was like a heavy version of the electron; and now Marty Perl finds a third one. It's got electric charge, just like the electron; it's got spin one-half; it does not interact strongly. Okay, it's not another meson, it's not another strongly interacting particle. It's truly a lepton, just like the electron and the muon. It's like a third generation, or a photocopy, of these other, lighter mesons. Of course, it's very radioactive, it decays very rapidly, leaving some characteristic signature; in fact it produces some new neutrinos—and that's a story that we'll come back and talk about. It's the third generation, so he gives it the Greek letter tau. It's the tau lepton; some people call it the tauon, but that's too hard to say, so we just call it the tau particle. And this was Nobel Prize material; Marty Perl got the Nobel Prize for the discovery.

But, in a certain sense, what is this telling us? I want you to try to imagine organizing the fundamental particles that we know about now in a little table, kind of like a periodic table of the elements; but now it's a periodic table of fundamental particles. So, the periodic table of the elements, you start with light ones and you work your way to heavier and heavier ones. Now, we have a slightly different scheme here. These aren't atoms we're talking about. I would say you've got quarks, on your one hand, and leptons on your other; and they're really quite different in many respects, so we'll keep them separate.

The lightest quarks are the up and the down, so I think they should get a row to themselves. They're distinct from each other; they have different charges. And then there are the two lightest leptons: the electron, and the electron neutrino, the electron-flavored neutrino. And those four particles really compose the world we live in. Right, your body is made of up and down quarks, which combine to form protons and neutrons—and maybe a few pions, which are still made of up and down quarks—and that's the nuclei; and then there's electrons; and the neutrinos, well, they're here, they're in the room. There's a hundred billion of them every square centimeter, every second passing through you, coming from the sun; although you don't feel them, or care about them, most likely, they are a part of ordinary, everyday life. It's just that that first generation, those lightest particles that seem to make up the ordinary world.

And then, we've got a kind of a photocopy, a second generation. Think about the quarks. We knew about the strange quark, which is like a heavy version of down: it's got the same charge and the same spin, but it's heavier and radioactive; and now, in 1974, we've discovered a heavy version of the up, the charm quark. So, we've got up and down, and then, charm and strange; and what about over in the lepton sector? Well, analogously, we've got the muon, and the muon-flavored neutrino. Now we have a complete second generation of quarks and leptons. And who ordered them? Well it's still kind of a mystery, although there are some arguments that we can make about why these particles are a necessary part of the world that we live in. They do appear naturally. They come raining down on us in cosmic rays—strange quarks for sure, muons for sure, and muon neutrinos—not so many charm quarks. They also exist, fleetingly, as virtual particles. They exist, perhaps, in certain unusual astronomical phenomena like massive neutron stars.

And now, what Marty Perl has discovered, is one more lepton. So, imagine, beneath the electron, and beneath the muon, now there's a new particle, the tau particle. So, you can think like a physicist now. If you found one new particle in a new generation, what do you expect? All right, I expect, by just sheer virtue of beauty and symmetry, that there should be a whole row. There should be a corresponding tau neutrino, right? Just like there was a previous generation of a heavy lepton and its corresponding neutrino. So I'm expecting that, if we go out and hunt, we should be able to find a tau-flavored neutrino. How about in the quark sector? Well, if nature is nice and symmetrical, if we've got a new generation of leptons, I'm thinking there's probably got to be a new generation of quarks: heavier still, even more radioactive. I think it's a natural guess, and everybody immediately started hunting. Okay, this was sort of a big, obvious project, starting right after this discovery in 1975 of the tau lepton.

It didn't take long. It was an experiment done by Leon Lederman and his team at Fermilab. Fermilab was, at that time, a fairly new nuclear particle-physics accelerator just outside of Chicago. It's still there; in fact, today it's the highest-energy facility in the United States. It's our premiere particle-physics facility, although we have others, including SLAC, which is still running and doing good stuff. Fermilab was not like SLAC. It was a little bit more like the Brookhaven facility: it was a giant ring. It's no longer a cyclotron. You, what you really have is a ring made up of a whole bunch of magnets, and the magnets are there to guide protons around in a circle; and you're whacking those protons all the time, giving them more and more and more energy as they go around. Fermilab is gigantic. It's four miles around, and if you every fly into the Chicago O'Hare airport, look out the airplane window and you'll probably see this big, very obvious from the sky, four-mile-round ring. When I first saw it, I was really psyched; it's very visible.

This was the accelerator facility where Leon Lederman was working. It's 1977, and the facility at that time could produce protons with an energy of 200 GeV—that's two hundred billion volt electrons. So, that's a lot of energy. And, if you dump that energy into a target, you should be able to produce new particles of nature. So, if you're visualizing this new periodic table, what are you going to see? You have an up and a down quark in the lightest generation, and the down quark, in a certain sense, is a little bit lighter than the up quark. And in the next generation, the strange quark is quite a bit lighter than the charm quark. You're always more likely to produce something that's lighter, because it's just easier to produce something that's lighter, and tougher to produce something that's heavier.

So, if there's going to be a new generation of quarks down there—first of all, what are we going to call them? People actually made up the name before they found these particles, and there were actually two competing schools of thought. The first thought was, "Well, we have up and down, how about calling them top and bottom—it's just a name, and it's kind of analogous and reminds us of the sort of symmetry of this picture of the world that we have—or T and B." And, as soon as people thought of T and B, the more whimsical characters wanted to call them truth quarks and beauty quarks, just like charm and strange. So there was a little bit of a dispute in the physics community—and I think today, top and bottom have won out. Everybody refers to the quarks in that way, and every now and again you'll somebody talking about a beauty quark. Mostly it's just T and B, or top and bottom.

If the pattern follows, the new quark that has charge negative one-third, the one that goes underneath the down quark and the strange quark, the bottom quark, is going to be the lighter of the two. And so, the guess is, that most likely, if this model of nature is correct, and if we're really going to find a third generation, that's what we are likely to find. And indeed, that's exactly what Leon Lederman found. It was completely analogous to the discovery of the J/psi, in the sense that Sam Ting had found evidence of a particle with definite energy by looking at protons smashing into a target, and investigating the debris that came out. The J/psi particle was a meson, a charm and an anti-charm quark bound together; and so, what Leon Lederman found was a bottom and an anti-bottom quark bound together.

And now, these quarks are so massive that the theory of quantum chromodynamics can really start to make some quantitative predictions about the lifetime, and the decay products, and the quantum numbers. Everything that you could imagine, that you could measure, about this bottom/anti-bottom meson is now theoretically predicted. So you might, you know, ask, "Well, did he get the Nobel Prize for that?" He did get the Nobel Prize, but for another experiment that he had done in his career. He didn't get the Nobel Prize for this one. It's kind of interesting: it's almost the same discovery as the J/psi, which got a Nobel Prize; but I think at this point, now you're just confirming what everybody's expecting. It's not such a big shock. And, you know, shoot, it's now quark number five—I mean, how many Nobel prizes are we going to give for new quarks? So, too bad.

Lederman named this quark the upsilon; it's another Greek letter, and at first I wondered why he picked upsilon; but I think, you know, upsilon sounds a little bit like Leon, which is his first name. And in fact, there's a humorous story: When they were looking for this particle—you know, what are you doing? You're collecting data, and you're looking for a large number of events that occur at one particular energy. And, at first, they begin to accumulate some events at one particular energy, so they said, "Aha! We found the upsilon." Of course they're good physicists, and they're cautious. It's like tossing that coin and you get ten heads in a row and you say, "I think we've got a weighted coin," but you should keep tossing for a while. If you discover that it was just a fluke, you have to apologize. And, in fact, it turned out that that first discovery was a mistake, it was just a statistical fluke; so, many people called it the Oops, Leon particle. And nonetheless, they very quickly did find the true upsilon.

There's another name for these particles, by the way. If you have a bottom and an anti-bottom quark in orbit around one another, like a little binary star system, you'd call it bottomonium. So, the charm/anti-charm quarks, which we call the J/psi, you could also call charmonium. Charmonium is sort of a generic name. You could imagine, if it's a little planetary system, that you could also have a particle where they spin a little faster—if you like, in a higher orbit. And in fact, you could predict the properties of that particle. And that would be an excited state, and that does exist. In fact, it was found at Stanford just within weeks of the discovery of the J/psi, exactly where you expected it. So a similar story happens with bottomonium, and in fact people have made electrons and anti-electrons go in orbit around one anther, and that's called positronium, just another naming convention.

So now, we have a situation where we've discovered a third generation of electron-like particles, a third lepton, which is called the tau; and we've discovered a third generation of quark. So now I think everybody's completely confident in the framework, in the reality of these objects; and so what do you predict? There's two particles left, in 1977. There should be a tau-flavored neutrino—which is going to be really tough to detect, because it's a neutrino. And you need lots of energy to produce a tau neutrino, because tau neutrinos are always produced as partners with the tau, which means you're not going to have very many of them; and if you don't have very many neutrinos, it's going to be a nightmare to try to detect them, because most neutrinos just go flying through detectors. So people knew right away: it's real, but it's going to take us a long time to find. And they were right. In fact, we only found the tau neutrino, directly, just a couple years ago. It's a 21st-century discovery.

So, the next particle that people are expecting to be able to find is the last, final quark, the one that sits underneath the up quark and the charm quark, the one with a charge of two-thirds. That would be what we would call the top quark. And everybody expects it, so—this is, you know, 1977, but everybody also expects that it should be heavier than the bottom quark; and nobody knows exactly what its mass should be. I want to remind you that the theory of quantum chromodynamics is a rigorous, mathematical framework; you can make calculations of almost anything, but it's a renormalizable quantum field theory. That's a fancy word that means that there's some infinities that arise, mathematically, and to fix them, there's a couple of things that you cannot compute. One of those couple of things is the mass of the fundamental particles themselves. You just can't predict that from the theory. So, you can't predict, from QCD, what the mass of a top quark is going to be. You can make some educated guesses. You can make some educated guesses, not by looking at the fundamental theory, but by looking at data. Okay, how do I mean that?

I mentioned in the previous lecture that people already had an idea for the mass of the charm quark, by considering events where a charm quark might appear virtually, very briefly, and then disappear again. The fact that it is real, and it exists in nature, means that you can sort of create a charm quark out of nothing, if you give it back again, even if you don't have enough energy to actually make one come free. And the pulling of the little charm quark, or any quark, out of the vacuum affects observables, even if you just scatter an electron against another electron and you look at the probability of them going off at some particular angle. If you calculate that with Feynman diagrams, you must include every possible intermediate state, every possible splitting of the vacuum into particle/anti-particle, and they will change the probability a little bit. So this is called precision particle physics. If you make a super-precise measurement, and then you make a very careful calculation, you can compare the measurement with the calculation; and you might be able to deduce the mass of this one, last, unknown quark, the top quark, by calculating the effects of—well, if it's there, then we should be able to produce a virtual top/anti-top. And if we do that, it's going to change any observable that you can think of, but by a very small amount.

So this was big business. Starting in 1977, people were looking at all of the data that they had, and at first it was very crude. People said, "Well, it's probably more than a hundred GeV and probably less than a thousand." It was that loose. As the years went by, I would go to conferences, and there was always a series of talks, "The Status of the Standard Model." And one of the things that they would always talk about is: "Okay, we've done these calculations, and we're honing in." And I remember, it was between 150 and 200 GeV, and then between 160 and 180 GeV; and this is now getting into the 1990s. I've gone through graduate school, and the standard model is now taught to graduate students. It's accepted pedagogy at this point, it's in textbooks; and you'll see a table of all the quarks and all the leptons with their masses; and then you get to the top quark, and there's simply a little question mark next to its mass. There's no questioning of the model—well, not much—but it's just assumed that this particle's got to be there and we just don't know quite where it is—yet. It was not until 1995 that the top quark was found. So that's a long time, almost 20 years, between the discovery of the bottom and the discovery of the top.

And yet, throughout that whole time, I think the physics community was not worried about it. We were beginning to expect where it should be, based on our precision measurements of unrelated experiments. It was finally found at Fermilab, once again. In the intervening years, Fermilab had been upgraded in energy. It's now called the Tevatron, because TeV is the acronym for trillion, or tera, electron-volts. So we've gone from billions to trillions now. The Tevatron now takes protons and anti-protons, accelerates them in this giant four-mile ring, and smashes them into one another. And so now you have matter/anti-matter annihilation of a proton with an anti-proton, a huge amount of energy, a trillion-volt collision. And that's the energy that's required to produce these ultra-massive top quarks and anti-top quarks.

I remember, in 1994, there was a colloquium at my institution. Some folks from Fermilab were claiming the first evidence for a top quark. So it was once again, kind of analogous to the discovery of the J/psi, this time at Stanford, where you had matter/anti-matter producing a particle and then you look at what comes out. They had a few events that were consistent with what you would expect if you produced a top/anti-top particle. And I remember playing my role as a physicist, and being skeptical. I thought, "Well, you know, it's a few events. Maybe they're tossing the coin enough times." And there's another thing that makes you nervous. Remember that these precision experiments have been homing in. We expect to find a particle somewhere between 160 and 180 GeV; and what they're claiming, now, is a particle at 170. That was the first evidence of the top quark. And I'm thinking, "Yeah, yeah; maybe they're finding what they want to find." And I think that's the right attitude. They needed to convince us, and they knew that too. It took another year of data collection before the folks at Fermilab would go from announcing hints of a discovery to actually writing a paper in which they discovered the top quark. So, you know, that's sort of part of the sociology of physics today. People are very much aware of the statistical nature of quantum mechanics, and there are criteria that you must satisfy. It's like tossing that coin; getting six heads in a row is definitely not going to be proof that a coin is unfair. It has to be repeatable; you have to be able to leave it and come back; it has to be discovered at other accelerator facilities.

The top quark requires so much energy to find that it's very difficult to duplicate this experiment. And, it's duplicated in a very interesting way. Let me tell you a little bit more about the sociology of modern particle physics. Okay, this is a huge machine. If you ever get to visit Fermilab, or SLAC, or any of these facilities, you've got to go. They're impressive machines, gigantic in scale. You walk into the area where the particle/anti-particles collide, and there's detectors the size of buildings all around the region. And there are teams of physicists: we're now talking about five hundred physicists working together on the discovery of the top quark. When they published the paper, you know, they give you the name of the collaboration—there was a CDF collaboration, and another group, the DZero collaboration. Each of them had five hundred names, so the next two pages of the paper just list everybody's names. And these are all Ph.D. physicists, or graduate students—and some technicians—who are working, essentially, full-time together. It's quite a spectacular collaborative effort, and it's also extremely international. People from throughout the world work on these big projects.

You know, physicists—some of them, have rather big egos. So you can imagine what it's like to try to get five hundred people to work together; and everybody has to agree before you can publish this paper. So, you know, you really have to have extraordinarily convincing evidence before these things get published. And, it's quite spectacular that not only does it work, it really works very well.

At Fermilab, you've got these two beams counter-rotating: the protons go one way, the anti-protons go the other way; and you have to get the beams very, very narrow—it's very high-tech—and then you have to make them run one into the other. Now, even though they're matter and anti-matter, they're still quantum-mechanical particles, and so they have a probability of interacting, but it's not 100 percent. Some of them will interact, and some of them will pass on by, and continue to rotate. So what they do is they design the rings so they're slightly off-kilter, so most of the time the protons and anti-protons just pass by one another. And, there are a couple of spots around the ring where they cross paths and those are the interaction regions; and that's where you set up your detectors.

There were two major detectors at the Fermilab ring—CDF and DZero were the names of the collaborations working to search for the top quark—and they were completely independent. It's like two teams, and they're fighting each other. Everybody wants to be first. If you're first you get the credit, and the glory; you get to name it, and everything. And they'll talk to each other, but you definitely have independent apparatus, completely independent ideas about what would be the best way to find a top quark, what sorts of detectors do you build; so these experiments really are independent, even though they're occurring at the same facility, and highly competitive. Both of them, in fact, came up with data, and in the end, the two publications agreed, within experimental uncertainties, on the mass of the top quark. Again, nice, confirming evidence that what we're finding is not just wishful thinking.

Today, what's the story now? We've passed through what you might call this golden age, and it's really now a period of consolidation. The top quark—and in fact, the tau neutrino was really the last; it came after 1995, it was the final discovery which really clarified and consolidated this theory of the strong interaction. The particles, we believe we know what they are. We think we have a complete list. At least there is no hint of any evidence, whatsoever, that there's a fourth generation, another new quark or lepton. It's an interesting question to ask. Could there be? You might say, "Well, maybe we just haven't built a high enough energy machine," and there are physicists who want to build higher-energy machines, still. But there's no, not even a clue right now, not even a hint, that there should be another generation. In fact, there are even some hints that there might *not* be another generation. That hint has to do with the neutrinos. We'll come back and talk about neutrinos in future lectures some more.

But, basically, although it's hard to detect neutrinos, you can actually at least learn about their existence, because it's not so hard to produce neutrinos in certain events; and the neutrinos, unlike everybody else, are extremely light. So even though everybody else is getting heavier and heavier, we're guessing that the neutrino, in a fourth generation, would be light; and if it was there, we would have already had evidence for it: at Stanford, and in European accelerators, and at Fermilab. So, there are even hints that there really isn't another generation. Maybe we really do know the fundamental building blocks of the world we live in. It's a lovely and compelling idea.

We also have a theory, quantum chromodynamics, which allows us to compute the properties of this top/anti-top pair: what are they going to decay into, what do you expect for the behavior of the system? Everything is fitting; everything is working well. We continue to discover new particles in the accelerators, but they're not exotic. Maybe we find a charm quark paired with an anti-bottom quark. That's something new, but we expect it. We know where to look for it. And so people continuing to look, to kind of verify this model—it's very important that, if you have a theory, you have to look for cracks, and breaks, and see if there's anything wrong with it— but, so far so good. We really do seem to have a well-established framework for understanding the world.

Now, we have six fundamental quarks: up and down, charmed and strange, top and bottom. And, we have six fundamental leptons: electron and electron neutrino, muon and its neutrino, tau and its neutrino. That's a lot of fundamental particles. It was nice when we had three—in fact, I think everybody really wants there to be one fundamental particle; that would be the most elegant of all. When we had three—you know, I think we can accept that maybe nature has three fundamental particles. When we have 6 and 6 is 12, plus the force-carriers, and then there's the anti-particles, and then there's all the different colors of the quarks; you know, it's starting to look like there's an awful lot of fundamental particles. And it's a perfectly legitimate question to ask, "How many fundamental particles can you have, before you start to wonder if there's some explanation, some deeper story?" And that's an idea that I want to revisit.

We don't know the answer to that. Right now, it looks like we have a very beautiful, very consistent mathematical theory. It describes all the subatomic data very well; and so, it's kind of just a feeling, a thought, that some people have, that maybe we might be able to go further beyond the standard model some day.

Lecture Seventeen
Weak Forces and the Standard Model

Scope: Progress in the 1960s and 1970s was not limited to understanding strong forces and quarks. We were also homing in on the weak force, the source of beta decays and the only force felt by the ghostlike neutrinos. It was a struggle at first. Scientists desperately wanted a field theory (like QED or QCD) for the weak force, but they couldn't make it work until the introduction of a novel particle called the *Higgs*. In this lecture, we cover the theory of Weinberg, Salam, and Glashow: the electroweak theory that unified the fundamental weak, electric, and magnetic forces. This has been one of the central goals of particle physics for a long time: to "unify" disparate laws into one single, simple, coherent whole. Just how unified are the forces now? At this point, we can summarize everything we know about particle physics: the players, the forces, and the rules! This is what we mean by the *standard model*. Do we now have the fundamental tools for understanding all microscopic phenomena?

> I will not try to define beauty, any more than I would try to define love or fear. You do not define these things; you know them when you feel them.
>
> —Steven Weinberg (*Dreams of a Final Theory*, p. 134)

Outline

I. At this point in the course, we have almost all the ingredients—the history, concepts, and terminology—to talk about the standard model of particle physics, which we believe explains the fundamental constituents of our world. One important ingredient that we are still missing is the weak force.

A. The theory of the weak force that we've talked about so far is an old one, dating back to the 1930s. Enrico Fermi had tried to explain beta decays, in which a nucleus spontaneously transforms itself into a different kind of nucleus and loses an electron and a neutrino.

B. Fermi's theory was not a particularly deep one. Although it served to make useful calculations, it began to break down if people tried to push it to make highly accurate predictions. Feynman taught us the importance of accounting for vacuum fluctuations when studying the electric forces of nature. If we attempt to account for these fluctuations in Fermi's theory, the result is infinity and nonsense.

C. Fermi's theory also makes a prediction about weak interactions in particle collisions. In collisions, just as in decays, particles can come together, interact weakly, and transform. However, as higher energies are achieved, the weak force becomes stronger. The strong force becomes easier to predict at higher energies, but the weak force is exactly the opposite.

D. In the late 1960s, it became clear that a better theory of the weak interaction was needed to go along with QED and QCD. Many important physicists were involved in the development of this theory, including Steven Weinberg, Abdus Salam, Sheldon Glashow, Gerard 't Hooft, and others.

II. The goal was to construct a relativistic quantum field theory for the weak force, and one of the most important ingredients for this theory would be the concept of gauge symmetry.

    A. Gauge symmetry relates to the fact that in working with electricity, the measure of voltage we call 0 is arbitrary. No matter what point we choose to call 0, once we've made that choice, the equations, the experimental results, and the laws of physics are the same.

    B. Also remember the idea of the force carrier in electricity. The classical idea is that a charged particle, such as an electron, has an electric field that reaches out into space. Imagine jiggling an electric charge, causing (electromagnetic) waves to propagate outward, similar to what you see in a pond. In quantum mechanics, we think of that propagating wave as a particle of electromagnetic energy, a photon.

        1. QED, particularly gauge symmetry, tells us everything about the photon, that it's massless, has no electric charge, and so on.

        2. QCD works in the same way. If we jiggle a quark that has color, it will propagate a gluon, and gauge symmetry tells us that the gluon is massless.

3. The single most important character in the story of electricity and magnetism was James Clerk Maxwell, a Scottish physicist in the 1800s. Maxwell took the idea of a classical field and formulated a mathematical field theory for electricity and magnetism, the precursor to QED. He recognized that electricity and magnetism are two facets of one universal underlying force of nature, which we now call the *electromagnetic force*.

C. As we said, in the late 1960s, physicists were trying to find an analog of this for the weak force, which is responsible for many radioactive decays, and neutrino physics. Every mathematician and physicist who tried working on this theory came to the conclusion that if gauge symmetry was a true principle of nature and they attempted to impose it on the weak force, then the weak force should have a carrier that is massless.

1. The carrier of the weak force, the particle that causes the weak force to travel from one place to another, is the *W particle*, or now, the *W* and *Z particles*.

2. The W particle is massive; it weighs almost 100 times as much as a proton. The mass of the W makes the likelihood of weak interactions improbable, because to have a weak interaction, we must, in a sense, excite the weak field. A W particle is an excitation of the weak field, a traveling wave of weakness. Thus, the weak interaction has its characteristics because its force carrier is so massive.

3. This leads to a dilemma. Gauge symmetry has helped us to understand electricity and magnetism. If we try to apply it to the weak force, however, we get results that are in direct contradiction with the data. Gauge symmetry says that the W boson should be massless, but nature says that the W boson is massive.

D. Even as QED and QCD became successful, it was difficult to make all the ideas fit together. A physicist named Peter Higgs came up with the idea that made it all work.

1. Higgs proposed that there might be a new particle of nature that nobody had thought of before. Those who were trying to understand the weak interaction adopted Higgs's idea.

2. The theory started with a what-if game. What if this new particle is everywhere? It could be a field that was present in the early universe and has spread everywhere.

3. Imagine a particle that is similar to a photon. It is a massless weak-force carrier, and it's propagating along through space that's filled with Higgs particles. The weak-force carrier bumps into Higgs particles all the time, which makes it effectively slow down. It seems to have mass because it is trying to move through this swamp of Higgs particles.

4. This idea could explain how a particle that, for other reasons, should be massless appears to be massive. In 1967, Weinberg produced a paper in which he put together a coherent mathematical framework for the weak force that took into account gauge symmetry and the Higgs particle.

E. The theory matched the data at low and high energies. As Maxwell unified electricity and magnetism, Weinberg, Salam, Glashow, and others now unified the weak force and electromagnetic forces. The weak force, which seemed so different in every respect, is really just like electricity and magnetism, viewed from a different perspective.

III. When Weinberg, Salam, and Glashow were looking at this theory, they noted that gauge invariance not only predicts the existence of the W particle, which is electrically charged and becomes massive in the presence of the Higgs field, but also the existence of another force carrier, the *Z particle*.

A. The Z particle is a consequence of the mathematics of the new theory. It is a new force carrier, which means that a new force exists. Although the masses of the W and Z particles could be predicted, the accelerators of 1967 were not capable of producing these particles.

B. The worldview at the time was as follows: We have gravity, electromagnetism, the strong nuclear force, and the weak force— four fundamental forces of nature. Then, electricity, magnetism, and the weak force were simplified; all of them were really one. However, we also seemed to have something brand new, because the weak force can be manifested in a way that is different from what we have seen before.

1. The Z boson is electrically neutral; it's a non-transformative weak force. Electrons can bounce off each other electrically, and now we have a way that particles can bounce off each other weakly.

2. Such an interaction could have all sorts of experimental signatures. For example, physicists could look for neutrinos bouncing off particles.

3. Such an event was seen at CERN, the European Center for Nuclear Research, in 1973. Using neutrino beams, researchers saw *weak neutral currents*. The term *weak* was used because the events involved the weak force, and the term *neutral* was used because the Z-zero particle was being transferred virtually between the neutrino and the target. This was the new force of nature.

C. The next hurdle was the discovery of the W and Z particles, which was made in 1983 at CERN with an Italian physicist, Carlo Rubbia, as the lead in the project.

D. The name for this theory of Weinberg, Salam, and Glashow is the *electroweak theory*; it is a mathematical field theory that is the analog to QCD. It unifies electricity and the weak force so that QED becomes part of the electroweak theory.

IV. We can now take stock of what we have been talking about throughout the course. We have in our hands all the ingredients of the standard model, even though a better title might be the standard *theory* of particle physics. Let's quickly reiterate these ingredients.

A. We have the fundamental particles of nature, which appear to be point-like, to the best of our knowledge even today. They come in two classes: quarks and leptons.

1. The six quarks are all strongly interacting. They are colored objects that get progressively heavier. We start with up and down quarks; then, heavier are the strange and charm quarks; then we have a third generation, heavier still, top and bottom.

2. In the category of leptons, we have the electron and its neutrino, a muon and the muon-neutrino, the tau and the tau-neutrino. Again, these get heavier and less stable, that is, more radioactive, as we go down in our table.

B. The forces are the electroweak force, which is described by the electroweak theory, and the strong force, which is described by

QCD. Those theories are consistent with everything we've ever learned in the last 400 years about science.

1. These theories include the principle of the field and all the logical consequences of having a field that also exists in a quantum mechanical world.

2. These theories predict new particles of nature, which we call force carriers. They predict the photon, the W+/W–, and the Z particle. These theories also predict the gluons, carriers of the strong force.

C. From these ingredients of the standard model, we can calculate the observable consequences of any imaginable subatomic physics experiment. In principle and in practice, we can calculate the properties and structure of fairly complicated particles.

D. The development of the standard model stands out as a grand human intellectual achievement, although certain interesting questions remain.

Essential Reading:

Schwarz, *A Tour of the Subatomic Zoo*, chapter 6.

Weinberg, *Dreams of a Final Theory*, Chapter V, pp. 116–end of chapter.

Recommended Reading:

Lederman, *The God Particle*, chapter 7, section on "The weak force revisited"; and chapter 8, up through the section titled "What are we talking about?"

Riordan, *The Hunting of the Quark*, chapter 9.

Taubes, *Nobel Dreams*. (Take your time and read the whole book. It's interesting and quite different from the usual, somewhat "drier," readings.)

Questions to Consider:

1. What is the difference between a "charged current" weak interaction, and a "neutral" weak interaction? Can you think of a specific example for each?

2. The mass of the W and Z particles is roughly (slightly under) 100 GeV. What can you say about the *range* of the weak interaction? (Either qualitatively or quantitatively, if you are so inclined)

3. Why do physicists make such a big deal about renormalizability? What does that mean again, and why does it matter?

THE STANDARD MODEL OF PARTICLE PHYSICS

FUNDAMENTAL PARTICLES

QUARKS   LEPTONS

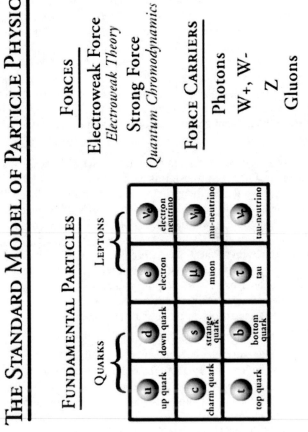

FORCES

Electroweak Force
Electroweak Theory

Strong Force
Quantum Chromodynamics

FORCE CARRIERS

Photons
W+, W-
Z
Gluons

Lecture Seventeen—Transcript
Weak Forces and the Standard Model

We are rapidly approaching a turning point in this course. We have almost all of the ingredients, the history, the concepts, the terminology, to put everything together and talk about the standard model of particle physics, the coherent framework that physicists have today which, we believe, explains the bottom layer, the fundamental constituents of the world that we live in. But there's one very important ingredient that's still missing, and that I want to talk about today; it is the weak force. Okay? We've been talking an awful lot about the strong force, and the quarks that feel the strong force. We talked about the weak force when we introduced the concept of neutrinos and beta decays. But the theory of the weak force that we've been talking about so far is an old one that dates all the way back to the 1930s.

Enrico Fermi had tried to explain the phenomena of beta decays, where a nucleus spontaneously just transforms itself into a different kind of nucleus, and an electron would go flying out, and a neutrino would go flying out, usually undetected. And Mr. Fermi's theory was a mathematical one; you could make predictions about what the probability of such decay, and there were lots and lots of data. Beta decays are very common in nature; many radioactive decays involve the weak force, as well as processes in the sun. It's quite a ubiquitous force of nature, even though we don't usually feel it on a daily, personal basis. Mr. Fermi's theory was okay, but it was not a particularly deep theory. Although it served to make useful calculations, if you tried to push the theory to make really accurate predictions it began to break down.

For example, Feynman taught us, when he studied the electric forces of nature, that you have to worry about the possibility that the electric field can spontaneously split, for a very short amount of time, into a particle-anti-particle pair. That vacuum fluctuation, those virtual matter-anti-matter pairs, can modify the electric forces between any objects. It's a modification If you're trying to do precision calculations, if you really want to do better than ten percent or something, you start to have to worry about those sorts of fine details, and you can't do that with Mr. Fermi's theory. If you try to look at, for example, how the possibility for a beta decay would change if we take into account the possibility of vacuum fluctuations, using his theory, you would get crazy answers - infinity and nonsense. It just doesn't make sense.

Also, Mr. Fermi's theory makes a prediction about weak interactions in particle collisions. Just as in decays, you can also have particles coming together, interacting weakly, transforming and turning into something else; the theory works reasonably well. As you go to higher and higher energies, starting in the 1930s and working your way up through the 1960s, at accelerators, the details of the weak interaction depend on the energy. That's not unusual. We've talked about how the strong force of nature actually gets weaker if the particles get higher and higher energy. That's why the strong force gets easier to predict when you go to higher and higher energies and more modern facilities. The weak force, according to Fermi's theory, works exactly the other way. As you go to higher energies, the weak force gets stronger. And at a certain energy, around, oh—it's a little bit hard to be precise, because it depends on the process—but around 100 GeV, the weak force goes off kind of crazy. It starts to go to infinity. You start to get calculations that tell you probabilities are greater than one of some weak interaction occurring. This is clearly nonsense. That didn't worry anybody for a long time because there were no accelerators anywhere near that energy until the 1970s. Until then, Fermi's theory was working fine, but it was clear to just about everybody that we needed a better theory of the weak interaction. It was a shame; we had this lovely quantum theory of electricity (in fact, between 1964 and the early 1970s, quantum chromodynamics, the mathematical theory of the strong force, was developed) and it, too, became very well established experimentally. It was clearly a good theory of nature, so we could explain the strong force, we could explain electricity and magnetism, but what about the weak force?

There were a lot of people who became involved in this business. It is hard to point to just one name. It's not as though there's an Einstein or a Fermi who we can say invented the theory. The Nobel Prize went to three characters: Steven Weinberg, Abdus Salam and Sheldon Glashow. I've mentioned Shelly Glashow before. He was the one who proposed the idea of the charm quark, and in some sense, Steven Weinberg was the one who wrote the paper that pretty much put it all together; but each of these three guys made many contributions, and there were many other people as well. There was a young graduate student, 't Hooft, in the Netherlands, who wrote a theoretical paper which really established the theory, ultimately, as a mathematically rigorous one. He and his advisor, Veltman, received a Nobel Prize at a later time. I think this is a story that occurs more and more frequently in physics today, that it's a little bit harder to point to one genius who came up with the theory all by himself. It may even be true, if you go back in history, that lots of people didn't get enough credit for the work they did.

What are people looking for? Quantum electrodynamics is a lovely theory that has quantum mechanics in it; it has relativity in it, and it has fundamental particles in it. We can do all of those things for any force, so let's try to build a relativistic quantum field theory for the weak force. One of the most important ingredients, for both QED and QCD, was the idea of gauge symmetry. Okay? It's a very abstract concept. We've talked about it briefly. Gauge symmetry has something to do with the fact that, in electricity anyway, it doesn't really matter where you choose to call zero volts. It's pretty much an arbitrary human convention. No matter what you choose, once you've made your choice, the equations are the same, the laws of physics are the same, and the experimental results are the same. It's an abstract mathematical symmetry, which is a little bit more complicated than that.

Gauge symmetry leads us, in fact, to the theory of QED, as well as to the theory of QCD. Not only does it help you write down the theory, it makes a clear prediction about the force carrier. Let me remind you about this idea of the force carrier in a field theory. Think about electricity. You have a little charged particle, which has an electric field that reaches out into space. That's a classical idea. Now, what is the photon? What is the carrier of electricity? Well, imagine jiggling that little electric charge, so the field lines get a little kink; and that kink propagates outwards like the rings of water when you disturb the water surface. It's a little wave that travels outwards, classically. In quantum mechanics we think of that propagating wave as a little particle of electromagnetic energy, a photon. So it's a wiggle in the field that we call the photon. The theory, in particular the gauge symmetry, tells you all about the photon. It says it's without mass, it says it has no electric charge, in the case of the photon; it tells you everything that you could ever want to know about the photon. We see the same thing in QCD. If you wiggle a quark that's colored, it will propagate a gluon. That's the particle, and the gauge symmetry predicts that the gluon should be without mass. In fact, it's very much like a photon, in a lot of ways, except that it itself has color.

Michael Faraday originated the idea of a field. I've mentioned Faraday, but I've never mentioned the most important character in the story of electricity and magnetism, James Clerk Maxwell. This is pre-quantum mechanics. James Clerk Maxwell was a Scottish physicist in the 1800s, and if I had to pick the most important physicist, far and away, between Isaac Newton in the 1600s and Albert Einstein in the early 1900s, it would definitely be Maxwell. Maxwell took the idea of the field, a classical field, and turned it into a mathematical field theory for electricity and magnetism. It was the precursor to QED, and in the process, he understood the principle of gauge symmetry. He wrote down all the mathematical equations discovered that electricity and magnetism are unified. It was a grand intellectual achievement. It's one that even today physicists look back, with just amazement, that these two forces of nature, electricity. You know, it didn't seem, to a physicist in the early 1800s, as though they had anything to do with one another. Electricity, as you know, sparks when you comb your hair; and magnetism moves compass needles, but, if you hold a compass needle near a spark, it doesn't do anything. There's no obvious connection between electricity and magnetism. People were finding clues about it in the 1800s, but Maxwell was the one who really unified the two and recognized that, in fact, electricity and magnetism are two facets of one universal, underlying force of nature which we now call the electromagnetic force. That's what I've been calling it all along. It's like one force of nature and you look at it from two different perspectives. You see electricity if you're moving this way and magnetism if you're moving another way. That's a kind of a simplistic way of thinking about it, but it really is one fundamental, unified force. That unification led to his prediction of light.

That's the classical story. Then the quantum analog of that is in QED; again, electricity and magnetism are unified, and the principle of gauge invariance now makes the statements about the carrier of the force. So people are now hunting for an analog of this. We would like to do the same thing for the weak force of nature, responsible for these esoteric radioactive decays and neutrino physics. Every mathematician physicist who tried working on this came to the annoying conclusion that, if you believe in gauge symmetry as a principle of nature, and try to impose it on the weak force, you should have, by consequence, a carrier of the weak force, which is without mass. We've talked a little bit about the carrier of the weak force, the little particle that causes the weak force to travel from one place to another, the quantum field of the weak force; we called it the W particle. Nowadays we have the W and the Z particles, and the W particle is massive - super-massive. It weighs almost 100 times as much as a proton. It's so massive that it is what makes the weak interaction so weak. It makes it very improbable for weak interactions to occur, because for it to occur, you have to excite this weak field. That's what a W is, the excitation of the weak field, a little traveling wave of weakness; and, despite its name, it's very massive. It's very hard to borrow that energy from nature, so the weak interaction has all of its characteristics because the force carrier is so massive.

You have this dilemma. On the one hand you really believe in gauge symmetry; it's helped you to understand the other two fundamental forces of nature. On the other hand, if you try to apply it to the weak force, it tells you something that's in direct contradiction with data. Gauge symmetry says the W boson should be without mass, but nature says the W boson is very massive. People were, for many years, kind thought it was hopeless. Even as QED became successful, and QCD was already successful, it was rather difficult to understand how to make this all work. The idea that made it work came from a physicist working on something else, as often happens. His name was Peter Higgs, and we'll talk a lot about Mr. Higgs's idea because it's a little bit complicated. We'll devote a whole lecture to the Higgs particle.

Mr. Higgs proposed that there was a new particle of nature, one nobody had thought of before. The folks who were trying to understand the weak interaction took Mr. Higgs's idea and said, "Okay, suppose there's this new particle of nature. Let's just do a 'what if'; it's the usual game. What if this new particle is everywhere? It's like a field that's present already in the early universe, and it spreads out throughout the entire universe; it's everywhere. It's inside you, it's out in deep, dark, empty space." The Higgs idea is everywhere. Now imagine that you have a particle that's kind of like a photon. It's a weak-force carrier without mass, and it's propagating along through space that's filled with Higgs. The weak-force carrier keeps bumping into Higgs all the time, and that makes it slow down, and get sluggish. It's as though it acts as if it had mass because it's trying to walk through this goo of Higgs's particles. It's a crazy idea, it's a mathematical, formal idea; but it could explain how a particle that, for other reasons, should be without mass, appears to be massive. People started working out the details. In 1967, Weinberg was able to produce a paper in which he put it all together and had a coherent, mathematical framework for the weak force. You could have gauge symmetry, and all the beautiful things that gauge symmetry gives you, and you introduce the Higgs particle. This theory reproduces Mr. Fermi's theory at low energies, so that's good; it agrees with all the data so far, but at high energies, no problems. Okay? Everything's finite, everything works, it makes sense, you can compare with the data; it all works. It's a quantum field theory. You can add many, many more complicated Feynman diagrams, and calculate to as an accurate, high precision as you want.

Now, I mentioned the unification of electricity and magnetism that Mr. Maxwell had discovered in the 1800s, because the same thing happened again when we were constructing this weak theory. This was discovered by all of the characters—Weinberg, and Salam, and Glashow, and 't Hooft— that there was a unification between the weak force and the electromagnetic forces. So what did that mean? It meant that this weak force, which seemed completely different in every respect, was really just like electricity and magnetism, viewed from a different perspective. It's always beautiful and wonderful, when different things actually turn out to be unified, different facets of one fundamental, deeper, underlying object. Physicists love that, and I think this unification of weak and electromagnetic forces was one of the compelling arguments for people to pay attention to this theory and take it seriously. At first the theory, although it was highly predictive, wasn't at first making predictions that changed any connections to data; but as the years went by, the theory was more and more sort of seriously verified mathematically. Nowadays it's incredibly well tested, but in the early days I think people paid more attention to it because of its aesthetic character than anything else.

When you have a gauge field theory, you predict the existence of force carriers. When Weinberg and Salam and Glashow were looking at their theory, they said, "Gauge invariance not only predicts the existence of the W particle, which is electrically charged and becomes very massive in the presence of the Higgs field, but our theory also predicts the existence of another force carrier, which I (Weinberg) am going to call the Z particle." Weinberg called it the Z particle maybe because it has zero charge, or maybe, he says, because it's the last letter of the English alphabet. He thought, "This has to be the last new particle we're going to have to propose." It turned out there have been others since then, but it was a nice dream. So, the Z particle is a consequence of the mathematics of this theory; and what is it saying? It's a new force carrier, which means there is a new force. Okay? This is really wild.

You are not only predicting a new particle of nature, you can state with some mathematical certainty what its mass is. Unlike the fundamental quarks and leptons, whose mass cannot be predicted from the gauge field theory, the mass of the force carrier *can* be predicted from the theory because it comes from gauge symmetry. So you could predict the mass of the Z particle, and the masses of the W particles, so we have numbers that are too high to look for right away in an accelerator. We need higher-energy machines than were available in 1967, so this just goes into the literature as a prediction about how nature is.

It's even a bigger deal than that. It's not just one more new particle; it's a new force. Okay? Think about this. Particle physicists have this worldview that—what do we have? There's gravity, there's electromagnetism, there's the strong nuclear force, and there's the weak force. That's it—four fundamental forces of nature. Now what we're seeing is—the story is in some ways more beautiful, because electricity, magnetism and the weak force are really all one. So there's a simplification; but also there is something brand new, because the weak force itself can be manifested, not just in the way we had always seen, transforming one nucleus into another; this Z boson is electrically neutral. So it's a non-transformative weak force. It's like a neutrino could just whack into an electron and bounce off of it. Electrons can bounce off of each other electrically, and now we have a way that particles could bounce off of each other weakly; and nobody had ever thought about this before. It would have all sorts of experimental signatures; you could look for neutrinos bouncing off of particles. "People immediately said, "If this is real, then we have to be able to see this in an experiment," and it didn't take very long.

It was in 1973 when, at the European facility, CERN - (the European Center for Nuclear Research; the acronym uses the French words for its name)—they had neutrino beams by then, at CERN - they could beam those neutrino beams right into a big bubble chamber and see the events. They saw what they called weak neutral currents. They called it weak, because it's the weak force; neutral, because it's the Z (zero) particle being transferred virtually back and forth between the neutrino and the target. And why was it called "currents?" I don't know; it was just a word. So, "weak neutral currents" were the new force of nature.

You have to appreciate both the sort of sweet success of this theory, and also the hubris of the theorists who are proposing new particles. It's not like the old days. All right? Dirac discovers anti-matter, but he was sort of timid about proposing it because it was too wild. Pauli, with the neutrino, was, also, too timid about proposing it. Even with quarks, in the early 1960s, after Murray Gell-Mann published his paper, then he gave talks and said, "I don't really believe they're real, I just think they're a mathematical framework." But, you know, Steve Weinberg went out there and said, "Oh, you're going to find weak neutral currents, you're going to find the Z particle; you're going to find the W right where we predict because the math looks good. He was pretty confident." It came together in just such an elegant way. The theory was unified, and it looked just like QCD, so everything really seemed to tie together in a beautiful way.

So what about the discovery of the W and the Z? It took a while. The Americans had kind of taken over from the Europeans. Europeans had dominated this physics story in the late 1800s and the early 1900s; we were just puppies back then. It kind of turned around after World War II. We started building the big accelerators and were putting big money into particle physics. We were making all the big discoveries. I think the Europeans, at a certain point in the 1970s, thought they should be big competitors again, and they pulled together. It was a brilliant stroke. Instead of the Germans just building a German facility, all of the Europeans got together; they built a European national center. The ring actually runs between France and Switzerland, so the particles go back and forth across the border as they run through the giant accelerating ring at CERN.

There was a physicist, an Italian physicist named Carlo Rubbia, who took the lead in the project to find the W and find the Z. What we're talking about is to put enough energy into a small region of space that you don't have a virtual force carrier, but you actually produce one. You see a little track—evidence, direct evidence, in some modern analog of a bubble chamber that this particle is real, and it has the charge and the mass and everything that you expected. Carlo Rubbia is quite an interesting character. There was a book written about this story called "Nobel Dreams," which I enjoyed. It's not a technical book at all; it's about the very strong personality of Carlo Rubbia. He had to step on a lot of toes to put together multimillion-dollar projects like this. He had to convince a lot of people that it was the right thing to do; in the end, he was quite successful. The Europeans, and Carlo Rubbia, got their Nobel Prizes. The discovery was made in the 1980s; 1983 was the year of discovery of the W, and just six months later they produced a Z boson. Some people don't like the book about Carlo Rubbia; I think it's somewhat critical, just on a personal level, of this character. You know, when you're a strong personality, you always step on people's toes, but I suggest reading those kinds of books as a way of seeing the human aspect of this story, which so often is just impersonal and formal.

The neutral current has been seen now many, many times; and the Z bosons, in fact, have been seen so many times I would call it probably one of the most well-studied particles ever in the history of particle physics. Shortly after its discovery, the people at Stanford said, "You know, we could build an upgrade. It's the same idea of taking electrons and positrons and annihilating them, but this time we're going to up the energy, so that the total energy of the e-plus/e-minus pair exactly matches the mass—$E = mc^2$, the mass-energy of a Z boson. And we'll have a little Z factory. We're just going to be producing nothing but Z's after Z's after Z's, and we can see how they decay." That was the experiment that told us there apparently were only three types of neutrinos. Because when Zs decay, they love to decay into neutrino-anti-neutrino pairs, and you can see how likely it is for the Z to decay. After that, actually shortly after that - CERN built a Z factory, and now they are producing millions upon millions of Z particles. So, extremely well verified, and matched with this theory, which I haven't named, is electroweak theory. The electroweak theory of Weinberg, Salam, and Glashow is the mathematical field theory that's like the analog of QCD. It put together electricity and the weak force unified them, so there was no more QED all by itself; it's a part of the electroweak theory.

So now we really are at a state where we can just take stock of what we've been talking about throughout this entire course. What we have in our hands now are all the ingredients of the standard model. I wish it weren't called the standard model. It's the standard theory of particle physics. A model, to me, implies kind of a crude collection of ideas that give you qualitative understanding. A theory, to me, is a mathematical framework, with well-defined constituents and a set of rules for computing observable facts. In particular, a theory, a tested theory, is exactly what we have with the standard model. Nobody calls it the standard theory, but that's really what it is.

Let me try to sort of reiterate the ingredients. We have fundamental particles of nature, which appear to be point like, to the best of our knowledge even today. They come in two classes: quarks and leptons. The quarks are all strongly interacting. They are colored objects, and there are six of them: up, down, and then heavier, strange, charm; and then a third generation, heavier still, top and bottom. Over in the leptons, you have the electron and its neutrino; a muon, and the muon neutrino; the tau and the tau neutrino. Again, getting heavier, and less stable, more radioactive, as you go down in the table, or up in mass. Those are the ingredients. Those are the fundamental particles of the world.

The forces are the electroweak force, which is described by the electroweak theory; and the strong force, which is described by quantum chromodynamics. Those theories are consistent with everything we've ever learned in the last three or four hundred years about science. They're consistent with Isaac Newton, although they go deeper; they're consistent with Maxwell and his theories of electricity and magnetism, but they go deeper. They're consistent with quantum mechanics, and Albert Einstein's relativity. They include the principle of the field and all of the logical consequences when you have a field that also exists in a quantum-mechanical world. These theories predict new particles of nature, which we call, force carriers. They predict the photon, the W plus and minus— there're both charges - and the Z particle, they're the electroweak's force carriers. Then, you have the gluons, which are the carriers of the strong force, and that is the standard model.

From those ingredients, you can calculate the observable consequences of any imaginable subatomic physics experiment. In principle, it's not just sub-atomic physics. For example, if you take two hydrogen atoms and combine them together, it's just electrons orbiting around up and down quarks. The forces are all described by this theory, the standard model. So, in principle, and in practice, you can calculate the properties and the structure of, let's say, a hydrogen molecule. We can keep building now. We can argue what the properties and structures are of a more complicated molecule. The more particles you put together, the tougher it gets to work out the math, and solve for the detailed, final answer. Chemists nowadays use quantum mechanics in order to describe—and quantum field theory—to describe chemical reactions, and to help us to design and understand novel chemical elements. So this story really is tying in to more complicated sciences.

As a reductionist, you can at least imagine, in principle, following this chain on up. Maybe we could understand DNA and its properties by examining the quarks that make up the nuclei, and the electrons that are orbiting around them. In practice it's just too formidable a problem. DNA is a very complicated molecule with lots and lots of constituents, and it's just too tough to really go directly from the bottom level up to such a complicated level, but in principle, we should be able to. People are making progress, and you can imagine that all of the biological properties that arise from DNA can be understood, at some level, from this fundamental, underlying theory, the standard model of particle physics.

So I think we're talking about what is really a grand human intellectual achievement. It stands out in the history of the evolution of human thinking, along with Isaac Newton's understanding about force, and the nature of gravity, or Albert Einstein's insight into the nature of space and time. The standard model of particle physics is such an achievement. It's really a very elegant and closed mathematical system. You know, it's a shame it doesn't fit on a T-shirt. If the standard model fit on a T-shirt, then I'd say we're done, you know, we've got the fundamental theory of nature. Alas, it takes several pages of a textbook to write out all the details; and in fact, there are still some issues; I won't call them problems. We're going to be talking in future lectures about why we don't just close the book; you know, publish it, set it on the shelf, and go off and study other stuff. In fact, particle physics is still very much alive. There are lots of interesting problems and questions.

We are now struggling to understand this theory itself, at some deeper level. I don't think we're trying to prove it wrong anymore. It might happen. You know, we're not trying to prove Newton wrong either, 300 or 400 years later. He's correct, but Newton's theory has been improved and deepened, and I think that's the state we're in now with the standard model of particle physics. There are still a large number of quantities, like the charge of the electron, and the mass of the strange quark, and so on—in fact, if you count it up, there are about 23 such numbers that don't come out of the theory. You have to go in the lab and measure them, and put them into the theory because it's a renormalizable theory, and that's the price you pay.

Twenty-three numbers, six quarks, six electrons; it is a little bit complicated. It's simple enough that I sort of view it as a lovely, simple theory; but I can understand this sort of sense that maybe, someday, we will be able to understand even this level of complexity in terms of something simpler still. It's a nice dream.

Lecture Eighteen
The Greatest Success Story in Physics

Scope: Make no mistake, the standard model of particle physics is an impressive accomplishment, an unprecedented artwork of mathematics and physics. It is a "minimalist" theory, composed in the 1960s with the simplest available framework, and guided only by the basic requirements of quantum mechanics, relativity, and the known data at the time on the fundamental particles. It is not fully unified, but it comes close, and it is an effectively complete description of subatomic particles and forces. The standard model has been the target of more than 30 years of intense efforts to find hidden cracks or problems in it—to no avail! The unparalleled success of the standard model includes qualitative and quantitative measurements, with years of increasingly precise tests. Still, many people are left with more than a slight discomfort with this theory. Could this really be the final story? Despite the descriptive successes, we are still missing a lot of explanations.

> The most incomprehensible thing about the universe is that it is comprehensible.
> —Albert Einstein

Outline

I. In the last lecture, we talked about the electroweak theory, which ties together electricity, magnetism, light, and the weak force of nature and all the associated particles and force carriers to form a coherent mathematical framework. This lecture consolidates everything we have learned thus far.

   A. Let's begin with a review of the standard model. We have a set of six quarks and a set of six leptons. Each one of these is considered to be a fundamental particle of nature. It has certain properties, including an electric charge, a weak charge, a mass, a spin, and not much else.

   B. Those are the fundamental building blocks of all the complexity in the world, everything from the hardness of a table to the softness of human skin; the colors that we see, the light that strikes our eyes, the sounds that we make—all these can be understood as coming from the rules and the players of the standard model.

C. The rules are the theories, QCD and the electroweak theory, or the Weinberg-Salam model as it's also called. The electroweak theory is a renormalizable relativistic quantum gauge field theory.

  1. *Renormalizable* is a mathematical statement that says that if we include all the subtle quantum mechanics, the result is some infinities, in principle. If we add progressively smaller and smaller corrections, they add up to be infinitely large. However, if we measure a few things that the standard model cannot predict, everything else is completely finite and mathematically consistent.

  2. *Relativistic* means that the theory must satisfy what Einstein has told us about the nature of space and time, including the symmetries that Einstein recognized.

  3. *Quantum* means that the theory must satisfy all the laws of quantum mechanics that were first postulated in the early 1900s, then rigorously tested and proven by Heisenberg, Schrödinger, Pauli, Fermi, and many others.

  4. *Gauge* refers to an abstract symmetry of nature that apparently all the fundamental forces have. Symmetry is observed when a change is made in an experiment or a theory, but all the answers come out the same. It is an invariance in a theory when viewed from different perspectives. Gauge symmetry was long observed to be true for electricity and magnetism, then later found to be true for the weak and strong forces.

  5. *Field* refers to our mental image of how action at a distance takes place. How can a particle in one place "communicate" with a particle in a different place? We visualize space as being filled with fields. Everything in these quantum field theories is a field.

D. The theory entails some fancy mathematics, but in the end, the story is one that we can think about in fairly concrete terms and has been successful aesthetically, mathematically, and physically.

II. First, let's examine the aesthetics.

A. This theory has built into it essentially every symmetry of nature that we have been able to think of and that we might hope exists in the universe.

B. It is also a minimal theory. It is extraordinarily simple in the sense of having a few ingredients and a few ideas behind it. We have observed, for example, the six quarks and six leptons in the laboratory, and we don't include anything else in the theory except the Higgs particle.

    1. This theory was essentially complete in 1967, and it hasn't been tweaked in any significant way since then.

    2. In biology, chemistry, geology, and other sciences, we often start with a simple approximation and make it richer as we gather more data and become more precise. In particle physics, the situation is different. We started off with the simplest theory and found that it just kept working. We built new accelerators, discovered new particles, learned a great deal since 1967, but it all fits in with the minimal electroweak theory.

C. The theory also unifies forces of nature. What we thought were completely different forces—electricity, magnetism, light, the weak force—all are just one force viewed from different angles in this mathematical theory.

D. Some symmetries of nature are broken in the standard model, resulting in mysteries, in some cases, and deeper understanding, in others.

    1. For example, left and right are no longer an exact symmetry of nature itself. The laws of physics are different for something spinning like a left hand spins and something spinning like a right hand spins, but they're only different when you consider weak interactions.

    2. The laws are the same for electrical forces, strong forces, and gravitational forces. Only the weak force appears to violate this symmetry of nature. That's in the standard model, but it can't be explained; it is just an observed fact of nature.

    3. Other broken symmetries can be explained. For example, symmetry is broken between two of the force carriers, the photon and the Z. In some ways, they seem similar. They both have spin 1, and they carry some aspect of the electroweak force, but the photon is massless, and the Z particle is massive. That's an electroweak symmetry breaking. We believe that this breakage arises from the presence of the Higgs particle.

4. The standard model, then, gives us an explanation for why these two forces, which we would think are completely symmetrical, seem, in fact, to be so different from each other in the real world.

III. Let's look at the mathematical success of this theory.

    A. The mathematics really boils down to the fact that this theory doesn't contradict itself in any obvious way. If we calculate something, we always get finite answers.

    B. There are no probabilities that can be computed in the standard model that come out greater than 1 or less than 0. Any other results would constitute a mathematical inconsistency. Earlier models had that problem, but the standard model does not.

IV. Finally, let's turn to the physical success of this theory. A theory is only good if the results it predicts match the results from the laboratory.

    A. Every year, the particle data group at Berkeley puts out a collection of all the information that has been gathered in particle physics.

        1. First, it shows a summary of the standard model, including the ingredients and the formulas. Then comes a list of particles and their properties. Every piece of data in this list includes a reference to a paper or a journal article, backed up by experiments that resulted in the numbers.

        2. In fact, this tabulation can be found published in a volume that runs to hundreds of pages. What's stunning about this mountain of information is that it all comes from the first couple of pages—the standard model and the formulas for doing these calculations.

        3. To date, I know of no significant discrepancies between the theoretical calculations and the experimental observables.

    B. Let's look at another success of the standard model that's a bit more qualitative but equally convincing.

        1. In 1967, Weinberg worked out a theory that predicted the existence of a new particle, the Z boson. Almost 20 years later, when we finally achieved the necessary energy to produce the Z particle, we detected its presence in a bubble chamber.

2. The marvel is that a human being made a prediction about a fundamental particle that would not be observed for 20 years. That kind of prediction of a new fundamental particle is a powerful indicator that our theory does, indeed, help us understand the world.

3. Of course, the W, the tau-flavored neutrino, and the top quark were all similarly predicted before their discoveries. In fact, the physics of the neutrino may be one of the few areas of the standard model that needs to be tweaked a bit for full understanding.

V. We should also note that even if we discover something odd about, for example, quarks, we're not going to throw out the standard model, because the model does have a bit of leeway in it.

    A. Imagine that we make a discovery of a fourth generation of quarks; such a discovery would modify the standard model, but it wouldn't shatter it. The framework would still exist and could be generalized to incorporate some new heavier particles.

    B. We might extend and deepen the standard model, as we have Isaac Newton's theory of gravity, but the basic model would still be accurate. No matter what we discover in the future, we're still going to have atoms and quarks, and they will continue to combine in certain ways.

    C. Consider also that the standard model makes both qualitative and quantitative predictions about the existence of new particles. We can predict, for example, exact numbers for the magnetic strength of an electron. As I mentioned earlier, this number is 2.00231930435.

    D. Thus, the electroweak theory may not be the ultimate theory of nature, but we know that it is extraordinarily accurate. If some new laws of physics take over beyond the standard model, we know that those laws will not affect an observable until the 13th digit, because we've already checked the accuracy of the first 12.

VI. In the future, you may read articles or hear news stories about discoveries that are touted as disproving the standard model.

    A. What these discoveries most likely will do is extend the standard model in an unexpected direction. We might be heading toward a deeper theory of nature, but the idea that the standard model would break down is almost inconceivable at this point.

B. That is not to say that the standard model doesn't have some bothersome aspects.

    1. For example, there are 23 properties that cannot be computed from the standard model; they must be measured in the laboratory.

    2. We also have a large number of particles, including quarks, leptons, force carriers, and the Higgs.

C. For the rest of this course, we will talk about the limitations and puzzles in the standard model, along with some even wilder speculations.

Essential Reading:

Kane, *The Particle Garden*, chapter 6.

't Hooft, *In Search of the Ultimate Building Blocks*, chapters 16 and 18.

Recommended Reading:

See recommended readings for Lecture Seventeen.

The central repository summarizing all standard model information is compiled by the Particle Data Group of Berkeley at http://pdg.lbl.gov/.

Information about the status of the standard model is usually quite technical. For example, Jens Erler and Paul Langacker at the University of Pennsylvania keep an up-to-date summary at http://dept.physics.upenn.edu/~erler/ electroweak/index.html. Some of it may be readable, but for the most part, it is designed for particle physicists.

Questions to Consider:

1. What other scientific theories can you think of that are as successful as the standard model? Are they "complete"? How well do they stack up when compared with the standard model?

2. Why do I say that the standard model can't be the "final story"? In what senses is it flawed or incomplete? (Your answer may depend on your personal philosophy about science!)

Lecture Eighteen—Transcript
The Greatest Success Story in Physics

Last time we talked about the electroweak theory; the theory of Weinberg and Salam that puts together electricity, magnetism, light, and the weak force of nature, and all the associated particles and force carrier that make a coherent mathematical framework. Today, I don't want to talk about anything new. We've learned an awful lot of particles, an awful lot of concepts, and I just want to try to consolidate everything that we've been talking about. We have now come to the stage where we really can talk about the standard model in its conceptual fullness. I want to talk about some of its successes, experimental, and a little bit more abstract successes. That's really the plan for today, just to take a step back and look at this intellectual achievement that the human race has made. You know, I view this as on a par with the collective works of Shakespeare, or something of that nature, where we really have come up with something that's very profound, and very deep, and, in principle, relatively easy to understand, although the details are some tough going.

Let me begin by reminding you of what I mean by the standard model. We have a set of quarks, six of them; and we have a set of leptons, six of them. Each one of these is considered to be a fundamental particle of nature. It has certain properties, which we just list; you can imagine tabulating these things. An electric charge, a weak charge—that's part of the electroweak theory—a mass, a spin, and not much else. That's pretty much the kinds of properties that you have to attribute to these points. Those are the fundamental building blocks of everything and anything. Think about this for a second. Think about the enormous complexity of the world around us. Everything from the hardness of a table, to the softness of human skin, to the almost the ethereal nothingness of the air around us; the colors that you see, the light that strikes your eyes, the sound that you make. All of that stuff can be understood, coming from just the rules and the players of the standard model.

The rules are these theories, quantum chromodynamics and the electroweak theory, or the Weinberg-Salam model, as it's also called. And let me throw out the, kind of, full-blown techno-babble at you; it's a renormalizable, relativistic, quantum gauge field theory. Okay? So, let's talk about those words. We can unpack those words because we really already have talked about all of them.

Renormalizable is a fancy, mathematical statement that says if you include all of the subtle quantum mechanics, you start to get, in principle, some infinities. You're adding a whole bunch of progressively smaller and smaller corrections that add up to be, not just big, but infinitely large, but renormalizable means we can sweep it all under a rug. We pay one little price, which is that there are a few things that the standard model cannot predict. We have to go out and measure them, but everything else is completely finite. We use Feynman's rules, and we can calculate anything we want. Renormalizability is a condition on any theory of nature; at least it's so far been an incredibly useful condition because it means that it's mathematically consistent.

Renormalizable, relativistic—it has to satisfy what Albert Einstein has told us about the nature of space and time, which includes the symmetries that Albert Einstein has recognized about space and time.

Relativistic quantum: it has to satisfy all of the laws of quantum mechanics which were first speculated upon in the early 1900s, and then turned rigorous by Heisenberg and Schrödinger, and ultimately Dirac and many other players—Pauli and Fermi and all those characters in the 1920s and 1930s. Quantum mechanics was quite well established by the early 1930s, and today really is unchanged. We have been testing it, studying it, and quantum mechanics hasn't continued to evolve in any way. So, a field theory, a quantum field theory like we're talking about, just has to incorporate all of those laws, such as the Heisenberg Uncertainty Principle, for example.

Quantum gauge field theory: a gauge that refers to symmetry of nature, a very abstract symmetry of nature that, apparently, all the forces, all the fundamental forces have. In the beginning, we knew that electricity and magnetism had this symmetry. That was observed from the mathematics of the 1800s, and it was such a nice symmetry. It's abstract; I can't talk about it as left-right symmetry, or sideways symmetry. It's a little bit more of a mathematical symmetry. Symmetry means you can make a change in what you call things. You can make a change, yet the answers all come out the same. There is an invariance in your theory, when you look at it from a different perspective. That's what symmetry is, in its most general sense. It's wonderful—and a little bit magical—that this gauge symmetry, which we observed to be true for electricity and magnetism, is also true for all the other forces of nature; at least the weak force, and the strong force, as well. How do I know that it's a good symmetry? Because we imposed it on the theory, and now we've been tested the consequences of that theory in the laboratory, and it keeps working. That's how it is that I can make the statement that apparently the strong force of nature also has this symmetry.

Gauge field theory means that we have this mental image of how it is that you can get action at a distance, how a particle over here can communicate with a particle over there. We visualize space as being filled with fields. Everything in these quantum field theories is a field. Normally I think of the forces as being force fields. But in fact, all of the particles are in a certain sense equally particles, the electron as much as the photon. There is an electron field in the universe, as well as a photon field. This is the theory: a renormalizable, relativistic quantum gauge field theory. It's quite a mouthful, and it's really what we've been spending a good part of all of the lectures so far trying to understand. It's fancy mathematics, but in the end, the story is really one that you can think about in fairly concrete terms. This theory has been successful on a lot of levels. I would call it successful aesthetically, mathematically, and physically; and I'd like to talk about all three of those dimensions.

First of all, I'll discuss "aesthetically." This theory has built into it, essentially, every symmetry of nature that we have been able to think of, that you might hope that the universe has; we've put it into the theory, and it makes it very compelling and very appealing. It's also a minimal theory. It may not look minimal, and it's a little bit hard, when you've been working with this for your whole career, to step back and ask if this really is simple. To me, it's extraordinarily simple, in the sense of just having a few ingredients and a few ideas behind it. And, minimal, meaning—I guess—that, for Mr. Weinberg in 1967, he had to write down a mathematical theory. In principle, you can write down a really complicated theory to describe the weak interaction, nut he chose to write the absolutely simplest one he could. Every time he had a choice, whether to make up a new particle, he chose to go with the fewest number possible. We've observed the six quarks and the six leptons in laboratory experiments, so we have to include them. We don't include anything else, except for the Higgs. The Higgs has not yet been seen, and we'll come back and talk about it. It's the sort of one mysterious piece, actually, of the story. With that one possible exception, it is a truly minimal theory.

You have to think about how amazing this is. In 1967, this theory was published, and it hasn't been tweaked in any significant way since then. You introduce a minimal theory, usually the physicists' first idea, "Let's just try the simplest one and see if it works," and if it doesn't, he can always add more. We can make it more complicated to try to match the data. That's the way biology, chemistry, geology, and so on, often work. You start with a simple approximation, then you make it richer and richer as you get more and more data, and become more and more precise. Here, we have a rather different situation. We started off with the most minimal theory, and then it just kept working, and it kept working even until today. We never had to make it more complicated. Even after we built new accelerators, we discovered new particles. We learned an awful lot after 1967, but it all fit in with the exact same minimal, electroweak theory. So that's a really lovely aesthetic idea that, I think, compels people to pay some attention to this theory, and believe in it as a statement about the scientific truth or nature of reality.

It unifies forces of nature. What we thought were completely different forces of nature: electricity, magnetism, light, if you like; the weak force, the weak neutral forces of nature, all of those are really one, in this mathematical theory. It's just one force viewed from different angles, abstract angles. And that's a really lovely idea. Again, there's an aesthetic appeal of this theory.

There are some symmetries of nature, which are broken in the standard model, and some of them are mysteries, and some of them are deeply understood now because of the theory. For example, left and right are no longer an exact symmetry of nature itself. The laws of physics are different for something spinning like a left hand spins, and for something spinning like a right hand spins. They're only different when you consider weak interactions. The laws are the same for electrical forces, strong forces, and gravitational forces; only the weak force appears to violate this symmetry of nature. That's in the standard model, and I can't explain it to you, because nobody knows why; it just is an observed fact of nature.

But there are other broken symmetries, which we can explain. For example, we have force carriers, the photon and the Z. They, in some ways, seem like similar things. I mean, they're spin one objects, and they carry some aspect of the electroweak force, but there's a broken symmetry between them. One of them is without mass, the photon; one of them is massive, the Z particle. That's an electroweak symmetry breaking. That's the fancy word for that observed fact of nature that one of these force carriers has mass and the other doesn't. We believe we understand that. It arises from the presence of the Higgs particle, which we're going to be talking about.

The standard model actually gives you an explanation of why it is that these two forces of nature, which, on the one hand seem to be completely symmetrical, but on the practical hand, in the real world, seem to be so different from one another. If you were to heat the universe, which you could do in a laboratory by smashing particles together, or by going back to the early universe like the Big Bang, you would discover that the weak and electromagnetic interactions are, in fact, identical. They merge together at super high energies because the standard model has unified them. Those are all sort of lovely mathematical, or aesthetic, arguments.

Mathematical arguments really boil down to the fact that this theory doesn't contradict itself in any obvious way. If you calculate something, you always get finite answers. There are no probabilities that can be computed in the standard model that come out bigger than one or less than zero; that would be a mathematical inconsistency. Earlier models had that problem, but the standard model doesn't. So, again, all of this stuff is very nice, and to physicists it kind of helps you to believe in the truth of the theory. But doggone it, you know, a theory is only good, in my mind, anyway, and this is a personal opinion, if you can go into the laboratory and make some measurements, calculate some numbers, and compare the numbers you predict with the numbers you get in the laboratory. That, to me, is really what it means to have a correct or successful theory of nature. I'll take quantitative over aesthetic any day, even if the theory gets a little ugly, if it's a good, predictive theory. That is certainly useful science, in any case. It may not be the most fundamental, or the end of the story, that's all.

Every year, the Particle Data Group at Berkley keeps up a table, a collection of all of the information that we've been gathering over the years. You can go on the Web and look up Particle Data Group, if you're interested, and you'll find tabulations. First, there's a summary of the standard model that tell you what the ingredients are and what the formulas are. Then there will be a list of the particles. First the fundamental ones: the electron has this mass, and this magnetic strength, and this electric charge, and so on; just a table. Every piece of data includes a reference to a paper, a journal article, or maybe several; so you can go and see what experiments were done to give us these numbers. They go on and have tabulations, not only of the fundamental particles, but of all the built-up particles, such as the the pions, and the kaons, and the protons, and the complicated particles in the zoo, which aren't really so complicated after all. They're just three quarks put together, or a quark and an anti-quark.

If you look at this—you can get a physical publication of it, and it's a big, thick, fat book. It's the collected information that we have, and each page is incredibly dense, because they're not really telling you everything, they're just telling you where to go to find out more details about that particular number, and that particular information. It's quite stunning, when you think about it. That big, fat book and all of the thousands and thousands of pages to which it refers, all comes from the first couple of pages, the standard model; and people have been spending years doing these calculations. To date, I know of no significant discrepancies between the theoretical calculations and the experimental observables.

Everybody's looking for it. Everybody would like to find some kind of a crack in the standard model, because that's the nature of physicists. It's no fun just going out and verifying what people are telling you you're going to see. It's much more fun if you get a surprise, a curiosity, something we don't understand; and, of course, these things still happen. They're often in subtle aspects of the standard model, but at some fundamental level, we keep hunting for cracks, and we're not finding them.

Let me talk about another success of the standard model that's a little bit more qualitative, but to me, equally convincing. Mr. Weinberg, in 1967, worked out a theory, and said, "I predict the existence of a new particle, the Z boson." Almost 20 years later, we go into a laboratory, and get to the energy, finally, that he said we need to get to; and there it is, in the bubble chamber, a new particle of nature. The Z particle is real; it leaves behind a track. You have to think about that for a second. How can a human being make a prediction about a fundamental particle that won't be observed for 20 years? Can he just be kind of making wild guesses? It's not as if Weinberg wrote paper after paper making a whole bunch of guesses, and one of them came true. He wrote one paper about this and argued convincingly that it should be correct, and it was. To me, I just find that kind of prediction, of a new fundamental force, or a new fundamental particle, to be a powerful indicator that we really do understand something. We're not just approximating or curve-fitting; we're really talking about some deeper level of understanding about the world. Of course it wasn't just the Z; it was the W, tau-flavored neutrino, and the top quark. All of these things were predicted in advance of their discoveries and later found where we were looking.

The tau neutrino, found in 2000, was a huge experimental effort, and it seems to be exactly like a muon neutrino, or an electron neutrino, except that it has tau-ness, whatever that might mean. It means that when it whacks into a proton, it produces a tau particle, rather than producing an electron or a muon, and you can segregate tau-neutrino beams, mu-neutrino beams, electron-neutrino beams. We'll be coming back, in future lectures, to talk about the current effort that people are making to understand the physics of these neutrinos. It's one of the few places, in fact, where there is already a hint that maybe the standard model does need to be tweaked up a little bit. Finally, after almost 30 years of attempts to find a crack, the neutrino sector is the one place where it seems to be that maybe you don't quite understand how the world works. There may be some interesting funny business, so we have to come back and talk about that.

Now, having said that, I also want to emphasize that, if we do discover some funny business with the quarks, we're not going to toss the standard model. At least, there would be no reason to, as long as the new data isn't in radical disagreement with the fundamental premises because the standard model does have a little bit of leeway in it. We could, for example, make a discovery of fourth generation of quarks, but that wouldn't shatter the standard model. It would be a wonderful surprise; everybody would be very excited, and we'd do lots of experiments to learn about it. That's already happened once. It would fit in to our framework of understanding. A fourth generation of quarks would, how should we phrase it? It would change the standard model, it would modify it; but it wouldn't shatter it. The framework would still be there, and we can generalize that framework to incorporate some new heavier particles. If we discovered a new light particle that didn't fit in, in any way, then we might have some serious worries. But that would be kind of like saying, "Yeah, maybe some day I'll hold this pen and it'll go flying up in the air, and that would ruin Isaac Newton's theory of gravity." Well it surly would, but I don't think it's ever likely to happen unless, you know, a little black hole goes flying over the top of the room, and then I guess we could understand it. In fact, it would be something beyond Isaac Newton's theory.

Today, we look back at Isaac Newton's theory and ponder if it is right or wrong. Some people would say, "Oh, it's wrong because Albert Einstein proved that F is not quite equal to ma when things get very relativistic," and I would disagree. I would say, "Sure, F=ma—for any object moving at ordinary speeds." It's a fabulously accurate law of nature; it describes the way the world works. The fact that a super-fast particle behaves slightly differently just means we have to stretch our understanding of F=ma; we have to deepen and broaden it. I can't imagine any scenario, ever, that's going to take away F=ma and Newton's law of gravity. All I can imagine is exciting new discoveries that deepen, or extend, the realm of applicability of the laws of physics.

I think the same thing has to be true of the standard model. No matter what we discover in the future, there are still going to be atoms in the world. We've seen direct evidence of them. There are still going to be quarks in the world; we've seen direct evidence of them. They're going to combine in the ways that the standard model says, and at low energies, or even high energies, like the highest that we can produce today, at the dawn of the 21st century. I think all of the laws that we have been testing, we've tested; and they're correct. So whatever wonderful things we discover in the future are going to certainly modify, almost surely modify our understanding of the standard model. But it's hard to imagine how they could break it.

Let me give you another example. The standard model not only makes these qualitative predictions about the existence of new particles, but you can predict numbers for experiments. What kind of numbers? Some of them are very exact specific numbers. For example, the standard model tells you what the magnetic strength of an electron is. I've mentioned that number before, because it is perhaps the greatest numerical success of the standard model. I'll repeat that number because I love it so much - 2.00231930435 - twelve digits that explain a fact of nature about this little particle, the electron, and how it behaves when you put it near a magnet. The fellow who did that calculation's name is Kinoshita, and here's a little quote from Mr. Kinoshita. "Unlike previous theories, QED"—and he actually meant the electroweak theory—has no obvious limitation since no theory is likely to be absolutely true, however"—I want you to think about that statement— "since no theory is likely to be absolutely true, physicists have been trying to discover where the limitations of QED might be found." There are a lot of deep ideas in that sentence. He doesn't believe that QED, really the electroweak theory, is necessarily the ultimate theory of nature, and neither do I. What he's saying is that it's an extraordinarily accurate one; and how do we find out its limits? Where does it begin to break down? Okay? If there's some new physics, beyond the standard model, some new particle or new idea that we've never thought of so far, what we know experimentally is, it's only going to affect that observable in the 13th decimal place because we've already checked the first 12. So the first place where the standard model can break down is off there in the parts-per-billion level. So it may be that that's going to happen, and it would certainly be interesting. It would change our worldview a little bit, but it's not going to change the first 12 decimal places of our understanding of the world.

That calculation is a lovely one, and people have done others like it. The calculation for the muon works almost equally well. The calculation for the proton failed because the proton wasn't a fundamental particle. But if you try to understand properties of the underlying quarks, then you can make more quantitative predictions. The energy of hydrogen atoms, and the transitions—for instance, the Lamb shift that was first observed in 1947, that led us on this path towards quantum electrodynamics, a little splitting of two lines of hydrogen from—wasn't really one color; it was really two closely spaced colors. That number has been calculated now to accuracy that approaches the accuracy of that magnetic number, the magnetic moment for the electron. All of these quantitative successes tell us that if you're going to find a crack in the standard model, it's probably going to be a subtle one. What that might mean is that it's going to occur at some very high energy. So we're probably going to find some very heavy particle, if there's something new beyond the standard model, and that's the likely place that you could imagine where the standard model might end up breaking down.

After you've listened to this course, you're going to hear lots of articles, hear stories, read articles about science, about physics —they appear from time to time in "The New York Times," in "Discover Magazine," and "Scientific American."—about the discoveries that are going on in particle-physics accelerators, and you'll hear all these buzzwords. I can almost guarantee you that, every time a big discovery is made, the press will make some sensational noises about, the standard model has been proven wrong, and you have to understand what they mean by that. The standard model was defined 30 years ago, and anything new, technically, could be argued to be proving the standard model wrong. If a neutrino is not without mass, we've proven the standard model wrong. If the magnetic moment of the electron disagrees in the 13th or 14th decimal place, we've proven the standard model wrong. Physicists will jump up and down because they want the press attention like anybody else; and the press is excited, because the physicists are excited. It is exciting, but bear in mind what it really means. It really means that we have extended the standard model in some unexpected way. We might be heading toward a deeper theory of nature, but the only way that I could imagine the standard model really breaking would be if gravity starts turning upside down some day. Or something—just sort of inconceivably radical—that all of a sudden, the new experiments disagree with the old ones. It's almost inconceivable at this point.

As a physicist, I of observe in my colleagues two reactions to these spectacular successes of the standard model. On the one hand, you have the physicists who are really pleased. They're a bit cocky and self-satisfied: I mean, look at it. We can write it down on a couple of pieces of paper; we can describe the world at a fundamental level—we've done it! All right? We understand the world. This has been a dream of human beings since 2,500 years ago, and probably longer. We know what Mr. Democritus's atoms are; we know what they're made of, we know the bottom line. It's a lovely thought, and maybe we are done. It's possible; it's at least conceivable. We've been peeling open this onion and we got to the nugget at the bottom. It's also quite conceivable. With the other group's reaction, I think probably the more common attitude among working physicists, is that there's still more layers to that onion. We have found a very lovely, round, smooth, elegant layer that explains everything up above it; and perhaps we can still look deeper. We can understand other aspects of the standard model. As a, sort of a cynical or aggressive physicist, you might start poking holes at it.

Nowadays, when I hear colloquia with especially young physicists, they always begin by saying well, we have the standard model, and it's very successful, but—then they start listing things that bother them about the standard model. For example, there are approximately 23 numbers, numbers that you have to go in the laboratory to measure and put into that particle data table. You can't compute them from scratch. What's that about? In fundamental physics, shouldn't we just be able to compute everything about the world? Well, maybe we can, maybe we can't. I don't know if that's the way the world works or not, but 23 is starting to be a big number. The same thing with the total number of particles: six quarks, six leptons, a whole bunch of force carriers, plus the Higgs; it's getting to be a big number. What's that about? Can't we understand the world in terms of something simpler? Even deeper still, what about these generations? Well, we've got the up, down, electron, and electron neutrino, the light generation; and they describe pretty much everything in the ordinary world. Who ordered the next generation? Why do we need them, why are they there, those heavier particles? It's a good question.

When we came to the third generation, in the 1970s, all of a sudden people are saying, "I don't know, this is starting to look like the periodic table of atoms all over again. There are rows and columns; the columns have very similar properties, the masses are increasing. Surely there's a story here." There's some underlying physics, and the standard model doesn't tell you that story. The standard model just says, this is the world, let's work from this, and now we can explain everything else. So it's certainly a popular idea nowadays that there's got to be something, and the buzzword is, beyond the standard model.

So for the rest of this course, we're going to be talking about the limitations, the present puzzles, and the curiosities in the standard model. In the very end, in the last couple of lectures, I want to really talk about some even wilder speculations. People have really started thinking about what, at the moment, are pretty radical ideas that are very elegant, they're mathematically compelling, and they really simplify the story at a conceptual level, although they complicate the math. It's very speculative and not yet tested.

What I want to leave you with, in this lecture, is the sense that the standard model really is successful, no matter if people say, "Well, it's a great theory, but..."It is a great theory. It really does describe the world, and every experiment I know of, in the atomic and subatomic realm, is described quantitatively by the standard model of particle physics.

Lecture Nineteen
The Higgs Particle

Scope: The mysterious Higgs particle is the least understood piece of our story so far and the *one* central part not yet directly verified. In this lecture, we ask a number of questions about the Higgs: Why should we believe in its existence? How can we picture it, and what role does it play in the standard model? Why is it so important to "find" a Higgs? How would we go about looking for such a thing, and what would happen if we looked for it and it wasn't there? We conclude by discussing the rise and fall of the superconducting supercollider (SSC), the machine specifically designed to answer our questions about the Higgs.

> No one can say whether any one accelerator will let us make the last step to a final theory. I do know that these machines are necessary successors to a historical progression of great scientific instruments ... Whether or not the final laws of nature are discovered in our lifetime, it is a great thing for us to carry on the tradition of holding nature up to examination, of asking again and again why it is the way it is.
> —Steven Weinberg (*Dreams of a Final Theory*, p. 275)

Outline

I. We have talked about all the fundamental particles that we currently know about, except for one—the Higgs boson.

   A. In the standard model, the Higgs is responsible for the defining characteristic of any particle, its mass. It also functions in distinguishing between the weak and electromagnetic forces, and it may have played a role in the evolution of the universe.

   B. The Higgs is the most poorly understood part of the standard model to date. Everything else we have talked about has been extremely well established for many years in the laboratory. The Higgs, however, has been established mathematically but not physically. We have no direct evidence of the existence of a Higgs particle.

C. Let's recall for a moment why the Higgs was proposed.

    1. In trying to understand the weak force of nature, physicists believed this force should have gauge symmetry, because the other forces in the universe had gauge symmetry.

    2. If gauge symmetry is observed in the weak force, as it is in electricity and magnetism, these forces will be completely symmetrical in many ways. Specifically, the weak force carrier will be massless. That idea, however, contradicted the data.

    3. Every piece of data says that the weak force has very short range. It is almost a contact force, which is why it's so weak. The force carriers—the W and Z particles—are massive. How could we reconcile gauge symmetry, which says that these carrier should be massless, with the fact that they are massive?

    4. The answer was a mathematical trick, the prediction or assumption of the existence of a new particle of nature.

        a. Remember, in quantum mechanics, particles and waves are two aspects of the same thing. Similarly, if we think about physics as a quantum field theory, particles and fields are two aspects of the same thing.

        b. This idea is most obvious in thinking about force carriers. How do I visualize a photon? First, I visualize an electric field, but that's not the photon. The photon is the traveling wave that results when the electric field is jiggled. The photon is also the particle of light.

        c. Every particle of nature can be thought of in this same abstract way as the ripple in a field.

        d. Thus, Higgs proposed a new field—not an electric field or a magnetic field, but the Higgs field, which permeates space. The ripple in that field would be called the Higgs particle.

D. The unusual thing about this Higgs field is that it is finite but everywhere. The Higgs field exists inside of atoms in the empty space between the electrons and quarks. It even exists in outer space. It is a uniform background in which we all live.

    1. Usually, we would think about outer space as being close to a vacuum. We would imagine that an electric field is strong when it's near an object but that it then fades away.

2. With the Higgs, however, to make sense of the weak interaction and to make the theory mathematically consistent, we were obligated to postulate that the Higgs field is everywhere.

3. When I walk through the room, then, I'm walking through a sea of the Higgs field. What would be the effect on me? That depends on the interaction of the particles in my body with the Higgs field.

4. One of the things that would happen as I'm walking, if the particles in my body are interacting with the Higgs field, would be some sort of resistance to my motion.

E. Before we talk about the search for direct evidence of the Higgs, which we have not yet found, I want to touch on one other aspect of it.

1. In nature, the weak force and the electric force seem completely different. One of them has to do with radioactivity; the other is everyday life. The carrier of electricity is massless, which means that generating electricity and electric forces is easy. But the carrier of the weak force is massive, which means that the likelihood of weak interactions is improbable.

2. A symmetry in the standard model says that these two forces of nature are the same. How, then, does that symmetry get broken? How does the universe change from one in which weak forces and electric forces are identical to the one we live in where the forces are so different?

3. The answer is that the Higgs is responsible for symmetry breaking. In the standard model, the W and Z bosons interact with the Higgs field very strongly and it slows them down a great deal. They become massive.

4. We call this phenomenon *electroweak symmetry breaking*. If we want to understand why the weak force and the electric force seem to be the same yet are so different, we must postulate something like the Higgs.

II. How do we look for a particle like the Higgs? With the neutrinos, a mathematical theory predicted their existence and the exact probability of their interaction. The situation is a little different with the Higgs.

A. The standard model cannot tell us how massive the Higgs field is. The phrase *how massive* refers to how much energy we have to pour into a small region of space to create a ripple in the Higgs field.

B. Again, imagine that we live in a sort of unusual fluid, and we want to create a wave. In water, the slightest perturbation will create a wave, but that is not true of the Higgs field. The Higgs is so massive that we need to add a lot of energy to create a ripple, which would be a particle and would leave direct evidence behind in a bubble chamber.

C. We could find this particle through a direct or an indirect search.

    1. The indirect search would be to look for the presence of a Higgs, because any particle of nature, if it's real, can be virtually created for a very short time, then would disappear again. This is how we began to search for the top quark before we had accelerators that were energetic enough to actually make one in the laboratory.

    2. This method is called *precision physics*. We make very accurate measurements and very accurate standard-model calculations. We have only one unknown—the mass of the Higgs—therefore, we can compare the measurements and the calculations and try to deduce the mass of the Higgs.

    3. Physicists have been trying this method for 30 years now and are beginning to hone in on the mass. Unfortunately, the Higgs effects on other processes tend to be extremely subtle compared to even the small effects of a top quark. The best guess for the mass of the Higgs today from precision data is somewhere between 100 times proton mass and 200 times proton mass.

    4. If we know this range, why don't we make a Higgs particle? At Fermilab, we can achieve a trillion eV. In principle, that's more than enough energy to create a Higgs. What happens, though, when we add all this energy to a small region of space? We would produce the whole zoo of particles, but the probability of producing a Higgs is so small that we would not expect to have seen one yet, even given the millions and millions of other quantum particle production events that we have observed.

    5. At Fermilab, the strategy to increase the likelihood of seeing an event has been to upgrade the intensity of the beam to increase the number of proton and anti-proton collisions. In Europe, at CERN, the strategy is to increase the energy.

Around 2007, CERN will be able to achieve 14 trillion eV, which will certainly be enough energy to see a Higgs if they exist.

6. In the 1980s, the United States began a project to build a truly gigantic accelerator—a superconducting supercollider. The project was abandoned in the 1990s because of lack of funding.

III. I believe that the demise of the superconducting supercollider was symbolic of a shift in U.S. priorities regarding particle physics compared to other areas of physics and other sciences.

A. From World War II until the 1990s, particle physics was at the top of the list for funding and facilities. This field was viewed as the most fundamental and, therefore, the most worthy of all the sciences.

B. That idea seems to have changed now. Particle physics is still very much alive, but now, it is one player among all the branches of physics and, indeed, among all the branches of science.

C. Let's finish this lecture by asking, "What if we had built this superconducting supercollider and hadn't found the Higgs?"

1. Mathematics tells us that we would expect to find this particle, but we could easily imagine that the Higgs might not show up the way we would expect it to. What would happen if we took the Higgs out of the standard model?

2. The answer is that some aspects of the model would survive, but in some ways, the standard model would collapse on itself. In particular, if we try to calculate events at a couple of trillion eV, then the theory would start to yield infinities. In other words, the theory is fine at low energies, but when we reach the energies of superconducting supercolliders, the theory might break down and yield answers that don't make sense.

3. That fact would tell us that something deeper is going on. The Higgs is a minimal theory. That is to say, it is the simplest possible mechanism we have come up with that is consistent with everything we know so far and makes the theory coherent.

4. I might almost guarantee that something will happen at a trillion eV that we don't know about today. The symmetry of nature between electric and weak forces is going to merge or break down at those energies, but we're not quite sure exactly how.

5. If we find something else, we would have to reconstruct the standard model to account for the Higgs sector.

D. The Higgs, then, is the last piece of the standard model and the most important in the sense of understanding mass, the fundamental property of everything, and understanding symmetry breaking, which is the key idea that unifies the weak and electromagnetic forces. And it's the one piece that we have not yet verified in the laboratory.

Essential Reading:

Barnett, Muhry, and Quinn, *The Charm of Strange Quarks*, chapter 9.2.

Kane, *The Particle Garden*, chapters 3 and 8.

't Hooft, *In Search of the Ultimate Building Blocks*, chapter 11.

Weinberg, *Dreams of a Final Theory*, chapter XII and afterward.

Recommended Reading:

Lederman, *The God Particle*, chapter 8, second half, starting from "The Standard Model is a shaky platform."

http://www.hep.ucl.ac.uk/~djm/higgsa.html. In 1993, the then-current Science Minister of the United Kingdom, William Waldegrave, issued a challenge to physicists to answer the questions "What is the Higgs boson, and why do we want to find it?" on one side of a single sheet of paper. This link is David J. Miller's prize-winning "qualitative lay-person's explanation." Other replies can be found at http://hepwww.ph.qmw.ac.uk/epp/higgs.html.

Questions to Consider:

1. Leon Lederman, former director of Fermilab and Nobel Prize winner, wrote a book about the Higgs called *The God Particle*. Few people in the physics community use that name for the Higgs. Why do you suppose he chose that title? Are you sympathetic with his choice? Why or why not?

2. If you could go back in time and cast the deciding vote in the Senate to end or continue construction of the superconducting supercollider, how would you vote? Why? What factors would influence your decision most strongly?

Lecture Nineteen—Transcript
The Higgs Particle

We've been talking about the standard model, trying to understand the conceptual framework and the players—the fundamental particles that make up the world that we live in. We've talked about all of the fundamental players that we know about today, except for one, which I've really only mentioned in passing. It's a very important player - very esoteric, but also quite relevant if you want to understand how things work, and that's the Higgs boson, the Higgs particle. The Higgs is named after a physicist named Peter Higgs, and its role in the standard model; although it's quite formal and mathematical, it really is the thing that's responsible for the defining characteristic of any particle, its mass. So the Higgs plays this key role and, also, plays a role in distinguishing between those forces that are presumably unified now: weak and electromagnetic. And it plays a role, potentially, in the evolution and development of the universe as a whole. It's quite unique in this respect: that something very, very large scale—cosmology, and the structure of the universe—may in fact depend, and be determined, by this subatomic particle.

So today I want to talk about the Higgs; what is it, why do we believe in it? It is definitely the most poorly understood part of the standard model to date. Everything else we've been talking about has been extremely well established for many, many years now in the laboratory. The Higgs has been established mathematically, but not physically. We have no direct evidence of the existence of a Higgs particle yet, and it's big business in the physics world. There are a lot of money, a lot of energy, a lot of time and a lot of effort being devoted to finding the Higgs particle. It will certainly be found. It has been in the news and will certainly be in the news again in the future. In fact, it has even played a role in U.S. politics, which I'll talk about today. So I think it makes it an interesting particle to learn about.

Let me remind you—we've mentioned it before—but let me tell you about why the Higgs was proposed, what it does. We wanted to understand the weak force of nature, and it was believed, pretty much just because the other forces had gauge symmetry, that the weak force should also have gauge symmetry. This is a very abstract idea; it's a kind of a mathematical property of a theory that says if you shift something, physics doesn't really care. The world doesn't really care what number you choose to define as zero volts in some electric circuit. So, if you believe that gauge symmetry is a symmetry of nature—and we've already had extremely good evidence that it was a symmetry of electricity and magnetism and the strong force, and maybe the weak force too; if you want that— then you discover that the electric forces and the weak forces will be completely symmetrical in a lot of ways. They will not only be unified; the weak force carrier would be without mass, and that's in direct contradiction with data. Every piece of data says that the weak force is very, very short range; it's like a contact. That's why it's so weak. A neutrino has to be right on top of another particle in order for you to have any kind of a chance of seeing a weak interaction, in a certain sense. The force carriers, the W and Z particles, are massive. So how do you reconcile gauge symmetry, which says they should be without mass, with the fact that they're massive? The answer is a mathematical trick. It's the prediction, or assumption, of the existence of this new particle of nature.

Now, when I talk about "particle"—and in quantum mechanics you can talk about particle or wave—it's kind of two aspects of the same thing. Similarly I can talk about particles or fields, now that we're thinking about physics as quantum field theory. The place where it's most obvious is when you're thinking about the force carriers. Okay, how do I visualize a photon, a particle of light? Well, first I visualize an electric field. I have a little charge, and there's this invisible, maybe even abstract or mathematical, electric field permeating space. That's not the photon; the photon is what you get if you wiggle that field. If you take the charge and make it move up and down, then the field will get a ripple in it, and the ripple will propagate through space; and it's that traveling wave, it's that ripple, that's the photon. That is the particle of light.

Every particle of nature can be thought of, in this same abstract way, as the ripple in a field, so Mr. Higgs proposes that there exists a new field of nature. It's not an electric field, and it's not a magnetic field; it's what we now call the Higgs field. If you have a ripple in that field then you would call it a Higgs particle. Now the really unusual thing about this Higgs field, the thing that makes it special, is that it asks if there is a field that permeates space. It was the sort of brainstorm of these folks like Weinberg, Salam, Glashow and Higgs all putting the story together. It's not zero; it's finite. It's everywhere - inside of atoms, in all that empty space between the electrons and the quarks, outside; even in deep, dark, outer space. There's a Higgs field everywhere, like a uniform background that we all live in. This is a wacky idea. I mean, normally we think about dark outer space as being pretty close to a vacuum. There are a few photons that travel through dark outer space, because they're starlight, but pretty much what we imagine when we think of fields is that there's a charge, and there's an electric field that's very strong near it, then that field fades away. So as you get far away from things, you expect that fields should get very small, but the Higgs does not.

In order to make sense of the weak interaction—in order to make the theory mathematically consistent—we were obligated to postulate that there's this Higgs field everywhere. So as I walk through the room, I'm walking through a sea of Higgs field, and what would the effect on me be? Well, that depends on the interaction of the particles in my body with the Higgs field. That's another of your assumptions. Right now, all we're doing is playing a what-if game. That's how theory often begins. You just say, "Well, what if there were a field of nature everywhere, and what if most of the particles, the other fundamental particles, interacted with it? Okay?

Well, one of the things that would happen is, as I walk, if the particles in my body were interacting with that Higgs field, there's going to be some sort of resistance to my motion. What I'm trying to do here is to describe some very formal mathematics and make it concrete. You have to appreciate that I'm always doing this in this course; and here I'm really stretching the metaphor about as far as I can. If what I was saying were literally correct, then you would imagine that all objects would come to a halt, because they are walking through this goo, but that's not what happens. An object in motion remains in motion. Newton's laws are still true, even in the presence of the Higgs, but an object in motion feels more sluggish than it would if there was no Higgs field around. That's probably the best way to think about it.

So we're kind of like fish underwater, swimming through this water; and the fish are always the last to notice the water. It's there, it's important for them, but they're going to notice other fish first; and they're going to notice the effects, maybe, of the water causing, you know, waves, causing things to move around. We live in this stuff. It interacts with our bodies. It's what creates the inertia, or in effect the mass, of all the particles; and yet, it took us a long time just to recognize the possibility of its existence.

It's a very formal idea, and, when I was first introduced, in graduate school to the Higgs mechanism, the mathematical framework, I learned that there's assumption after assumption being piled on. We're assuming a new field; we're assuming that it's not zero; we're assuming that the particles interact with it in specific ways. Then, you start to draw some conclusions, and those conclusions match all the weak data. In fact, lots more than just the weak data; the masses of the particles, everything begins to tie in, in this lovely way that we've seen. The standard model is a beautiful, self-contained theory, which describes the world. So on the one hand, everything fits together, and the standard model is a spectacular quantitative success. On the other hand, this Higgs business just seems pretty ad hoc. Now, maybe it's just because it's a more recent discovery. You know, way back in the early 1930s Mr. Dirac proposed something very mathematical and formal, namely, anti-matter and could hardly believe it himself. We see the same thing with the neutrino. Many of the particles that came out of mathematics were at first sort of viewed with suspicion; it seemed so formal and abstract. The quarks themselves didn't seem like real things. What we want is evidence, direct evidence, for the existence of the Higgs.

Before I get to our search for the direct evidence—because we don't have it yet, okay, this is still a hunt—I want to say another word about why the Higgs is such an important thing in this theory. In nature, the weak force and the electric force seem completely different. They are as different as night and day. One of them has to do with radioactivity, the other with everyday life. The carrier of electricity is without mass, so it's really easy to generate electricity, and electric forces; but the carrier of the weak force is very massive, so it's highly improbable to create—to have weak interactions occur. There is, in fact, symmetry in the standard model that says that these two forces of nature are one and the same, viewed from a different perspective. So how does that symmetry get broken? How do you go from a universe where weak forces and electric forces are identical—that has some sort of mathematical or high-energy universe—to the one we live in, where the forces are so different? And the answer is the Higgs. It's the Higgs that's responsible for symmetry breaking.

In the standard model, the W and Z bosons interact with this Higgs field that's everywhere. As they're traveling along they, in fact, interact very strongly with it, and so it slows them down a lot. They become very massive. So it's the Higgs field that breaks the symmetry. We call this electroweak symmetry breaking; that's the fancy expression for it in the world of physics. And, if you want to understand that phenomenon of nature—why does the weak force and the electric force seem both to be the same, and to be so different—you have to understand the Higgs. There either has to be one, or there has to be something else. If it's not the Higgs, if we're wrong in our guess, it has to be something. At least, we believe there should be some reason why this symmetry gets broken.

I think we should be glad that the Higgs exists, if it does, because symmetry breaking is not a bad thing. It's a good thing. The world is much more boring when everything is completely symmetrical. Artists know this. When they create paintings, it's very rare that they'll draw perfect geometrical symmetry. It's the breaking of the symmetry that gets your interest. Imagine that we lived in a sphere, a very symmetrical universe, and it was filled with some gas, some water vapor. Okay? It would be a beautifully symmetrical universe. There's water vapor in all directions, the pressure's the same everywhere, and nothing interesting is going on. Life would get more interesting if you cooled down that sphere, cooled it down to a certain critical temperature where the water vapor begins to condense into liquid water. Okay, at that lower temperature, you have some physics going on. You've got a little pool of liquid water here, and some vapor over there, and you can swim in the water and jump out of the water and, there are boundaries; that means there are physics phenomena going on. Life gets interesting, and that's sort of what's happening in our universe. We could imagine a universe. In fact, in an accelerator at ultra-high energies—higher energies than we've reached so far—or, in the early universe, very close to the Big Bang; when the universe is, in effect, very hot, the electric force and the weak force become like this uniform water vapor. It's all the same. The two forces are identical, and indistinguishable, and it's lovely and symmetric, but not so interesting and not so much fun to live there. In our universe, which has cooled off, the two forces separated, and so we get the two different, distinct phenomena. So this is all kind of motivation for why we believe in it, and why it might be important to our understanding of the world. So let me talk now about how you go hunting for a particle like this.

It's, in a certain sense, like looking for neutrinos. You believe in them because there's a mathematical theory that predicts their existence. With the neutrinos, it's pretty clear from the theory how you should go looking for them, because the theory itself tells you exactly what the probability of interaction is, so you can figure out how to design a big detector. It's a little tougher with the Higgs because, unfortunately, the one thing that the standard model cannot tell us is how massive this Higgs field itself is. When I say how massive I mean, how much energy has to pour into a small region of space to create a ripple in that Higgs field. So, I'm sort of imagining an unusual fluid, right? We're walking through this universe that's filled with, not water, but Higgs; and what we want to do is create a wave. It's not like water, where just the slightest little perturbation will create a wave; that's a particle without mass that's easy to make. The Higgs is massive, so you need to pour a lot of energy in to create a ripple, which would—once you've created it, it's a real particle, the Higgs particle—leave a track in a bubble chamber; it would decay into other particles. It would be as obvious as any of the fundamental particles, if we could create one.

There are a couple of ways you could imagine doing this, either a direct search or an indirect search. The indirect search would be to look for the presence of a Higgs because any particle of nature if it's real, can be virtually created, and then disappear again, for a very short amount of time. Any particle, including the Higgs, can do that. This is how we begin to search for the top quark before we had accelerators that were energetic enough to actually make one in the laboratory. It's called precision physics. You make a very, very accurate measurement of something. You make a very, very accurate standard-model calculation, and there's only one unknown—the mass of the Higgs—and so you can compare the two, and try to deduce the mass of the Higgs.

We've been trying this for 30 years now, ever since the Higgs mechanism was proposed, and we're beginning to hone in on it. Unfortunately, the Higgs effects on other processes tend to be rather wimpy compared to, say even the small effects of a top quark. The effects are very, very subtle effects, so they're hard to pin down exactly. We had predicted the mass of the top quark to within ten per cent before we actually found one. The Higgs is, right now I think a mass somewhere between 100 times a proton mass and 200 times a proton mass, so that's the best guess today from this precision data. We continue to collect precision data, so presumably those limits are going to get narrower and narrower.

Why not just make one? Between 100 and 200 proton masses is how much energy you have to put into a small region of space, so we ought to be able to do that. At Fermilab we have a TeV, or a trillion electron-volts which is plenty enough energy, in principle, to create a Higgs. The problem is, if you pour all this energy into a small region of space, something happens. Most of the time you get protons and anti-protons, electrons and anti-electrons, pions, and kaons or all the zoo of particles, including top quarks and anti-top quarks. Everything you can produce, you will produce. All right? That's sort of the way quantum mechanics works. Everything is probabilistic. The more probable an event is, the more you will see coming out in your detector, and we worked out the probability of creating a Higgs particle in the standard model. The one unknown is its mass, so you just sort of take a guess. Suppose it's 200 times the mass of a proton, what would the probability be? The answer is, very small. So we're going to Fermilab and looking for these things, but the probability is so small, in fact, that it's so far not even expected that we would have seen even one such event, compared to the millions upon millions of other quantum-particle-production events. So you're truly looking a needle in a haystack. If we make one, it'll be pretty clear. The signature of a Higgs particle in a bubble chamber, or a modern version of a bubble chamber, is going to be quite distinctive. We believe that once we make one we'll know, but we just haven't made one yet.

The American strategy at Fermilab has been to upgrade the intensity of the beam. It's very difficult for us to increase the energy. We've already done the upgrade to a trillion electron-volts; it was a big expensive project, and upping it even more would require stronger magnets, bigger accelerator. We just don't have the funding at the moment for that kind of a project, so we're just trying to make more and more and more protons and anti-protons collide together. If you have a low-probability event, it's like rolling the dice—or, more appropriately, buying lottery tickets. If you have only got a one-in-a-million chance, that's all you've have. You might as well buy more and more and more tickets; actually, that's a very bad strategy with the lottery, because you have to pay money for each of those tickets. Well, with the Higgs, that's really the idea at Fermilab. Increase the luminosity, or intensity, and hope that we get enough events that one or two of them will be a Higgs.

The Europeans decided to take the brute-force approach. There was an accelerator in Europe called CERN (the European Center for Nuclear Research), which they have just recently shut down. They're revamping the energy, the magnets, and the whole system, so they can go up to much higher energy. I believe their new energy, in 2007 if all goes well, will be 17 trillion electron-volts, compared to our one. They're going to go for high energies, which also increases, it turns out, the probability of producing the Higgs, and they will certainly be able to see them if they're there, if they're where we're expecting them to be.

The United States made a big effort to find the Higgs a long time ago. In the 1980s, a bunch of physicists got together. The standard model was in existence, and we knew almost all the pieces that we know today, except for the Higgs. So in the early 1980s it was proposed that we build a truly gigantic accelerator. It was going to be called the Superconducting Supercollider. It's a little ridiculous; there are a lot of supers. Physicists need some new adjectives, I think. The term "superconducting" was considered because, if you really want ultra-high-energy protons to go in a circle, and not fly away, you have to have a big magnet to make them curve around in a circle. This machine was of mammoth proportions. It was going to be something like, boy, 53 miles around; that is, fifty-three miles around this ring. Magnets were to be used; if they just used a conventional magnet, it would not have been strong enough. If they had used a very powerful electromagnet, it would not have been strong enough. They needed superconducting metal, which can withstand enormously high electric currents, to create the magnet. That's why it was superconducting, and supercollider because it was to be of extraordinarily high energy. We're talking 20 TeV in each particle beam that was going to smash into each other, so that means forty-trillion-electron-volt collisions. That would have been plenty. Had this gone as was planned at the time, it would already be running today. In fact, by the late 1990s this machine would have turned on.

It was proposed in the 1980s and went through a lot of reviews and then— first in the physics community and in the United States political community— Congress approved the project. The original estimates were eight billion dollars, and it was funded. This was big science, biggest particle-physics experiment since the Manhattan Project - which really doesn't belong in the field of particle-physics research; that was a U.S. military project. That eight-billion-dollar figure started climbing, and they started building the tunnels. They started designing and building the magnets. Everything was going, and every year, of course, Congress would revisit and engage in more debates. Finally, in the early 1990s, times were tougher, and the project was canned. It was pretty much, I think, a political decision.

In fact, it was completely a political decision, not a scientific decision. They had already spent, by that time, two billion dollars digging part of the tunnel and designing some of the system. There were many tragic aspects of this Superconducting Supercollider's demise. You know, was very sad, in a certain sense, for the physics community, that we lost this opportunity to find a new particle of nature; but, it's also tragic when the U.S. government essentially threw away two billion dollars. You can argue about whether it's worth the money to design such a facility; it's an interesting and important argument, and physicists were also arguing this just as much as the politicians were. Nobody wants to start a project and throw it away, so I think there were many lessons learned; and one of them is, if you're going to do a project like this, we really are going to have to make it an international collaboration. We're going to have to make sure that the funding gets spread out in some way that it's not going to disappear in the middle.

Just to set the stage, I guess it's worth looking at the financial scale of some other big science projects. We have a space station in orbit right now, and that project, I mean, it's hard to estimate costs of these things, but it's roughly a twenty-five-billion-dollar project. I think most scientists, even astronomers and physicists, agree that the science coming out of the space station is probably less obvious and less exciting then the physics that was aimed for in the Superconducting Supercollider. The motivations for the space station are somewhat different, I think. The Human Genome Project is a biology experiment that was cost three billion dollars, somewhat cheaper than the Superconducting Supercollidert, and it was a great success. Its interesting to compare these projects. Another useful number to look at is what fraction of the U.S. federal budget would the Superconducting Supercollider have been? The answer is one one-hundredth of one percent of the U.S. federal budget. So, on the one hand, it's a huge amount of money; on the other hand, it's small compared to all the other monies that we're spending. You can imagine the political and social debates about where we should be spending our money. Even within the physics community, some people are nervous that if you put too much money into one pot, it's going to get taken away from another, but others say, "No, it's really just federal funding that's adding to our scientific effort." In any case, I believe that the demise of the Superconducting Supercollider was symbolic of a shift in U.S. priorities regarding particle physics, compared to the rest of physics and the rest of science.

I think from World War II until the Superconducting Supercollider, particle physics was kind of at the top of the heap. They were getting the big money, the big facilities, and there was a certain sense in which it was viewed as the most fundamental, therefore, the most worthy of all the fields. I think that idea has changed now. Particle physics is alive, very much alive, even though this big project disappeared. In future lectures we're going to be talking about all the exciting stuff that's still going on in other accelerators and in other types of particle-physics experiments. But I think, now, particle physics is one player among many, among all the branches of physics, and indeed among all the branches of science. That's probably a healthy thing as well.

I'd like to finish with this question: "What if we had built this multi-billion-dollar machine but didn't find the Higgs? I mean, this is a mathematician effectively telling you that we expect this particle of nature to exist. It's quite compelling, but it's not proof. You have to see it in the laboratory. This is science. You could always imagine—in fact you could easily imagine—that the Higgs wouldn't show up the way we'd expect it. I think the argument at the time was that that would have been really interesting, too.

Okay, we have this extraordinarily successful model. You may ask yourself what would happen if you took the Higgs out of the model. The answer is that some aspects of the model survive, but the standard model begins to collapse in on itself. In particular, if you try to calculate what goes on at a couple of trillion electron-volts of energy, then the theory would start to give infinities. In other words, the theory is fine; still, at low energies. But, when you reach these Superconducting Supercollider kind of energies, or the European (I believe their acronym is LHC, Large Hadron Collider.), which is what they're building at CERN, because they're going to smash big hadrons – really big, strongly interacting particles - together. The theory would break down and give you some crazy answers, which would tell you that something is going on. It might be the Higgs. The Higgs is called a minimal theory. That is to say, it is the simplest possible mechanism we've been able to come up with, which is consistent with everything we know so far, and it makes the theory coherent. It's minimal. It doesn't mean it's true.

What I can almost guarantee you is that something has to happen at a trillion electron-volts that we don't know about today. It's either the Higgs, or something else. The symmetry of nature between electric and weak forces is going to merge, or break down, at those energies. We're not quite sure exactly how. We have a theory that makes a prediction, and we'd like to verify it; that's sort of the beauty of finding the Higgs, but if you find something else, well that's great, too. Then we go back to the drawing boards, and we reconstruct our standard model to fix it up in the Higgs sector. That would be the way we would express that.

So, the Higgs is the last piece of the standard model. It's he most important in the sense of understanding mass, the fundamental property of everything. Understanding symmetry breaking, is sort of the key idea, the unification of weak and electromagnetic forces. It's the one piece we've never really verified in the laboratory. We haven't seen one. We don't have any direct evidence yet. There was actually a hint at CERN just before they shut down for their upgrade. There were one or two events that were consistent, not gold plated, but they were consistent with being Higgs. There was a big hoop-de-doo at that laboratory, because the physicists said, "Gosh, let's keep running, let's keep running, and, you know, keep rolling those dice. If we have one or two, we ought to find some more." Then the laboratory director had a terrible decision to make: to continue running at this lower energy, with a low probability or to shut down for eight years, build the brand new machine, spend all that money, and be guaranteed of finding the thing and learning all about it. It was an agonizing decision, but in the end they decided to shut down to do the upgrade, and that's what they're doing.

Meanwhile we, in the United States, are continuing to look at Fermilab. So if they were correct—actually, that evidence has since been essentially retracted—there may have been other explanations for those events. It would be great if we could find evidence for the Higgs, with our existing facility; but most likely, we're going to have to go to a higher-energy machine in order to do it.

We really won't be able to say with confidence that the standard model is complete until we've either found the Higgs or found what else there is that plays the role of a Higgs. At that point, I think we will be able to argue that the standard model is truly a complete theory of nature. So this is, as you can understand, a pretty big deal in the world of particle physics. It's one of the reasons why so much money and effort is being spent. I could almost guarantee you that, in the next few years—certainly in the next dozen years—you will be seeing newspaper articles about either the discovery, which would be spectacular, or the non-discovery, which would be equally spectacular, of the Higgs particle, either in the United States or at this new facility in Europe.

Lecture Twenty
The Solar Neutrino Puzzle

Scope: Neutrinos are among the most mysterious and intriguing particles in the standard model. There remains a great deal that we do not know about them. We have always assumed that they are massless, but is that really the case? What would happen if they did have mass? We might also ask about solar neutrinos. Evidence suggests that far fewer neutrinos are coming from the sun than we would expect. Is there a way to make sense of that? Could the answer be something novel about the way neutrinos behave? In this lecture, we discuss physics deep in mine shafts, an alternative to accelerator physics. We also talk about neutrinos changing "flavor." What does that term mean, and how could we ever tell? We already have a small but growing body of evidence for "neutrino physics beyond the standard model." If we establish this science, why won't it "shatter" the standard model?

> You could argue that all the experiments are simply wrong, but this is highly unlikely.
> —Sudbury Neutrino Observatory Web page

Outline

I. The neutrino is an intriguing particle and still not fully understood. Just recently, some data have hinted at the possibility that the standard model may not be completely correct concerning neutrinos.

   A. Remember that the neutrino is a lepton. It has no mass, no charge, and no color. The only force of nature it feels is the weak force.

   B. The massless aspect of the neutrino is part of its mystery. If a particle is massless, it is incredibly easy to produce.

      1. The other massless particle with which we're familiar is the photon. A photon is a traveling electromagnetic energy bundle, but it has no rest mass. It is always moving at the speed of light.

      2. A neutrino is quite different. It is not a force carrier. It is itself one of the fundamental constituents. Yet, relativity tells us that if it is massless, it will always be moving at the speed of light.

3. We might ask: Why is it massless? The photon is massless because of gauge symmetry, but there is no such principle associated with the neutrino.

4. Suppose that we measured the mass of the neutrino and it turned out to be .00001 in some unit system, a tiny but non-zero number. Would that break the standard model? The answer is no. We would have to adjust the standard model in some places, but finding mass in a neutrino would not break the model. In fact, we now have some hints that the neutrino may have mass.

C. Another important aspect of neutrinos is that they seem to exist in three flavors: the electron flavor, the muon flavor, and the tau flavor.

1. We talk about electron number as a property of particles. An electron or an electron-neutrino both have electron number 1. Apparently, electron number is a conserved quantum number.

2. As particles move and interact, an electron could just bounce, preserving electron number, or it might undergo a weak interaction. A W boson might be exchanged, and the particles would transform. The electron will transform but only into an electron-type neutrino to conserve total electron number.

3. Again, until recently, it was believed that electron number, as well as muon number and tau number, were absolute exact conserved numbers. We have not discovered any symmetry of nature that would correspond to conserving electron number; it seems to be an accidental conservation.

II. Let's leave the world of neutrinos for a moment and talk about the sun. The two subjects seem disconnected, but they are closely tied together.

A. For many years, physicists and astronomers have been trying to understand what the sun is and how it works. These scientists have constructed what we might call a standard model of the sun.

1. This truly is a model, not a fundamental theory of nature. It is a collection of many branches of physics—heat, thermodynamics, light, sound, vibration, and materials—put together to describe the sun. This model is very successful and, to a certain extent, gives us a good understanding of the sun.

2. One aspect of that understanding involves the nuclear reactions at the core that power the sun. We know how many such nuclear reactions are occurring, and we know that they involve protons smashing together with other protons at very high temperature and energy to form nuclei.

3. When protons bind to protons, in order to stabilize, one of the protons must turn into a neutron; that is a weak decay that releases a neutrino. According to the laws of nuclear physics, if we know how many nuclear reactions are going on, we know exactly how many neutrinos are coming out of the sun.

4. The standard solar model makes a definite prediction about how many neutrinos flow out of the sun every second, which is 60 billion neutrinos per square centimeter per second everywhere in the orbit of earth.

B. In the 1960s, a physicist named Ray Davis thought that with a big enough detector, we should be able to see neutrinos from the sun.

1. Davis looked at nuclear reactions in which a neutrino hits a nucleus and makes a transformation occur. One such reaction was in a chlorine atom. If a chlorine atom is hit by a neutrino, the neutrino will convert into an electron because it's a weak interaction, and the chlorine nucleus will convert into an argon nucleus, which forms a radioactive gas and should be easy to detect.

2. Davis filled a railroad car with carbon tetrachloride, a cleaning fluid, then buried it in a mine. He buried the material, because otherwise, he would get too many other interactions from cosmic rays.

3. The idea was that a couple of times a day, one or two atoms in the railroad car would transform from chlorine into argon. The argon would bubble up to the top of the tanker and be collected, then the radioactivity would be detected by a Geiger counter.

4. Davis's experiment worked, but he collected only about a third of the neutrinos that the standard model predicted he would. This was the first direct evidence that neutrinos come from the sun, but the fact of the missing two-thirds of the neutrinos was bothersome.

C. Davis and others worked to solve the solar neutrino puzzle.

1. Davis placed a highly radioactive source next to his tank and collected the number of neutrinos he expected. This result points to the idea that there really is something special about the neutrinos coming from the sun.

2. In the 1980s, the United States, the Soviet Union, and the Europeans conducted similar experiments using tanks of gallium, which is liquid metal and transforms into germanium. These results showed that the number of neutrinos coming from the sun was less than predicted by a factor of two.

3. Japan entered the story with a giant water detector, a tank of very pure water, called Kamiokande, deep in a mine in Japan. This experiment found something extra.

   a. When a neutrino hits the nucleus or the electrons in the water tank, the transformation occurs. Some energy is released, and the outgoing particles carry the momentum of the neutrino. If the neutrino comes in from a certain direction, we see a spray of particles going out in that direction.

   b. Using light detectors, scientists at Kamiokande saw for the first time that the neutrinos really were coming directly from the sun. In fact, Kamiokande and its second-generation upgrade, Super Kamiokande, have produced images of the sun.

4. Despite all these efforts, we were still left with the solar neutrino puzzle: too few neutrinos arrive from the sun.

D. In working on the puzzle, scientists saw two obvious points at which our understanding might be incorrect.

1. First, the standard solar model may be wrong; the sun may produce fewer neutrinos than we thought.

2. The other problem could be with the standard model of particle physics. Suppose that an electron-flavored neutrino, which is produced in the nuclear reaction in the sun, travels the long path from the sun to the earth, and as it's moving along, imagine that it could oscillate in flavor from an electron-type neutrino to a muon-type neutrino.

3. The laws of quantum mechanics say that if neutrinos are not truly massless particles, then this can happen. In fact, we've

seen exactly the same thing happen before with quarks. (A down quark can oscillate into a strange quark.)

4. How do we visualize this phenomenon? Imagine that a number of neutrinos leave the sun; they are all electron types because that's what the nuclear reactions in the sun produce. As they travel, they oscillate and, by the time they reach earth, half might be electron-flavored and half, muon-flavored. Unfortunately, our detectors have been designed to look only for electron-type neutrinos.

E. What experiment would verify that the total number of neutrinos predicted is coming out of the sun, but they are changing flavor before they reach earth?

1. The Canadians built an underground neutrino detector called the Sudbury Neutrino Observatory (SNO). SNO is a water detector and can see transformations, the same as all the other detectors. SNO also has the possibility of measuring a weak neutral current.

2. Remember that the weak force can manifest itself in two ways according to the standard model. First, the weak force can be seen when a neutrino changes into an electron and, at the same time, transforms whatever it is hitting from one nucleus into another. That's the usual beta decay weak interaction.

3. The weak force can also be mediated by the exchange of Z bosons. This interaction is neutral; there is no transformation. A neutrino comes in, exchanges a virtual Z boson, and hits the nucleus, but no transformation occurs. In detecting the neutrino in this weak interaction, all that can be observed is a recoil of the nucleus.

4. SNO is capable of detecting any flavor of neutrino and is beginning to publish results. Scientists there are seeing the full solar neutrino flux when they look in this weak neutral sector. Thus, the sun is behaving just as we thought it should.

III. We seem to now have evidence that neutrinos can change flavor and, therefore, according to quantum mechanics, neutrinos are massive. What is the significance of this discovery?

A. The most important theoretical argument for the significance of this finding would be that it is a break in the standard model. Such a break would mean that our understanding of the world is incomplete, which is exciting. The discovery that the neutrino is not massless would also enhance the standard model by making it more symmetrical.

B. The practical aspect of this discovery lies in an astrophysical use that no one would have dreamed of even 20 years ago. We have now formed images of the core of the sun using neutrinos that other scientists may be able to use as tools for gathering information.

C. The third consequence of this discovery relates to cosmology, the study of the universe.

    1. If we consider that astronomically huge numbers of neutrinos exist in the galaxy and that they have mass, we must ask whether they will have a gravitational effect on the universe.

    2. As the universe is expanding, all these neutrinos are pulling it back, and they may produce enough gravity to help halt the expansion and cause the opposite of the Big Bang, the Big Crunch. The fate of the universe may rest in this exotic particle.

Essential Reading:

Barnett, Muhry, and Quinn, *The Charm of Strange Quarks*, pp. 143–145.

Kane, *The Particle Garden*, chapter 5, pp. 86–90.

't Hooft, *In Search of the Ultimate Building Blocks*, chapter 19.

Recommended Reading:

See John Bahcall's Web site, http://www.sns.ias.edu/~jnb/ (check the link on the left for "popular accounts").

The SNO collaboration's Web site, http://www.sno.phy.queensu.ca/, is a good place to find its latest news and summaries of other experimental efforts.

Questions to Consider:

1. Until SNO, all the solar neutrino experiments agreed that there was a *deficit* in the number of electron neutrinos from the sun. What did SNO do that was new and helped resolve whether the deficit comes from the *solar* physics (e.g., lower temperature in the sun's center than we thought) as opposed to *neutrino* physics (e.g., flavor oscillations)?

2. Most physicists have speculated that the electron neutrinos from the sun might oscillate into mu or tau neutrinos. Others have claimed that the electron neutrinos might oscillate into yet another kind of neutrino, a *sterile* type that would not interact in any way with ordinary detectors. What would be the *difference* in what SNO would see in these two scenarios?

Lecture Twenty—Transcript
The Solar Neutrino Puzzle

Despite the mystery and the importance of the Higgs particle to the standard model, my favorite is still the neutrino. Far and away, it's the most intriguing and appealing of all the fundamental particles to me. I'm not quite sure why; it's partly that it's such a mysterious little object, and even today, there are still lots of aspects of neutrinos that we don't really understand fully.

The standard model is a minimal theory. That is to say, we take what data we have, we write down the simplest possible model that's consistent with it, and then we go out and test and see how well we've done. So far it's worked very well, but just recently there have been some hints that perhaps we haven't quite got the story completely correct with the neutrinos. So today I want to talk about where we're at, and what we're learning about neutrinos.

Remember what these particles are. This is a particle which is fundamental, it's one of the leptons; it has no mass, no charge, no color, and no strong force. The only force of nature that it can feel, really, is the electroweak, truly the weak force of nature - not the electric part, just the weak part. This massless business is part of the mystery of the neutrino. What does it mean for a particle to be without mass, "massless?" It's incredibly easy to produce. The other massless particle we're most familiar with is the photon. If you take a photon, it's a traveling electromagnetic energy bundle, but it has no rest mass. You can't take a photon and make it sit still. It's always running at the speed of light. A neutrino is quite different from a photon. It is not a force carrier; it is itself one of the fundamental constituents, yet, what we're saying is you can't take a neutrino and hold it still. If it's massless, relativity says it's always going to travel at the speed of light.

Now, a deep question that you could ask is why is it massless? See, that's a little bit of a funny question. In the standard model, don't we just collect the masses and write them down? We fill out tables by looking at the data. So the answer is, it's massless because it appears in experiments to be massless, but sometimes, in the standard model, there is a reason. For example, there is a deep, theoretical reason why the photon is massless; namely, gauge symmetry. Gauge symmetry is a principle. It's symmetry of the universe. It has to do with the nature of electricity and magnetism; and if gauge symmetry is correct, then the photon must be massless. It works out from the mathematics. It shouldn't necessarily be obvious to the casual viewer, but it is a definite consequence, and there is no such principle associated with the neutrino. There is no fundamental or deep mathematical reason why it needs to be massless. We have just said that it is because, so far, there's no experimental evidence for sluggishness, or inertia, associated with the neutrino.

Suppose we made a very delicate little experiment and measured the mass of the neutrino, and it turned out not to be zero after all. Maybe it's .00001 in some unit system, a tiny but non-zero number. Would that break the standard model? The answer is definitely not. You could just scratch-out the zero in the table and replace it with .00001. Actually you'd have to do a little bit more than that. The standard model, as it's constructed, really makes the neutrino special. There are certain characteristics of the neutrino; for example, the fact that every neutrino in the universe, as it's running away from you, has a spin that points counterclockwise. It's a left-handed particle. That would also have to be modified. That statement about the world would get fixed up, if we discovered that the neutrino has mass.

People have been looking for many, many years, making very delicate measurements of the production and detection of neutrinos, trying to see if they have some mass. We have no direct evidence yet, so whatever it is, it's a very, very small number. But we do have some hints now that maybe it isn't zero.

There's another aspect of neutrinos that's very important in our present understanding of them. There seem to be three different kinds—we call them flavors—of neutrinos. It's just a kind of a whimsical name. There are the electron flavor, the muon flavor, and the tau flavor, that seem to be distinct particles. In fact, now we talk about electron number as a property of particles. An electron, or an electron neutrino, each has electron number one. So, electron number apparently is a conserved, quantum number in nature. As particles move and interact, an electron could just bounce, preserving electron number; or it might undergo a weak interaction, so a little W boson might be exchanged, and particles transform. The electron will transform, but only into an electron-type neutrino, in order to conserve total electron number.

Again, until recently it was believed that electron number—and separately, muon number, and separately, tau number—were absolute, exact, conserved numbers. You know, Emmy Noether, a mathematician in the early 1900s, said that when you have symmetry of nature, it's associated with a conservation law. We don't have any symmetry of nature, that we've discovered, that would correspond to conserving electron number, so, it's sort of an accidentally conserved thing. When that happens in nature you're sort of looking and asking if we can find an experiment where we see lepton number changing. Again, this is something we're hunting for, and I'm going to tell you about the way that we hunt for these things shortly.

Let me leave the world of neutrinos for a moment, and tell you about the sun. It seems disconnected, but we'll see there's a very close connection. For many, many years, physicists and astronomers have been trying to understand what the sun is, and how it works. It is obviously very important for an astronomer to know what the sun is; because, after all, all the other stars are suns as well, and this is the closest one. We can study it very easily, so people have constructed what you might call the standard model of the sun. I use the word "standard" now in completely the generic English sense; just as an ordinary, everyday model of the sun that people can accept and believe that describes how it behaves. How bright it is, the color, what it's made of and all of the things that you associate with the sun. This truly is a model. It's not a fundamental theory of nature. We just have a big ball of gas, and we're trying to describe it as accurately as we can. There are certain things about the sun that are very difficult to measure, such as how hot it is in the middle. We can't get in there with a thermometer, so we have to make deductions. Based on what you see on the outside, and based on the laws of physics and nuclear physics, as we know them, what should the temperature be in the middle?

So the standard model is a kind of a collection of all the branches of physics—heat, and thermodynamics, and light, and sound, and vibration, and materials—all put together to describe this object, and it's a very successful model. It can predict all of the features of the sun down to details such as sunspots and solar flares. To a certain extent, we have a good understanding of the sun.

One of the aspects of that understanding is nuclear reaction at the core, which powers it. We know how bright it is, so we know how many such nuclear reactions are occurring very, very precisely. We know what they are; it involves protons smashing together with other protons at very, very high temperature and energy, forming nuclei. It's a strong interaction of nature that binds protons and neutrons together. Here you've got protons binding to protons, and in order to stabilize, one of the protons has to turn into a neutron, effectively; and that is a transformation that's a weak decay, and so a neutrino is released. According to the laws of nuclear physics, if you know how many nuclear reactions are going on, you know exactly how many neutrinos are coming out of the sun. So the standard solar model makes a very definite prediction for how many neutrinos flow out of the sun every second, and—since we know how far away we are—how many are passing through us, our bodies, our detectors.

A long time ago, in the 1960s, a physicist named Ray Davis decided this is a big number. It is, to be exact, 60 billion neutrinos per square centimeter, per second, everywhere, and here at the orbit of planet Earth, almost 100 billion. That's a big number. Even though these neutrinos are very low energy—they're coming from low-energy nuclear reactions in the sun—and even though neutrinos are very wimpy particles, if we built a big enough detector, thought Ray Davis, we should be able to find evidence that there really are these neutrinos coming from the sun.

After all, any individual neutrino has a tiny probability of interaction, but you've got an awful lot of them. So you're buying a lot of lottery tickets, and if you could build a big detector, you would just be increasing the number of nuclei available for those little neutrinos to bump into. So what are you going to do? Ray Davis looks at all the various nuclear reactions where a neutrino comes in, hits a nucleus and makes a transformation occur. And his favorite was to take a chlorine atom, and if a neutrino hits it, the neutrino will convert into an electron—because it's a weak interaction—and the chlorine nucleus will convert into an argon nucleus. That's what was known from nuclear physics. Argon is a gas, but it's not only a gas. This argon that you produce is a radioactive gas, so it should be fairly easy to detect. A little Geiger counter would click if an argon atom was nearby and decayed. So what we need is a big tank filled with chlorine.

What he used was cleaning fluid. A very clever idea because it was carbon tetrachloride, so it had lots of chlorine atoms in it. It's also quite innocuous stuff, not a big biohazard. He took a railroad car, a tanker, and filled it with carbon tetrachloride then buried it deep in a mine, the Homestake Mine in South Dakota.

It was buried in a mine because we're looking for a very, very tiny signal. Only once or twice a day is such an event likely to happen, according to the laws of particle physics, which he knew about in the 1960s when he built this device. If we put this detector here in this room—remember there are cosmic rays raining down on us all the time, and cosmic rays can also make nuclear reactions happen, and at a much higher rate because the cosmic rays interact electrically as well as weakly, and strongly as well as weakly—we would just be completely swamped out with other stuff going on. It's as if you're looking for a really dim light, you're not going to go into a bright room and look for the light; you're going to go into the darkest room you can possibly find.

So they went deep down into the Homestake Mine in South Dakota and just started waiting. The idea was that every now and again, a couple of events a day, a single atom (I mean you have to realize the spectacular scale of this experiment.) or one or two atoms, in this railroad car is going to transform from chlorine into argon. Argon is a gas, so it will bubble up to the top; you have some collectors up at the top, and then you hold them in front of a Geiger counter for a while to see if a radioactive argon atom will appear. He would sweep through the tank with some other gas to collect the individual argons, and it worked. It was an amazing experiment, and indeed, every month he would collect a couple of dozen atoms.

Now, remember, Ray Davis knows what the prediction is for the total number of neutrinos; and he knows all of the nuclear and particle physics of the day, which is pretty much the same as what we know today, almost, so he could make a quantitative prediction. The standard solar model, combined with the standard model of particle physics, says you should get so many neutrino events per month. What he got was about a third of that.

At first people were just very proud of him. He did a great job, which was, in a certain sense, the first direct proof that neutrinos come from the sun. You might even argue it's the first direct proof that the sun is really undergoing nuclear reactions. This was very cool, good stuff. But after a while this factor of three started to bug some people. They wanted to know why they were seeing fewer neutrinos, by a big factor, coming from the sun than were predicted.

At first, it was known in the literature and began to sort of grow and be called the solar neutrino puzzle. So, where are these solar neutrinos? I think that even when I was a graduate student, I remember reading about the solar neutrino puzzle, and most people said, "Ah, come on. Maybe he's, well, missing a couple. I mean, these are atoms in a giant tank! Maybe he doesn't collect them all." But he was a great physicist—actually still is, he's continuing work on this experiment—and he did a lot of careful checks. For example, he would take a highly radioactive source, and put it right next to his tank. He had a source, and he knew how many neutrinos were coming out of it; and then he *did* collect what he expected. So there really did seem to be something special about the neutrinos from the sun; he really seemed to understand his tank.

This kept going. He was collecting more and more data and improving the accuracy. Finally other people began to get interested. There were several new experiments in the years after this. In the 1980s, the United States got together with the Soviets and built something kind of like it down deep in a mine, but instead of using chlorine we used gallium, which is a big tank of liquid metal. The gallium would transform; the neutrinos would turn into electrons, and the gallium would turn into germanium, then we could count and collect the germanium. There was a European experiment, also, that was done with the same idea, with gallium; both of those independent experiments agreed that the number of neutrinos coming from the sun is fewer, by a factor of two in their case, or three in the chlorine case, than what we expected. So the solar neutrino puzzle is just getting bigger. It's not going away.

The Japanese entered the story with a giant water detector, a giant tank of very, very pure water. It was called Kamiokanda, deep in a mine in Japan; and again, it was a neutrino coming in, whacking into the water and making a transformation. But this time you have something extra. When the neutrino whacked into the nucleus, or the electrons, in the water tank and the transformation occurred, some energy was dumped; and that energy carried the momentum of the neutrino. If the neutrino came in from a certain direction, you had a little spray of particles going out in that direction. If the tanks are lined with detectors—actually just light detectors, big photo-multiplier tubes is the name for a big light detector—it was possible, actually, to see which direction the neutrino was coming from. So, for the first time ever, not only did we have evidence that there were neutrinos, but there was direct evidence that they really were coming from the sun.

In fact, Kamiokanda, and its second-generation upgrade, Super Kamiokanda, which is even bigger—this is gigantic, okay? It is many, many stories tall. It's like a ten-story building filled with water, and they have produced images of the sun. The images were beautiful, beautiful images, because you actually seeing the neutrinos that were coming right from the center of the sun. Then, instead of seeing the sun in light, they were seeing the sun in neutrinos. So we started to do some astrophysics. But again, they saw fewer neutrino events than they had expected, by about a factor of two or so. So everybody agreed that there was a solar neutrino puzzle. By the way, this is a little footnote. There was a tragic accident at the Super Kamiokanda facility just a year or so ago. As they were filling it back up again with water after a routine cleaning, one of the photo-multiplier tubes imploded, and the shock wave traveled through the water, created a chain reaction, and they lost thousands of these photo-multiplier tubes. It costs millions of dollars of damage, and so this wonderful detector has been shut down. Hopefully it will be rebuilt because it was a great facility, doing all sorts of good neutrino physics.

How are we going to understand this puzzle? I think that by the time all of these experiments have been put together, and they all agree with one another, no longer will they be dismissed any one as being just experimental uncertainty. There are two obvious places where we might not understand the world correctly. Maybe we don't understand the sun; maybe the standard solar model is just wrong, and there are fewer neutrinos produced in the sun than we thought.

There are various reasons why you would believe the standard solar model; it's a good, robust model. And of course, those folks were paying careful attention to these experiments. As the years went by they refined their model and came up with new tests. In recent years they came up with a lovely test. If you understand the sun, including the detailed structure of temperature as a function of distance in the sun, you can predict the resulting vibrations of the surface of the sun. It's called "helioseismography," and it's like measuring seismic vibrations on planet Earth, but we're actually watching the sun vibrate. The pattern of vibrations is very sensitively dependent on the solar model. If you change the temperature in the middle of the sun a little bit, you change the pattern of vibrations a lot, and it's working very, very well. It really seems as though we understand the sun, and we certainly know the total number of neutrinos, because we know the total brightness of the sun. So I think the standard-solar-model folks defended themselves very well, and we're almost obligated to believe what they're telling us because the data is very robust.

So what else could be wrong? It could be particle physics. Here's the wild idea that people came up with, actually, fairly early on in the story. Suppose that what I said in the beginning of this lecture, that what the minimal standard model says is wrong. Suppose that an electron-flavored neutrino, which is produced in the nuclear reaction in the sun, is traveling the long travel path from sun to earth, and as it's moving along, imagine if it could oscillate in flavor, from an electron-type neutrino to a muon-type neutrino.

Now, why would this happen? Turns out the laws of quantum mechanics say that, if you have massive neutrinos—if they're not truly massless particles—this can happen. It's one of those weird quantum things that is really hard to understand, but it comes out of the math. In fact, we've seen exactly the same thing happen before with quarks. A down quark can oscillate in flavor, if you like, into a strange quark. We have observed this phenomena. It's a fascinating phenomena in the quark world, which we haven't had time to talk about. It works in the quark world because quarks have mass.

The neutrinos need to be massive for this story to explain what's going on, and if they're massive, it's allowed to happen. The rules of quantum mechanics allow an electron-flavored neutrino to turn, spontaneously, into a muon type. In fact, it'll turn into a muon type, or a tau type, and then back again. It's a kind of a cyclic oscillation.

How am I visualizing this? You have a bunch of neutrinos leaving the sun, and they're all electron types because that's what the nuclear reactions in the sun produce. It's beta decay; beta decays produce electrons, and therefore electron neutrinos. Now they might oscillate on their way, and by the time they get to Earth half of them might be electron-flavored, and half might be muon-flavored. Every detector that I've described so far was specifically designed to look for the transformation where an electron-type neutrino produced an electron, and some other transformation. If you had a muon-type neutrino coming in, it wouldn't do a darn thing in your detector, because it doesn't have enough energy to produce the muon that it needs to produce. Muons are very, very heavy particles. So muon neutrinos would just go cruising right through your detector and the electron neutrino are the ones you would detect. That might explain why you're seeing fewer than you expected, because you're only measuring one flavor.

Okay, it's as if you go to the store and buy a pint of vanilla ice cream. As you're walking home, quantum mechanics takes over, and some of your vanilla ice cream transforms into chocolate and some into strawberry, so by the time you get home you have Neapolitan—I think it's Neapolitan—and if you only like vanilla ice cream. You say, "Hey, I bought a pint, but now I only have a third of a pint because I don't want to eat that chocolate and strawberry." It's a slightly weak analogy, but that's the basic idea here. If that's true, how could we find out? What experiment can you think of that would verify that there really are the total number of neutrinos coming out of the sun, but they're simply changing flavor?

The Canadians came up with a great idea, and they built another one of these underground neutrino detectors. It's in a little town called Sudbury; it's the Sudbury Neutrino Observatory and has the wonderful Canadian acronym SNO. SNO is designed to make the following detection. First of all, it does the same thing as all the other detectors did, it can see transformations; it's a water detector. But they also have the possibility of measuring a weak neutral current. Remember that the weak force can manifest itself in two different ways, according to the standard model. The obvious way, the way that it's been seen since the 1930s, is where a neutrino changes, transforms, into an electron; and at the same time transforms whatever it's hitting from one nucleus into another. That's the usual beta decay, weak interaction.

But Weinberg and company discovered a new force of nature. It's the weak neutral force mediated by the exchange of Z bosons. It's neutral, so there's no transformation. A neutrino just comes in, exchanges a little virtual Z boson, and whacks the nucleus. No transformation happens, just neutrino in and neutrino out. So what do you see in the laboratory? You see a nucleus just happily sitting there. Then, all of a sudden, boom! It was knocked into, by something invisible. You can detect the presence of the neutrino by this weak interaction, and all you see is just a little recoiling nucleus, which will show up as a little light signal in your detector.

So they built this thing capable of detecting, and it doesn't care what flavor neutrino you have. An electron neutrino can bump into you, a muon, or a tau-type neutrino, it doesn't matter; they will bump equally as strong. It's a low-probability event, so you have to be patient. They've been running for over a year; and, indeed, they have begun to publish their results. They're seeing the full solar neutrino flux when they look in this weak neutral sector. They're seeing them all. So the sun is, after all, behaving just as we thought. There is the full, total number of neutrinos. And, it really does look as though some of them are transforming into other types because when they only look at the electron type, they're seeing the usual reduction; but when they look at them all, with this neutral current, they see everything.

So that's big news. Okay, we seem to now have evidence that neutrinos can change flavor and therefore, according to quantum mechanics, neutrinos are massive. So now, it's a big experimental effort. There are lots of new facilities being designed, coming online, to try to pin the story down. I'm sure that in the near and distant future, you will be reading newspaper articles about neutrinos, as we discover what kinds there are, and what's the likelihood that one will transform into the other, and what's their mass. It's fun stuff, and it's also pretty esoteric stuff. I mean, I know that a perfectly legitimate reaction to all this is to say, well, it's interesting, but do we really care? I mean, why does it matter if the neutrino has mass, or doesn't have mass? So let me sort of finish up here with a discussion of a number of reasons why you might care, ranging from the, kind of, theoretical to the more applied.

The most important theoretical reason that I can argue for why you care is we have this standard model of particle physics, a framework for understanding the world. It's been around for, basically, thirty-plus years now. It says the neutrino is massless, and, once an electron neutrino, always an electron neutrino. This would be the first time—in my professional career, and in many physicists' lifetimes nowadays—that we would have a breaking of the standard model. That's a pretty big deal. If you break the standard model, then it means that our understanding of the world is really not complete, and we expect that it's not complete. So that's, you know, that's exciting. It's a shift in our worldview. It's not a complete destruction of the standard model, by any means. The whole picture still holds together. After all, in the quark sector we had up and down quarks; they were both massive. Then, in the lepton sector we had the electron and the electron neutrino, and one of them was massive. Shoot, why isn't the other one massive? It almost seems as though it should be. It seems desirable that neutrinos should have mass. It's lovely and symmetrical and would make the standard model nicer. It was a slight artifice of the physicists of the 1960s to say, "Well, the neutrino is massless." They were trying to keep the theory as simple as they possibly could do, and by making it simple, they had to make the neutrino this oddball, the massless particle. So now, what we're suggesting is maybe the neutrino isn't such an oddball; it just happens to be quite light, and it took a while for us to figure that out. So there's this breaking-of-the-standard-model aspect that makes it interesting to physicists.

Then there's the practical aspect and it's not practical in the sense of American consumers, at least, not yet. I'm not talking about neutrino TV. I am talking about an astrophysical use of neutrinos that nobody had ever even dreamed of even 20 or 30 years ago. We have now formed images of the core of the sun using neutrinos. That's a wonderful tool. It's a tool for other scientists. Now, you can start to think of all the possibilities. We can measure the inside of the sun; there's no other way to get information directly from the inside of the sun. The only reason this works is, neutrinos are formed in the middle of the sun, and then they have a tiny probability of interacting, even though the sun is hot and dense. Most of them just make it right to our detectors without hitting anything along the way. In fact, most of them go right through our detectors; that's why we need such big neutrino detectors. Light is produced in the center of the sun, and it immediately bounces around and gets absorbed by other atoms. It takes a million years for light to reach from the center of the sun to the outside. And by then, it's undergone so many interactions that we're not seeing the middle of the sun; we're seeing the edge of the sun.

You could imagine looking at a supernova, and the core of the supernova— something that's unimaginable by regular astronomy—or the center of our galaxy. So it's a tool, neutrino astrophysics. This is just now beginning. If you want to do neutrino astrophysics, though, you really better understand neutrinos. You can't do neutrino astrophysics and say, "Well, we're not sure if it has mass, or if it oscillates, because you're going to be off in your predictions about the center of some galactic phenomenon by factors of two or three, unless we figure out what the story is with the neutrinos." So this is a motivation for the particle physicists to get their story straight and figure out what's going on.

The third consequence—kind of a very large-scale consequence is cosmological. Cosmology is the study of the universe as a whole. How did it begin, the Big Bang; how did it evolve to where it is today, and where is it headed in the future? It's the big question, in a certain sense, where is the universe headed?

Remember the sun is producing these, just gigantic numbers of neutrinos every second, and it always has been; they're all flying out. Most of them just continue to fly along their merry way in straight-line path essentially forever; and every star in the universe is doing that. So the universe is just filled with neutrinos. There are an astronomically huge number of neutrinos passing through us, and there are neutrinos other than the ones coming from the sun. There are neutrinos coming all the way back from the Big Bang itself.

Now imagine, for a moment, that these neutrinos have mass. Okay? That's a new idea that we're just beginning to wrestle with. If all of those neutrinos have a tiny mass; add up a whole lot of tiny numbers, and you get a big number. Think about the effects on, say, a galaxy. It there are a whole bunch of massive neutrinos floating around, they're going to gravitationally attract. All right? Mass creates gravity. As the universe is expanding—and we are observing that it's expanding—all these massive neutrinos are pulling it back again. So if they're massive enough—we don't know yet, but if they're massive enough—they may produce enough gravity to turn around that expansion, halt it; and then we might end up with a big crunch, the opposite of the Big Bang. So the fate of the universe may rest in the hands of this exotic little wisp of a particle.

Lecture Twenty-One
Back to the Future (1)—Experiments to Come

Scope: This lecture highlights some of the current issues and directions in experimental particle physics. Many physicists continue to look for small deviations in the standard model, but we are also looking farther afield, for physics beyond the expected. One of the big efforts today is a search for violations of matter/antimatter symmetry. Some such searches take place at "bottom factories," high-energy accelerator facilities redesigned to produce heavy quarks. Novel neutrino beams that travel vast distances through the earth are also under construction, as are new high-energy machines, including RHIC, designed to study the conditions of extreme density and temperature that existed a fraction of a second after the Big Bang. Plans are underway for next-generation accelerators. We'll close with the increasing role of non-accelerator physics, including the giant arrays of detectors built in the desert, tundra, and under the Antarctic ice sheet to look for ultra-high-energy cosmic rays.

> It is a capital mistake to theorize before one has data. Insensibly one begins to twist facts to suit theories instead of theories to suit facts.
> —Sherlock Holmes (*Scandal in Bohemia*)

Outline

I. In the last couple of lectures, we've talked about puzzles that physicists are investigating today, such as the puzzle of the Higgs boson or the mass of neutrinos. This lecture looks at some of the other directions that particle physics is taking by focusing on current programs led by experimentalists.

II. The first project we'll look at is called *CP violation*, which might be thought of as the difference between matter and antimatter.

    A. *CP violation* refers to a breaking of a symmetry that for years, many scientists believed should be present. This symmetry is between particles and anti-particles.

B. We've talked about how the weak interaction violates the symmetry of handedness. If you have a particle that's spinning in a left-handed sense and it interacts weakly, it will have a completely different interaction than a particle that is spinning in a right-handed sense. Only the weak interaction violates parity or mirror symmetry.

C. Even after the discovery of parity violation, physicists thought that matter/antimatter symmetry was an elegant symmetry of nature.

1. Imagine that you have some heavy particle that can decay, according to quantum mechanics, in various ways. If you're working with high energy, there might be various pathways or possibilities for how the particle might fall apart.

2. We would expect that if matter and antimatter are symmetrical, the way a particle would decay would match with the way its anti-partner would decay. Of course, the particle will decay into certain particles and the anti-particle might decay into the anti-particles, but the relative probabilities should be the same.

D. In 1964, Cronin and Fitch performed an experiment in which they looked at the decay of one of the heavy mesons now called a *kaon*.

1. The kaon is a strange particle. Microscopically, it consists of a strange quark and an anti-down quark.

2. Remember that a strange particle has charge $-1/3$, and a down quark has charge $-1/3$. An anti-down is $+1/3$. This kaon, then, is a neutral particle.

3. The kaon also has various decay possibilities. A kaon decay might result in an electron and a pion. The electron would be negative and the pion would be positive so that total electric charge is conserved. The decay must also result in some electron-type neutrino to cancel the electron number that is lost. We will see the electron and the pi+, or we might see an anti-electron and a pi –, the anti-particles.

4. We might also expect, given that we started with a neutral object, that these two results would be equally likely: matter and antimatter are symmetrical.

E. Cronin and Fitch discovered, however, a violation of symmetry. For every 1,000 times the kaon might decay in one path, it would decay 1,003 times in the other path. A great deal of data had to be collected before this tiny break in matter/antimatter symmetry was recognized.

F. How was this break built into the standard model?

   1. The answer is rather subtle and mathematical, but the general idea is as follows: The standard model includes an experimental number that must be measured, and this experiment did that measurement. You might think of this number that belongs in the standard model as telling us how much CP violation, or matter/antimatter discrepancy, there is.

   2. The number can range from 0 to 1. Zero would mean no breaking, perfect symmetry. One would mean a 100 percent breaking, absolutely no symmetry. In principle, the number could have been anything, but in practice, it's .003.

G. Physicists were amazed by this breaking of symmetry and the expansion it offered in our understanding of antimatter.

   1. We might use this knowledge to ask very basic questions, such as why the world is made of matter. Why isn't it made of antimatter or, at least, why isn't it half antimatter?

   2. If matter and antimatter are completely mirror images, then every time an event created a particle, it would be equally likely to create an anti-particle. Go back all the way to the Big Bang when all matter was originally produced from a hot bundle of energy. We should have equal numbers of electrons and anti-electrons, protons and anti-protons, and so on.

   3. Most of these would have been annihilated, and the universe today would basically just be energy. Yet the universe that we live in appears to have matter everywhere and almost no antimatter.

   4. How do we know that the sun isn't made of antimatter and we're made of matter and we're just lucky that we haven't yet touched the sun? The sun spews out particles constantly—mostly protons—in the solar wind. If they were anti-protons, there would be antimatter explosions all the time, but we don't see that. We know the sun is not made of antimatter.

   5. How about other stars in our galaxy? Not only our galaxy, but even the space between galaxies, is filled with a gas, mostly hydrogen, a little helium and lithium, and other light elements. Astronomers can see light being absorbed and emitted by this gas. It also appears to be matter because cosmic rays pass through it and we don't see matter/antimatter explosions.

6. We return to the original question: Why are we here? Apparently, there's something special about matter in relation to antimatter. This is really the motivation for studying fundamental physics. An understanding of the difference between particles and anti-particles can help explain the cosmic fact that we are here.

H. Scientists have made measurements in these kaon systems starting in the 1960s and ever since and found that there is an asymmetry in the universe between matter and antimatter. That finding goes in the right direction to explaining why there is a little bit more matter than antimatter, but it doesn't go far enough. The numbers don't work out.

I. If we work out how much matter and antimatter should be left over today from the Big Bang, based on the standard model, we cannot account for the amount of matter that exists so it's still something of a puzzle. Physicists are setting up accelerator experiments to look at modern versions of this kaon decay.

III. Let's look at some other experiments that are going on today that give a sense of the kinds of topics physicists are interested in.

    A. I've talked a lot about the solar neutrino puzzle and the desire of particle physicists to understand whether neutrinos have mass and whether they oscillate.

        1. The accelerator at Fermilab is used in experiments that are different from those involving big tanks buried in mines. In this research, protons are smashed into a target and the resulting spray of particles is examined. Many of these particles are pions and kaons and particle/anti-particle pairs.

        2. Next, magnets are set up to sweep most of the particles out of the way and leave behind a beam of, for example, positively charged pions. The energy of that beam can be controlled by tweaking the energy of the accelerator and the strength of the magnet.

        3. Pions decay, and one of the results of that decay is neutrinos. The pion beam creates a beam of neutrinos, as well as electrons, muons, and other particles. If you smash these particles into a target—a wall that is 200 feet thick of steel and concrete—they are absorbed, and they come to a halt. Nothing comes out the other side of the wall except a beam of neutrinos.

4. The direction and energy of this beam can also be controlled, as can the flavor of the neutrinos. Instead of waiting for low-energy neutrinos from the sun to come and hit our detectors, we are actually able to produce many neutrino events.

B. Another topic of interest in particle physics is the ability to build higher energy machines. This has been a quest since the origins of quantum mechanics, because the higher the energy that can be achieved, the finer the microscope that results.

1. Some scientists are beginning to do research on the "next linear collider." The idea is to accelerate electrons and anti-electrons or muons and anti-muons in a giant linear machine.

2. Brookhaven has also recently built a machine, the *relativistic heavy ion collider* (RHIC), used to study both nuclear and particle physics. The idea of RHIC is to accelerate large particles, such as lead nuclei, into each other. The result is something like a pancake filled with quarks and gluons. The prediction of the standard model is that we may actually form a new state of matter, a quark-gluon plasma with certain fairly simple properties.

C. We also see developments in physics that do not involve accelerators. Some researchers are returning to the physics of the 1920s and looking at cosmic rays.

1. Cosmic rays come from a variety of sources, not just the sun. In fact, we're not even sure where all the cosmic rays come from. They might be from supernovas; they might be from some very energetic, exotic phenomena at the center of galaxies.

2. We do know that cosmic rays are high-energy particles, mostly protons, that strike the earth's upper atmosphere, producing some kind of nuclear reaction. A proton strikes against a nucleus and produces some particles, which in turn, strike other nuclei in the atmosphere, which in turn, strike others, resulting in a shower of particles.

3. We've observed that the higher the energy you go to in looking for cosmic rays, the fewer events you get. If we wait long enough, however, we occasionally see a cosmic ray event of stupendous energy, along the lines of 10^{20} eV—billions of times more energy than we can produce.

4. Nature gives us this particle accelerator, but the bad news is that we have to wait for it and we don't know when or where it will occur. Detectors for these occurrences are built in out of the way places, such as in the desert or under the Antarctic ice sheet.

5. These ultra-high-energy cosmic rays are interesting both to cosmologists, who want to learn about the astrophysical phenomenon that produces them, and to particle physicists, who want to know what will happen in a particle physics event with 10^{20} eV of energy.

IV. We could go on about particle physics and astrophysics and experiments being performed today that boil down to looking for high-energy physics, trying to solve the puzzles in the standard model. Your job in the future will be to connect the exotic experiments and results you hear about with the elements of the standard model that we've constructed in this course.

Essential Reading:

Barnett, Muhry, and Quinn, *The Charm of Strange Quarks*, chapter 2.11.

Kane, *The Particle Garden*, Appendix C.

Recommended Reading:

Lederman, *The God Particle*, chapter 7, section on "Slightly broken symmetry."

Riordan, *The Hunting of the Quark*, chapter 16.

Wilczek and Devine, *Longing for the Harmonies*, chapter 27.

Questions to Consider:

1. If an astronomer looks at a distant star, is there any way she can tell whether the star is made of matter as opposed to *entirely* of antimatter? (If not, how could we conclude that the universe is mostly matter and very little antimatter?)

2. If a physicist is interested in learning about neutrino oscillations, what are the primary advantages of using an accelerator-generated neutrino beam, rather than just looking at the naturally produced (and enormously abundant) solar neutrinos?

Lecture Twenty-One—Transcript
Back to the Future (1)—Experiments to Come

In the last couple of lectures, we've been talking about puzzles and phenomena that people are investigating today concerning the standard model; for example, the puzzle of the Higgs boson, or trying to understand the mass of neutrinos. These are hot topics, and the physics community, the particle-physics community, is actively investigating, both in the laboratory and in theory calculations. It is by no means the end of the story. There are an awful lot of topics that are very hot right now, in the sense of either puzzles, or curiosities, or something that's connected with this world of particle physics that we're trying to understand. In today's lecture I want to try to just give you a sense for some of these other directions that particle physicists are heading in, and I'm going to focus this lecture on the experiments and the experimental-led programs. The theorists are kind of off now on their own, looking at more exotic theories, theories beyond or beneath the standard model. And in the next lecture we'll come back and talk about where the theorists are thinking about, and where the action is. There are a whole bunch of projects going on, and I can't possibly explain all of them. What I'd like to do is begin with one that's a pretty big research program. There isn't just one group, there are many, many groups at many different accelerator facilities studying this phenomena. I'll tell you about it, and then we'll go on and talk just a little bit about a wider number of other experiments of different types.

The experiment that I want to first tell you about, or the project, goes by a really dry name. It's called, CP violation. It's a shame because physicists are usually pretty savvy about coming up with amusing or interesting names for things, part of which, I suppose, is that it makes it better when you're trying to talk to the general public. People can get more interested in a gluon than they can in CP violation. I believe that CP violation could just be called the difference between matter and anti-matter, although that's not quite technically right, and physicists shy away because it's not exactly right. CP violation refers to a breaking of symmetry that many people, for many years, believed should be present. It's symmetry we haven't talked much about, but it comes down basically to the symmetry between particles and anti-particles.

We've talked about how the weak interaction violates the symmetry of handedness. If you have a particle that's spinning in a left-handed sense, and it interacts weakly, it will have a completely different interaction than a particle that's spinning in a right-handed sense, as a right-handed, threaded, screw is different from a left-hand threaded screw. One of them will fit in the hole and the other one won't. Only the weak interaction violates parity, or mirror, symmetry. For a long time, even after the discovery of parity violation, people thought that matter-antimatter symmetry was a beautiful, elegant symmetry of nature, and there were a variety of experiments done to test this. I think it's one of those things where you ask physicists and they say, "Oh sure, it's obvious that there's no difference between matter and anti-matter". Yet, it's so obvious, but if it's not tested you really should go and look, and see whether nature does the obvious thing. Nature often has big surprises up its sleeves for us.

Let me try to get a little bit more specific about what I'm saying. Imagine that you have some particle, one of the members of the zoo, some heavy particle that can decay, and it has various, different possible decays because it's quantum mechanics. So if you have lots of energy, there might be various, different pathways or possibilities for how you might fall apart. Now what you would imagine, if matter and anti-matter were completely symmetrical, is that they way a particle decays would match the way its anti-partner decays. Of course, the particle will decay into certain particles, and the anti-particle might decay into anti-particles, but the relative probabilities should be exactly the same. There should be a deep natural symmetry between matter and anti-matter. There was an experiment done in 1964, by Cronin and Fitch, in which one looked at the decay of one of those mesons, one of those heavy mesons, called the kaon. They received the Nobel Prize for this experiment.

The kaon is a strange particle. In fact, microscopically it consists of a strange quark and an anti-down quark inside of it. Now let's just think about that for a second. Strange particle has charge minus one-third; down has minus one-third, so anti-down is plus one-third. This is a neutral particle. The kaon has total charge zero, and it has various different decay possibilities. Let me just look them up, because you never remember these things. You might see an electron and a pion come out. The electron would be negative, and the pion would be positive, so that total electric charge is conserved. There must also be some electron-type neutrino coming out, to cancel the electron number that's running away, but you won't see that in the laboratory because it's a neutrino, so it has very low probability of interacting. You see the electron and the pi plus; or, you might see an anti-electron and a pi minus, the anti-particles. You might expect, since you started with a neutral object, that it would be equally likely to go to either. Matter and anti-matter should be symmetrical. You might expect it, as everybody did, but when they did the experiment they discovered there was a violation of symmetry. It was a little bit different. It wasn't like parity, where the symmetry was broken completely. In parity we saw 100 percent left-handed neutrinos and no right-handed neutrinos. It was as broken as could be. Here, it was more like, for every thousand times it chose to decay in one path, it would decay a thousand and three times in the other path. As you can imagine, it took a long time, and a lot of statistics, before they could convince themselves that there was really a difference. Because it's quantum mechanics, you're always tossing a coin. One thing might happen or another thing might happen. You have really have to collect a lot of data before you can see such a tiny breaking of matter-anti-matter symmetry. CP, the term, comes from C, which is charge reversal and P for parity. It's a technical word, and it really doesn't matter why we call it CP, but it amounts to something that's very closely related to matter-anti-matter symmetry.

You might ask, if this phenomenon, which was observed in 1964 already, was consistent with the standard model as we know it today. The answer is yes, it's a little bit of a cheat to say that because the standard model didn't exist yet in 1964, and as we built the standard model, we made sure that it was consistent with all the data known at the time. How do you get this matter-antimatter symmetry to be broken in the standard model? The answer is rather subtle and mathematical, so let me not try to go into the details; but the point is, there is a number, an experimental number, in the standard model—it's a parameter—and you simply have to go and measure it. This experiment did that measurement, so we didn't know in advance that the standard model didn't predict this phenomenon; we simply observed it, and we wrote it down and tabulated it. You might think of this number, that belongs in the standard model, which tells you how much CP violation, or matter-anti-matter discrepancy, symmetry breaking, there is. It's a number that could range from zero to one. Zero would mean no breaking, perfect symmetry. One would mean 100 percent breaking, absolutely no symmetry between matter and anti-matter. In principle it could have been anything. In practice it's .003, a really, really tiny number, but not quite zero. It's a funny story.

You know, most people don't really come into this business with any expectations. If I told you that there's matter-anti-matter symmetry, you would have accepted it; and if I said there's not matter-anti-matter symmetry, you would have accepted. To a physicist, though, matter and anti-matter—going all the way back to the 1920s—really seemed to be mirror images. Until the breaking of mirror symmetry, it surely seemed that matter and antimatter really should be identical. To the physics community it really was kind of a shock. Maybe not quite as much of a shock as the breaking of parity, but still one that made people shake their heads and say, "This is quite amazing." We use anti-matter all the time. I don't want you to think that this is completely esoteric stuff. We make anti-matter in the laboratory; we use it in medical technology. PET scans involve the use of interaction of anti-matter with human beings, and I can easily imagine a future where our technological society will make even more use of anti-matter, so I think it is useful to understand what anti-matter is and how it works. It may even become practically useful. There's a story that 150 years ago the prime minister of England was visiting a laboratory where they were studying electricity and asked—I believe it was Michael Faraday—"Of what use is this electricity stuff?" and Michael Faraday said, "I'm not sure, sir, but someday you may tax it." That was an interesting response, kind of a techno-response that seems more contemporary than from back in the 1800s. He was quite farsighted.

We want to understand CP violation. Let me also motivate it with an even deeper question. Suppose I ask you why we are here. I'm not a religion professor, so when I ask why are we here, I'm not asking some moral or ethical question or some metaphysical question. I mean literally, why are we here? Okay? Think about it for a second. What I'm proposing to you, throughout this course, is that we have a world made of matter, particles, but why isn't it made of anti-matter? Or at least, why isn't it fifty-fifty? If matter and anti-matter are complete mirror images, if there were a perfect symmetry in nature, then every time an event created a particle, it would be equally likely to create an anti-particle. Go back all the way to the Big Bang, when all the matter was originally produced from this hot bundle of energy. You should have equal numbers of electrons and anti-electrons, protons and anti-protons, hydrogen atoms and anti-hydrogen atoms. Most of them would have then annihilated with one another, and so the universe today would basically just be a bunch of energy, a bunch of photons running around. Maybe, here and there would be a few smatterings of atoms, and a few smatterings of anti-atoms that haven't yet met. It doesn't appear to be the universe that we live in. The universe that we live in looks as though there is matter everywhere, but there's almost no anti-matter. You really have to struggle to find a few rare bits of it in cosmic rays, or in our accelerators. It would be a fair question to ask how we know that. Maybe the sun is made of anti-matter, and we're made of matter, and we just are lucky that we haven't yet touched the sun. But we have touched the sun; we do it all the time, because the sun is spewing out particles. It's called the solar wind. It consists mostly of protons, some neutrons, and some electrons, and they're all flying out and they hit our atmosphere. If they were anti-protons, there would be little anti-matter explosions going on all the time, but we don't see that. So we know the sun is not made of anti-matter, and when you look at the star, it's glowing. You can't tell, *a priori*, whether it's matter that's glowing or anti-matter; because, if matter and anti-matter are symmetrical, they glow the same.

We have this direct physical evidence that the sun is made of matter. How about other stars in our galaxy? Well, it turns out that not only our galaxy is filled with a gas, even the space between galaxies is filled with a gas. It's mostly hydrogen, a little helium and lithium and other light elements. It's tenuous, but it spreads out throughout the entire observed universe; the astronomers can see light being absorbed and emitted by this gas. It also appears to be matter because cosmic rays pass through it, and we aren't seeing little matter-anti-matter explosions anywhere. There are no boundaries anywhere in the universe, so if some stars were matter and some stars were anti-matter, there should be places in between them where you see these little explosions. Matter-anti-matter explosions are very easy to detect. If an electron and an anti-electron annihilate, they produce radiation of a very definite energy. So we would know, and it surely looks as though the universe really is made of matter. So now I ask the question again, why are we here? Why isn't it fifty-fifty? Apparently, there's something special, or better, about matter than about anti-matter. I think this is really the motivation for studying the fundamental physics. Let's look in our laboratories and understand why, and what, is the difference between particles and anti-particles so that we can explain this cosmic fact that we are here.

People have made the measurements in these kaon systems, starting in the 1960s and working their way forward, and they had this very small number. Here is the puzzle. If you work it out, there is a difference between the decays of particle and the decays—and therefore, working backwards—the production of particles and anti-particles. So there is a difference; there is asymmetry in the universe between matter and anti-matter. That goes in the right direction toward explaining why there is a little bit more matter than anti-matter, but it doesn't go far enough. The numbers don't work out. If you do the calculation, you think you understand the basic physics of the

Big Bang. We think we understand the standard model, and we work out how much matter should there be today, and how much anti-matter should there be left over today from the Big Bang, based on just the standard model. What we would discover is, it should be more matter than antimatter, but not the overwhelming amount of matter that we have today. It's not a qualitative problem, but it is a quantitative problem with the standard model. People are busy setting up accelerator experiments to look at modern versions of this kaon decay. For example, instead of having a strange quark and an anti-up quark, how about if we use a bottom quark? Use bottom particles in order to study CP violation, matter-anti-matter violation, at higher energies, which begins to take us a little bit closer to the conditions of the Big Bang, so I can't tell you the answer; we don't know. We're working on it. There are lots of accelerators and lots of groups of physicists who are studying this. I imagine that when they make their discoveries, especially if it's something that disagrees with the standard model, you will hear about it. This will be one of those things that the popular press will surely pick up.

Let me leave CP violation and just spend some time telling you about a bunch of other experiments, not quite in so much detail, that are going on today, to give you a sense about the kinds of new stuff that folks are curious about and are thinking about. I've talked a lot about the solar-neutrino puzzle and the desire of particle physicists to understand whether neutrinos have mass and whether they oscillate. It's not the case that the only experiments that we've been doing involve building a big tank, of water or cleaning fluid, burying it in a mine, and sitting and waiting patiently. We have also decided to take a more proactive approach. If you have a particle accelerator, such as Fermilab, and you take that proton beam, normally you smash the protons and anti-protons together. You have a lot of energy, and you produce a whole bunch of particles. Typically, you analyze all those particles to learn some particle physics with them. Another thing you might do is just to take those protons and smack them into a wall. You're going to put a lot of energy into the wall; you're going to produce a whole spray of particles. Many of them will be pions, kaons, and particle/antiparticle pairs, and you could start setting up some magnets. With magnets, different particles follow different paths, so you can sweep most of the particles out of the way and leave behind one little beam of, let's say, positively charged pions. Now you have a beam of positively charged pions, and you can control the energy of that beam by tweaking the energy of your accelerator and of the magnet. Pions decay, and one of the things that they decay into is neutrinos.

So imagine that you run this pion beam for a while; it's creating a beam of neutrinos as well as other stuff, electrons, for example; muons. Then you put another wall in front of them and this time it's thick - maybe 100 or 200 feet of steel and concrete. The particles smash into the wall become absorbed and dump their energy. They heat the wall up just a little bit, and they come to a halt, you still have another couple hundred feet of concrete to go. What's going to come out the back? Absolutely nothing; this is the biggest, most secure radiation-prevention system that you could imagine, except the neutrinos will make it through. Now you have a beam of neutrinos, and you can control the direction of the beam of neutrinos by controlling the direction of your original pion beam. You can aim those neutrinos at something, and you can change their energy. You can even, if you're clever, decide whether you want a beam that's all electron-flavored neutrinos or all muon-flavored neutrinos, by controlling the original particles that you started with. So this is great stuff. Instead of just kind of waiting for these low-energy neutrinos from the sun to come and hit our detector—first of all, these are very high-energy neutrinos, and the higher the energy, the stronger the weak force of nature appears; it interacts with more probability—these neutrino detectors are now getting many, many, many events. So, it's actually getting easier to detect neutrinos in this way because they have higher energy and, therefore, higher probability.

This program is just starting; we have some preliminary data. There are a whole bunch of accelerators around the world that are trying to do this. If you want to learn about the properties of neutrinos, if you want to learn whether they change from one flavor to another, you have this very beautifully controlled experiment now. You can start to figure out exactly how it's all working, and what the energy dependence is, and all that good stuff. In fact, there's an experiment that they're working on, where they take that beam of neutrinos and they just beam it right down into the earth. There's a beam at the end of Fermilab, in Chicago, that's just pointing down into the earth at an angle, and so everything gets absorbed except the neutrinos, which are now traveling into the earth at an angle. They're cutting a chord through planet Earth, and they're aimed carefully so that they come out in Minnesota where they've set up a big detector. You have got this huge distance, right? One of the things we would like to mock up is the huge distance between the sun and the earth, to explain the solar-neutrino puzzle. That's a really long distance. If we can get that one to work, we might even be able to beam them all the way through to Japan. It's kind of wild stuff; and this is now physics of the future.

Another possibility, of course, in the world of particle physics which people are very interested in is the ability to build higher energy machines. It's always been a quest, ever since the first origins of quantum mechanics. The higher the energy you go to, the finer microscope you have. You're learning about new particles; you're learning about the substructure of the existing particles. People always want to go to higher energies. In the United States, Fermilab is the biggest, highest-energy machine we have, and there are no immediate plans for a higher-energy-still machine. We are starting to do research on what some people are calling the "next linear collider." That would be the next-generation experiment or facility. It's—nobody's working on the Superconducting Supercollider anymore, because Congress definitely ended that one. So one of the key ingredients is not to make it quite so expensive. The next linear collider; the word linear refers to a straight line instead of a big circle, such as Stanford. Stanford was such a successful machine, the idea here was to accelerate either electrons or anti-electrons in some giant linear machine, or maybe even muons and anti-muons, which would give you even more wallop. This would be a great way to produce ultra-high energies in a very small region of space.

This is a very international collaboration; I think there's a lot more political savvy, now, about what it takes to get a big, high-energy facility constructed all the way through to completion; we will see. I'm sure that as this makes progress beyond the research and development stage, and people are actually seeking funding in the hundreds-of-millions to billions-of-dollars range in Congress, there will probably be articles, and you'll have to decide if you want your local senator to vote for or against this machine. So it's nice to have some sense of what's coming up.

We have just recently built a new machine. It's not quite a particle physics machine; it's kind of a machine that studies both nuclear and particle physics, so it's not trying to go to ultra-high-energies, exactly. It's called RHIC, the Relativistic Heavy Ion Collider. It's in Brookhaven, Long Island, in the existing Brookhaven National Lab facility. The idea of RHIC was, instead of accelerating an electron, or a proton, in a big accelerator to take the biggest thing possible, such as the lead nucleus then accelerate two lead nuclei and smash them one into another. It's an extraordinarily high-energy machine, but the real big deal is not to go to higher energies, but to go to high density. It's another regime in physics that we really haven't explored. It's walking away from simplicity and heading more in the direction of complexity. Lots of physicists are interested in what happens when complicated systems interact and how they behave. In fact, the purpose of most of physics, and most of science is to understand complicated things. One of the lucky and beautiful things about particle physics, in general, is that we get to study the absolutely simplest things. We can all be simpletons about our science. At RHIC, they're going to take, basically, a highly relativistic lead nucleus, and another highly relativistic lead nucleus, or gold on gold; a variety of things can be accelerated then smack them together; you'll get a little pancake. It will be a pancake filled with quarks and gluons. The prediction of the standard model is that we may actually form a new state of matter. You have solid, liquid, and gas. This is a quark-gluon plasma, which would have properties that the standard model says are actually fairly simple. When you have that much stuff that close together—remember, the strong force gets weaker when particles are close together—it should actually be almost like a free-particle gas. We think that we have some ideas of what to expect, but there will surely be many surprises. That's another direction of high-energy physics in the near future. They're just online now. They've been collecting data for a year or so, and I expect to see news coming out of these facilities.

There's also physics going on that you might call non-accelerator. The accelerator is a wonderful laboratory, a tool for studying high-energy physics. We migrated towards the accelerator starting in the 1930s, all the way up through today, because of the control and the high energy that it gives us. It's like a giant magnifying glass that magnifies; the higher the energy, the finer the magnification, but you can also just go back to the old physics of the 1920s, and look at cosmic rays. In a certain sense, that's what those neutrino folks are doing. They build a detector deep in a mine, and they just wait for the solar neutrinos to come through them.

People are also thinking about just good old cosmic rays, the stuff that rains down on us all the time. Cosmic rays come from a variety of sources, not just from the sun. In fact, we're not even sure where all the cosmic rays are coming from. It might be from supernovas; it might be from some very energetic, exotic phenomena at the center of galaxies. What we do know is that there are high-energy particles, mostly protons, which strike the earth's upper atmosphere, and when they hit the upper atmosphere, there's some kind of nuclear reaction; it's called a shower. The proton strikes against a nucleus and produces some particles, which in turn strike other nuclei up in the atmosphere. They, in turn strike, others and others, and you get a shower of many, many particles. The higher energy you start with, the more particles you will end up with here on planet Earth. That's the origin of the cosmic rays that were discovered and understood way back in the 1930s.

In those days, they would send up an emulsion in a balloon. They would go up to a certain height and see some small event. Typically the energies associated with those events were rather small: MeVs, millions of electron-volts, small by modern standards of accelerators. But cosmic rays are random, and we've been looking ever since. We haven't completely abandoned cosmic-ray research, and what we've observed are a whole bunch of cosmic rays at low energy. There are some cosmic rays at higher energy, and a few at even higher energy. The higher energy you have, the fewer events result. We haven't found a cutoff yet, so if you're patient, and you wait long enough, every now and again you can see a cosmic-ray event of stupendous energy. We're talking you know, 10^{20} electron-volts; billions upon billions of times greater energy than the energy that we can produce at Fermilab, or even with the Superconducting Supercollider. It's as if nature had produced this particle accelerator for us, and that's the good news; it's free, but you have to wait; and that's the bad news. At Fermilab you get millions per second, and in the cosmic accelerator you may have to wait for a year, and then you don't know where it's going to be. Where are you going to build your detector? The trick is, just to pick some out-of-the-way place, where land is cheap. You go out to the desert, or the tundra, or under the ocean, or under the Antarctic ice sheet, and you build a big detector, and you just wait.

This has been going on for some time and there are lots of new projects being constructed. Frequently these projects are collaborative with Third World countries; sometimes that's where the facilities are actually being built. It's a nice opportunity for a country that doesn't have the financial resources to build a giant accelerator or detector, to nonetheless develop really contemporary high-energy physics research. These ultra-high-energy cosmic rays are interesting both to the cosmologists, who want to know where the heck these things are coming from—"What kind of astrophysical phenomenon is producing these high-energy rays?"—and to the particle physicists who say, "Hey, what's going to happen if you have a particle physics event with 10^{20} electron-volts of energy? One of the experiments being designed is under the Antarctic ice sheet. You just basically put detectors down there, and it's as though the ice sheet itself were a giant bubble chamber. If the particles strike the ice sheet and they produce a shower, they will produce a little bit of light; and you can spread this thing out over hundreds of square miles. It turns out that beneath the snow pack, the ice itself is very optically transparent, so once you create a photon, it will travel a long distance. That's really what you want. If you want to see a high-energy event, it's going to spread out over very large distances. So, kind of wild stuff is going on there, too.

I could just keep going. I could go for entire lectures about all the particle physics, and astrophysics, and connection experiments that are going on today, that boil down to looking for high-energy physics, trying to pin down the details and the puzzles that still exist in the standard model. I wish I could tell you about all of them. I think what I have to leave you with is the sense that, when you go out, you will certainly see these things and it may be something I've never mentioned. Something such as a magnetic monopole search, or a dark-matter search, or a search for gravity waves or a search for large extra dimensions. These are exotic names. I guess what I would like for you to walk away from this course with is the understanding that we do have this framework, the standard model, that you can hang this exotic new physics on. Whatever we find is going to fit into this picture that we've developed, most likely. It's a very robust picture, the standard model. It's extremely successful, quantitatively. And so, no matter what kind of cool, exotic, new stuff we find, your job, then, is to hook it in and say, "Ah yes, this connects with the Higgs, or the neutrino mass, or the zoo of fundamental particles. I think that's part of the fun because there are going to be all sorts of new experiments coming out that I can't even dream of today.

Lecture Twenty-Two
Back to the Future (2)—Puzzles and Progress

Scope: If the standard model is such a great success, why are many physicists looking beyond it for some deeper and more fundamental theory of nature? Their reasons fall into the categories of practicality, aesthetics, and mathematics. None of these is, by itself, convincing proof that such a deeper theory exists, but taken together, they make it hard to believe that we're at the end of the line. Where are we headed next? We must begin with the missing link of gravity. We'll talk about issues of simplicity, unification, and grand unification. Then, we'll look at two conceptual developments that, to many physicists, seem to be the best candidates for new physics. We'll look at both supersymmetry and the new and highly promising theory of strings.

> I know that this defies the law of gravity, but, you see, I never studied law.
>
> —Bugs Bunny

Outline

I. In the last lecture, we talked about the directions that particle physics is taking in the world of experiments. In this lecture, we'll see that theorists are also looking to go beyond or beneath the standard model. We have several theoretical hints that the standard model, beautiful and successful as it is, is not the final story. In fact, for a number of reasons, it can't be.

    A. You may have noticed that I have left one topic—gravity—out of this story almost throughout the course. Gravity is a fundamental force of nature and the one that is most obvious to us.

        1. Newton's description of the force of gravity was fairly simple and, over the years, quite successful. In 1915, Newton's theory was deepened by Einstein with the general theory of relativity.

        2. Newton's theory left open the question of why the force of gravity is present between two massive objects, but Einstein gave us a geometrical picture of space-time that allowed us to understand this attraction.

B. You might think that we could merge these two theories—the standard model of particle physics, which explains all the forces of nature except gravity, and the general theory of relativity, which includes gravity. The result would be a complete description of the world.

C. The problem, though, is that Einstein's general theory of relativity is inconsistent with quantum mechanics. It is not a quantum field theory; it's a classical theory of nature.

1. Gravity does obey relativity, which is necessary for any correct theory of nature as far as we know today, but it doesn't, for example, satisfy the Heisenberg uncertainty principle. There is some mathematical contradiction between general relativity and quantum physics.

2. Gravity hasn't played a significant role in our study, because it affects only big objects. In fact, despite the attraction that we feel on earth, gravity is the weakest force. The attraction of individual atoms is miniscule, but it adds up so that we feel the force.

3. Think of a hydrogen atom. It has a proton and an orbiting electron. The electron is held in orbit by an electrical attraction. Electricity is not a particularly strong force of nature, but if we compare the electrical attraction of the electron to the proton with the gravitational attraction, the ratio is 10^{36}.

4. That tiny effect is almost unnoticeable, and it's also why physicists have not really addressed the idea that gravity is inconsistent with quantum mechanics.

5. Some particle physics theorists, however, have begun to wonder if a fundamental theory of nature can be found that would be consistent with both quantum mechanics and general relativity.

6. The study of black holes, for example, requires such a connection. Black holes are particle-like, involving quantum mechanics, but they're also heavy, which brings in gravity. The same is true for the study of cosmology. Once again, the distance scales are incredibly small, but the energy density is very high.

II. Let's turn to another topic, called *supersymmetry*, that theorists are exploring to try to come up with a deeper theory of nature.

    A. We have found and worked out all the possible geometrical symmetries that are available in space-time, such as translational symmetry, parity, and matter/antimatter symmetry. They are all either realized or broken in some way, but they're all relevant in the theory of gravity or particle physics, except for one, supersymmetry.

    B. We can think of supersymmetry as analogous to matter/antimatter symmetry, that is, as a "pretty good" symmetry of nature.

        1. Imagine a symmetry between fermions and bosons. Remember the difference between these two: Fermions— electrons, protons, and quarks—are like particles; they don't sit on top of each other. Bosons, such as photons and the other force carriers, will sit on top of each other.

        2. What if these two kinds of particles were completely symmetrical? What if there was some intimate connection between them, just as there is an intimate connection between protons and anti-protons?

        3. This is the question of supersymmetry: Does every particle have a partner that has a different spin but is otherwise almost the same? We know that a top quark has an anti-partner, the anti-top quark, but does it have a supersymmetric partner? We currently have no evidence for super partners for any of the existing particles.

        4. The names for these partner particles, if they're discovered, will be whimsical. For the fermions, the partner of a quark will be a *squark* and the partner of a top quark will be a *stop squark*. For the bosons, the partner of a photon will be a *photino*, and so on.

    C. Theorists have begun working out the mathematical consequences of supersymmetry and made some interesting discoveries.

        1. They've discovered, for example, that if supersymmetry exists, we can add it to the standard model. In fact, the model tells us a little bit about the behavior and properties of these particles.

        2. For example, we would expect to produce these particles at an energy of around a trillion eV. Our next generation of particle accelerators at Fermilab and the large hadron collider in

Europe should be able to produce supersymmetric particles if they exist.

III. Probably the most exciting explorations in the world of theoretical particle physics are in the complex field of *string theory*.

A. Let me try to give you an idea about string theory by asking: What is the single most fundamental idea of particle physics? Your answer might be that the world is made of particles.

B. That's what particle physics is based on. QCD and the electroweak theory postulate a world filled with point-like objects. We can't visualize a point because it is a mathematical entity, but we can attribute certain properties to it, such as charge, mass, and spin. We can then make calculations about these particles according to the rules of quantum field theory.

C. String theory asserts that perhaps the world is not made of points. Maybe the fundamental constituents, the building blocks, are actually little lines or strings. A constituent might be a little line with two ends, or a closed loop, like a doughnut.

D. The most significant aspect of this theory is that it would allow the whole universe to be understood in terms of just one string. In string theory, we don't have many different kinds of fundamental particles. There is only the string, and all objects are made of string. The theory is as unified as it could possibly be.

   1. An up quark might be thought of as a string that vibrates in a particular pattern. A different, more energetic pattern might be a charm quark.

   2. The forces, too, would be understood to come from the string.

E. String theory also addresses the problem of renormalizability in the standard model.

   1. Remember that some calculations in particle physics result in mathematical infinities that must be renormalized. In string theory, there are no such infinities.

   2. The infinities in the standard model spring from the assumed existence of physically point-like particles. Points can get arbitrarily close to each other, and the closer they get, the stronger the fields become. If the particles are infinitely close, the calculations can go all the way to the limit of infinite fields. Strings, however, spread out over some distance, which can never get any smaller than the size of the string itself.

F. What is a string made of? String theorists postulate the existence of a string, but it's not made of anything. It is not, for example, made of "string atoms," which would be fundamental particles.

G. How do strings interact?

    1. In the current standard model, we can imagine two particles interacting as they move along and creating, spontaneously, a virtual photon that travels between them.

    2. In string theory, two strings move along and, at a certain point, they come together. Perhaps they sort of fold together to form one larger string for a brief moment, then that one string might split apart again into two strings. That might be the interaction of two fundamental particles.

    3. In our contemporary language, the intermediate state describes particles and virtual particles, but in string theory, there was never anything but strings interacting. Further, there are never any infinities because the size of the string is limiting.

H. What is the size of a string? String theorists can actually put a number on it, a number that comes from another idea that is bothersome in particle physics.

    1. Remember that the electric and weak forces are unified at a high energy scale—trillions of eV. But also recall that we have not yet determined whether the strong force is unified with electricity and the weak force. Right now, we assert that the forces are different, but might they, in fact, merge together at some ultra-high energy?

    2. Once scientists thought about this *grand unification*, they also wondered about the inclusion of gravity. We can even estimate the energy at which all these forces might come together, which is about 10^{19} GeV.

    3. Quantum mechanics says that an ultra-high-energy scale corresponds to a tiny distance, and that's the size of the string.

IV. String theory offers the possibility of an ultimate unified theory in which all physical observables can be computed from a mathematical framework.

    A. In working with the mathematics, we have found that supersymmetry is a consequence of string theory. The theorists are, in a sense, making a prediction about the existence of supersymmetric particles, which begins to place the theory in the realm of physics.

B. String theory also predicts the theory of general relativity as a low-energy effective theory. That is to say that the mathematics of strings predicts Einstein's general theory of relativity.

C. Finally, string theory has some strange aspects. For example, in order to make the theory mathematically consistent, it has to be true that strings live in more than three spatial dimensions. That would mean that other dimensions exist that we haven't noticed yet.

D. String theory has the potential to be the greatest revolution in physics since the discovery of relativity and quantum mechanics. It is not yet fully developed, but it is definitely a topic to take note of.

Essential Reading:

Barnett, Muhry, and Quinn, *The Charm of Strange Quarks*, chapters 9.3–9.5.

Kane, *The Particle Garden*, chapters 9–11.

't Hooft, *In Search of the Ultimate Building Blocks*, from chapter 22 on.

Weinberg, *Dreams of a Final Theory*, chapter IX.

Recommended Reading:

Calle, *Superstrings and Other Things*, chapter 25, sections on "GUTS" and "Superstrings."

Greene, *The Elegant Universe*, Part III. (A great book—take your time and read the whole thing!)

Krauss, *Fear of Physics*, chapter 6.

Lederman, *The God Particle*, first half of chapter 9, through the section on superstrings.

Wilczek and Devine, *Longing for the Harmonies*, Tenth Theme.

Questions to Consider:

1. Given the successes of the standard model and the arguments that something may exist "beyond" the standard model, what are your feelings about looking for such new and exotic physics, requiring

significant commitments of taxpayer-supported funding and physicists' time?

2. Look up the size (length) of a superstring. How many times smaller than a proton is that? Can you make up some kind of analogy in the form "String size is to a proton as proton size is to _____"?

3. What would you call the supersymmetric partner of each of the quarks? (Remember the top's partner is the *stop*) Can you pronounce them all?

Lecture Twenty-Two—Transcript
Back to the Future (2)—Puzzles and Progress

In the last lecture, we talked about the direction in which particle physics is going in the world of experiments, and about the exotic experiments that are trying to probe the puzzles and corners of the standard model. Theorists are also very busy, and the big question here is, what's beyond, or beneath, the standard model? There are lots of hints—no direct evidence whatsoever, but lots of theoretical hints—that perhaps the standard model, beautiful and successful as it is, is not the final story. It's not the last chapter in our understanding of the world. So today I want to talk about some of these, both the motivations and the wild ideas, the speculative ideas of the particle physicists today.

There is one, perhaps the most, compelling reason for me to argue to you that the standard model not only is not the final story, but it can't be. Okay. It cannot be the ultimate theory of nature, and, you may have actually noticed that there's one thing I've left off this whole story, almost throughout the entire course, and that is gravity. Where's gravity in this story? It's a force of nature that's very fundamental; we live with it every day. In fact, of all the forces of nature, everything I've been talking about, the one that's the most obvious to all of us, the one that we live with, is gravity. I haven't said word one about it. I use it as a metaphor, when I talk about the old physics of planetary systems, and Isaac Newton. Well, Isaac Newton did figure out gravity to a certain extent. He wrote down a formula, which explained it—I shouldn't say explained—described how gravity works, in a fairly simple way. Over the years it was actually very successful. It was successful in explaining planetary motion, and it was successful enough for us to get people to land on the moon, and for probes to land on Mars and Venus. The theory of gravity developed by Newton is a great and successful theory; and in 1915, it was modified, deepened, incredibly and radically, by Albert Einstein. This post-special theory of relativity is called the General Theory of Relativity, so Albert Einstein, you could argue, fixed up the theory of gravity. Not that there was really anything wrong with it, until you pushed the theory of gravity off to some sort of extreme, as near a black hole or in the early universe.

Albert Einstein's theory didn't just push the extremes; it also went a level deeper. You could always ask the question, why? Why is the force of gravity present between two massive objects? Isaac Newton doesn't tell you why; he just tells you how it is. He describes the world. Albert Einstein went much further in explaining why. He gave us a geometrical picture of space-time, in which we could really understand why massive objects attract one another. They warp the geometry of space-time then follow the path of least resistance, in a certain sense. You can always ask, but why? You can always think that there might be another layer deeper. Nobody found anything beyond Albert Einstein's General Theory of Relativity, at least until some of the modern speculations that I'm going to discuss.

So, what's the situation here? We've got the standard model of particle physics I've been telling you all about, on the one hand, which explains all forces of nature, except gravity; then, on the other hand, there's Albert Einstein's General Theory of Relativity, which explains gravity. Now, you might say, "Well, that's fine. We have these two theories, and when we put them together and we call that the standard model. Now we've described everything because after all, the standard model tells you what all the particles are. Einstein doesn't change that; there are still quarks and leptons. We have a theory of all the other forces; we have a theory of gravity, and everything is successful. The theory of gravity describes our solar system; Isaac Newton's laws are the correct approximation for Albert Einstein's laws of gravity in the regime of ordinary masses. You can also test general relativity by going to some super extremes, or by going to very, very high precision. You discover that Newton's laws just begin to break down in the nth decimal place, then you need general relativity.

There's a deep problem still. You can't just tack the two theories together and say, "That's our model of the world because Albert Einstein's General Theory of Relativity is inconsistent with quantum mechanics." It is not a quantum field theory; it's a classical theory of nature. It does obey relativity, which is important and necessary for any correct theory of nature, as far as we know today; but it doesn't, for example, satisfy the Heisenberg Uncertainty Principle, which is experimentally verified. There's something contradictory, mathematically contradictory, between general relativity and quantum physics, so it cannot be the case that we completely understand the world.

So far it hasn't mattered a hoot, because when do you worry about gravity? You worry about gravity when you're looking at big things. I mean, the theory of gravity describes the weakest force of all. Gravity is truly a weak, weak, weak force. You're probably thinking, I don't understand: gravity's a pain in the neck. Right? Gravity is the strongest force I can think of, and the only reason you think that is because you just happen to be born right next to a really big planet. Okay? The only reason you think gravity is strong, is because, just by coincidence, we all live next to this gigantic ball of rock. Every atom in the planet Earth attracts us. And, that attraction is really miniscule. It is almost immeasurably miniscule. It's only because there are so many of them, and they all add up, that we've ever even noticed gravity. If we had been born and raised on some starship floating around in outer space, we wouldn't, maybe, even know about gravity unless we'd started looking at planets going in orbit. Certainly, in particle physics experiments, there is no way we would have ever noticed gravity.

Think of a hydrogen atom, a proton and an electron, and the electron's going around in orbit. Why? Why doesn't it fly away? It doesn't fly away because of an electrical attraction. Now, electricity is not a particularly strong force of nature; the strong nuclear force is much stronger. But, compare the electrical attraction of the electron to the proton with the gravitational attraction, and the difference is 10^{36}. Now, I can't even think of that number. It's a billion, billion, billion, billion times less gravitational force than electrical force. There's such a tiny effect that nobody would ever even think of worrying about it, or noticing it. There's not even a conceivable particle-physics experiment, to date, that comes anywhere close to caring about gravity. This is why physicists have not really worried about the fact that gravity is inconsistent with quantum mechanics. You see, quantum mechanics is only important when you're looking at little things; when you look at little things you don't care about gravity. They have been disconnected in the laboratory, so the fact that they're mathematically inconsistent hasn't worried people. Well, it hasn't worried too many people, but the particle theorists have begun to wonder whether there is, in fact, a fundamental theory of nature that would be consistent with both.

It's partly just mathematical, or aesthetic, argument compelling us to look for a truly unified theory. It's also the case that if, for instance, you're interested in black holes, very exotic, but real—at least we believe they're real, there is some astronomical evidence that these things are real objects—black holes really require that you know both quantum mechanics and general relativity, or gravity and the other forces. Everything is put together. They're tiny, they're particle like; you need to worry about quantum mechanics, but they're heavy, so you also have to worry about gravity. It's the same thing if you're interested in cosmology, the study of the universe as a whole, and in particular its origin, the Big Bang because once again, there is this incredibly small distance scale, with very high energy density. Putting those two things together means you need gravity and quantum physics.

These are motivations for looking beyond the standard model. The standard model cannot be consistent in this way, so we need some new theory of nature. It's also true, just forgetting about gravity, as I've already talked about some of the reasons why you might be curious. For instance, there are so many unknowns, 23 numbers that you have to go out into the laboratory and measure, before you can start making predictions. There are so many particles, six quarks and six leptons; and they even start to form these little lovely patterns of increasing mass, and are duplicating generation two and generation three of the particles, which are like photocopies of the lighter ones. What's all that about? Okay? All of these things lead us to try to come up with some other, deeper theory of nature. Let me tell you about some of the ideas - two, in particular, that are being bandied about right now. I say it loosely; actually these are intense fields of theoretical research and lots of particle physicists are spending their careers trying to think about what this new physics might look like.

The first new idea—and it's not really a complete idea, it's a part of the second idea I want to tell you about—goes by the name of supersymmetry. I have to laugh a little bit, because as I said, physicists are over using this word "super", but, when we were studying the nature of space-time and particle physics, we were discovering all these symmetries of nature. Some of them were geometrical, like translational symmetry, which says the laws of physics are the same in different places; and there is time-translation symmetry, which says that the laws of physics are symmetric when you look at different times. Then there were these more abstract symmetries, such as mirror or parity symmetry, or matter-anti-matter symmetry. The mathematicians have been able to actually work out all the possible geometrical symmetries that are available in space-time and found them all. They are all either realized or broken in some way, but they're all relevant in the theory of gravity and/or particle physics—except for one. One symmetry remained that had never played any role in the theory; and it's now called "supersymmetry."

The easiest way—it's very mathematical, I can't, sort of, give you a geometrical explanation of supersymmetry—the way I think about it is an analogy to matter-anti-matter, just as that's a pretty good symmetry of nature. Imagine, imagine if there was symmetry between particles that we call fermions and particles we call bosons. That is to say, spin one-half, the fermions, and an integer such as zero, or one, or two, the bosons. Remember the difference between fermions and bosons. Fermions are kind of particle-y. They are the particle-y particles. They are the ones that don't like to sit on top of each other. Electrons, and protons, and quarks - pretty much anything you think of as a particle is a fermion. The bosons are the ones that do like to sit on top of each other. For example, light is perfectly happy—you can have two photons sitting in the same place at the same time. In fact, all of the force-carrying particles are bosons. The gluons, the photons, even the W and Z particles. Whether they're massive or not.

You have these two kinds of particles in the world. The particles and the force carriers, and what if they were completely symmetrical? What if there was some intimate connection, just as there's an intimate connection between a proton and an anti-proton? That's supersymmetry. It's an idea, just an idea. Is it real? Is there really a partner, of every particle out there that has a different spin, but is otherwise almost the same? Let's go look. Suppose that you took a quark, like a top quark, and asked, does it have a supersymmetric partner? It has an anti-partner, the anti-top quark, but does it have something with a different spin that is otherwise almost the same? The answer is no, there's no evidence for super-partners of any of the existing particles.

I have to tell you about the naming convention, because it's kind of whimsical, and you may be hearing about super-particles in the future, if these things are real, if we do discover them; so here's the convention. If you have a particle of spin one-half, and you're talking about its super-partner—which would have an integer spin, like spin zero—then you get the name by just adding an S on the front. It's the super-partner, so you add an S on the front. So the partner of a quark is a squark. The partner of a top quark is a stop squark. It's embarrassing, and when you start thinking about all of them, some of them are pretty goofy. The partner of a neutrino would be a sneutrino, but that's the convention for all the existing fermions. If you have a boson, there's a different naming convention; we add an "ino" at the end. I kind of like that because I become a Steverino, assuming I'm a boson. A photon becomes a photino. A gluon's partner would be a gluino. So what am I saying? It's like having a proton and an anti-proton. They're not the same thing, but they're connected, they're partners, they're symmetrical. Now we're saying a photon would have a partner in the universe, a photino, which has a different spin. It's a different particle, but they're somehow connected. So people are looking for super-particles.

The theorists have begun working out the mathematical consequences. What if supersymmetry really is symmetry of nature? They've discovered some very interesting things. First of all, they've discovered that, if this is true, we can fix up the standard model. We don't toss it. Instead, we just add the super-partners, and in fact, the model almost already tells you how they're going to behave, and what their properties are. Not completely, so the theorists are still struggling, and working to try to understand, what would a photino be like? One of the things that they have concluded is, in order for the mathematics to all be consistent, these super-partners can't be super-duper heavy, to use the technical word, but they can't be as light as the existing particles, because otherwise we would have already seen them. So it's a kind of a Goldilocks thing, they can't be too heavy and they can't be too light; and in fact the region of energy where we expect them is around a trillion electron-volts. So our next generation of particle accelerators: Fermilab, and in particular the laboratory being built in Europe, the LHC, the Large Hadron Collider, will be able to produce supersymmetric particles if they exist. So that's, you know, another one of those things that you can expect to see in the newspaper. It'll be a big deal. It would be a whole new family of particles. If we found them, it would be a radical, radical revision, but still not a scrapping of the standard model.

Now, this is all kind of fun and interesting stuff for particle physics, but that's not really where the action is right now. The big action, in the world of theoretical particle physics, is in a field which is called string theory. I think string theory has actually reached the popular media. There have been some nice books that try to present string theory in terms like in this course that are intuitive. It's a pretty difficult mathematical theory. In fact, compared to relativistic quantum field theory, that's kindergarten stuff compared to string theory. The mathematics is really horrendous. It would take me—I would have to dedicate many years of my life, just to learn the math that's required to become a string theorist. So I haven't done that; so what I'm telling you about is what I know from reading colloquia, but not from actually doing the calculations myself, unlike many of the topics in particle physics, which I have studied myself.

So what is the idea of string theory? Let me ask you, what is the basic, single, most fundamental idea of particle physics? I would say, it's that the world is made of particles. Okay? That's what particle physics says; everything else is detail. The world is made of particles. It goes all the way back to Mr. Democritus's idea of how the world is built. That's what particle physics is based on. Quantum chromodynamics and the electroweak theory postulate a world filled with little point-like objects. You can't visualize a point, it's a mathematical entity. You attribute properties to the points: charge, mass, spin; and then you start calculating according to the rules of quantum field theory, which is a theory of points. So that's the basic idea of particle physics as I have been describing it to you.

String theory says, maybe the world is not made of little points. Maybe the fundamental constituents, the building blocks, are actually somehow little lines, little strings. They might be a little line with two N's, or they might actually be a little loop, like a little doughnut of some kind. It's a, kind of a wild idea, but if you can believe that the world is made of points, I suppose you can also believe that the world is made of little strings, running around interacting with one another. Now why is this such a great idea? First of all, it's generalizable. Once you start thinking about little strings, you can start asking about little sheets, or little balls, in higher and higher dimensions. And in fact, string theorists are heading in that direction, and that generalization is called M theory. That's the buzzword. I don't want to go into any of those details. Let's just talk about the basic idea here.

The beautiful thing, the amazing thing about string theory is, there is just one string. That's it. The whole universe can be understood in terms of the string. Okay? It's not like you have an up quark particle, and a down quark particle, and an electron particle, and each one of those is a completely distinct kind of point—so that's why we're a little bit unhappy with the standard model, there's so many different kinds of fundamental particles. Now we have this theory, which is as unified as it could possibly be. There is only the string, and all objects are made of string. So, for example, what's an up quark? Imagine a little string that vibrates in one particular pattern; one pattern might be an up quark. A different, more energetic pattern might be a charm quark. A different pattern might be an electron, or even a W boson. Everything, not just the particles, but the forces, are all understood in string theory to come from the string. It's just, you know, as a mathematician, or from an aesthetic point of view, you can see the appeal of this idea. And now the question is, is it physics? Can you really make mathematical equations? Can you work it out, and decide whether this is in fact a correct, accurate, quantitative description of nature?

At the moment, string theory really is a mathematical theory. It is not, yet, in my personal opinion, physics; in the sense that there are very few direct physical predictions that string theory can make. People are still working on the basic structure of the theory. It's kind of, to my mind, a little bit like Niels Bohr in 19—early 1910s, who is trying to piece together a theory of quantum physics, but he doesn't know quantum mechanics yet. He's trying to create quantum mechanics, but he doesn't have all the tools, he doesn't have all the ideas, he doesn't have all the data yet. And, so he's just forming a tentative theory, the beginnings of quantum mechanics. And I see string theory, now, as a formative science, and it may pan out, or it may not. It— we're going to have to check it out, and see whether it does. But let me tell you some of the great things about string theory, because there's been a lot of effort.

It's quite recent business. Okay? People started thinking about strings, I don't know, in the '70s and '80s; and there wasn't really a whole lot of activity until really the 1990s. So this is really a very recent theoretical phenomenon. Let me tell you about a few of the amazing things. First of all, one of the problems that we all have with the standard model, one of the things we don't really like about the standard model, is this renormalizability thing. Remember what renormalizability meant. If you have a theory of point particles, and you work out, what is the effect of virtual particle-antiparticle creation, which is a part of quantum mechanics, you discover that you get infinite contributions to your theory, mathematical infinities which you have to sweep under the rug. It's not quite a cheat, but it's almost a cheat. There are few things that you cannot compute in the standard model, like the mass of an electron. There's just no way, in the model, to compute that physical quantity. You've just got to go out and measure it.

In string theory, there are no such infinities. It turns out that the origin of the infinities, the reason why you have to worry about renormalizability, is because you've got points, and points can get arbitrarily close to one another. And the closer they get, the stronger the fields get; and you can go all the way to the limit of infinite fields, if the particles are infinitely close. That's the kind of physical explanation for that mathematical infinity. But strings are different. They spread out over some distance, and you can never get smaller distances than the size of the string itself. Okay? Let me say some more words about this string, because you might be asking well, what is the string made of? And that's not a question that you can, or should, ask a string theorist. They are postulating the existence of a string. It's not made of anything. It's not made of little string atoms, otherwise those would be your fundamental particles. So that's a little bit hard for me to wrap my head around, but it's just one of the mathematical inputs to this theory.

Imagine the interaction of two electrons. Okay? In the present model, I imagine two particles cruising along, and then they create—spontaneously out of nowhere—say a virtual photon, which travels between them; and that's how they interact with one another. How would I picture this with string theory? I would imagine two little strings moving along and at a certain point those strings come together. They might even sort of fold together to form one big string for a brief moment, and then that one big string might split back apart again and turn into two little strings. That might be the interaction of two fundamental particles. And the intermediate state would be describing particles plus virtual particles in our contemporary language. But in the string-theory language, there was never anything but strings interacting. There were never any infinities because there's this limiting size.

Now what is that size? The string theorists can actually put a number on it. And that number comes from another idea, which is something that we've been worrying about in particle physics. I've talked about how the electric and the weak forces are unified. And they unify at a rather high-energy scale, trillions of electron-volts; and in fact, in some sense, even higher still. We haven't really reached the true electroweak unification in experiments yet. The question that many physicists have today, which the standard model does not answer, is, how about the strong force? Is it unified with electricity and the weak force? Right now we just say no, we've got the strong force, and we've got the electroweak force, and they're different. The question is, might they, in fact, at some ultra-high-energy, merge together and truly unify? After all, as you go to higher and higher energies, remember that the strong force gets weaker, but the electroweak force gets stronger. So it's possible that they merge together at a certain very high energy, and in fact, we can even estimate that energy. By looking at how they depend on energy, we can estimate how high up you would have to go. And it's astronomically high; we've come nowhere close to even being able to dream of a particle accelerator that could reach this grand unification scale.

Once people thought about grand unification, then they started thinking about—I don't even know what to call it—a complete unification, which would include gravity as well. Gravity is ultra-weak down in our ordinary world, but as we go to higher and higher energies, it gets stronger. And again, you can make a crude estimate; and the energy at which all the forces of nature seem to come together is about 10^{19} GeV. Remember, the highest energy accelerator we have right now is 10^3 GeV. This is 16 orders of magnitude higher energy than any machine we can imagine building right now. It's really the realm of theorists that we're talking about, not the realm of experiment; but at least in principle you could imagine that everything unifies at that ultra-high-energy scale. Quantum mechanics say an ultra-high-energy scale corresponds to a very tiny distance, and that's the size of the string, basically.

So we've got these strings, and if this is the theory of nature, one of the nice things about it is, no infinities pop up. So that means that you have at least the hope that you can compute everything. Okay? There are no quantities that you say sorry, the theory isn't going to tell you about that. We should be able to compute the masses, and the charges, and all the forces, and everything. And it's the ultimate unified theory where all physical quantities, all physical observables and phenomena, can be computed just from this mathematical framework. Very, very elegant.

Here's another amazing thing about string theory. If you suppose that the world is made of strings, and you start working out the mathematics, and you insist that you have a coherent theory—that is to say, you insist that you can't have any infinities cropping up—that actually puts some limitations on how you write down the mathematics. And you discover that supersymmetry is present; that is to say, supersymmetry is a consequence of string theory. So the string theorists are, in a sense, making an experimental prediction that they believe there should be supersymmetric particles out there. So that begins to make it physics, if you make a prediction that we can test. Although I imagine, if we didn't find them, they would be able to sort of change their theory a little bit, because it's still in a pretty nebulous state, and say, well no, they're a little bit too heavy, and that's why you haven't seen them yet.

Not only does the theory of strings predict supersymmetry, it also predicts the theory of general relativity as a low-energy-effective theory. That is to say, the mathematics of strings predicts Einstein's General Theory of Relativity. Okay? That to me is the most mind-blowing single aspect of string theory of them all. And I think that has sold many physicists enough on the theory to spend their careers studying it some more. And think of that. We have the mathematics to describe fundamental particles, and out pops the theory of gravity as a mathematical consequence. A theory of gravity, that has been with us since 1915 and that is extremely well verified and tested. Okay? So this is—not proof, but it's certainly a compelling argument that this is a theory of nature. So it's wonderful! It includes gravity correctly, and it includes quantum mechanics correctly, and it includes the standard model. It's all there; it's all buried in this one lovely little idea of the string.

There are also some very weird aspects of string theory, almost scary weird. For example, in order to make the theory coherent, mathematically consistent, it has to be the case that strings live in more than three spatial dimensions. Okay, what does that mean? I have three dimensions that I live in, and so do you. I can go forwards and backwards, I can go left and right or I can go up and down. And that's it. Three different, perpendicular directions are available to me. Any other motion is a combination of those. I can go a little forward, and a little left, and a little up; but those are the three independent dimensions that we live in. It's the degrees of freedom of an object in quantum field theory.

In string theory, to make everything work out, we live not in three spatial dimensions, but, maybe, ten dimensions or even 26 dimensions. It depends on the details of the string theory. Okay, what does that mean? How do we picture that? This is one of the wild parts of string theory; and they even have an explanation for why we haven't noticed it so far. And that stuff is just so wild that I would need a whole other course, just on string theory, to start to convey all of these ideas about the theory. And it's well worth learning about, I think.

Let me finish off with my own personal opinion about string theory. Fascinating stuff: it has the potential to be the greatest revolution in physics since the discovery of relativity and quantum mechanics 100 years ago. It's very mathematical. It's not quite a theory yet so we kind of have to just sit and wait for these theorists to tell us whether they can come up with some quantitative predictions, whether they can really conclude that the standard model is the low-energy-effective theory of their string ideas. Okay? It may turn out to be the correct theory of nature—and it may just be completely wrong. It may just be a bunch of mathematics. So I don't think we know yet, but it's certainly something to be paying attention to. Very wild stuff.

Lecture Twenty-Three
Really Big Stuff—The Origin of the Universe

Scope: Cosmology, the study of the universe as a whole, relates to particle physics, because matter at the very largest scales requires understanding of matter at the very tiniest. In this lecture, we examine the Big Bang scenario and see how it fits into the standard model. We also explore the relationship of *inflation* to the cosmos and define the latest buzzwords in physics, *dark matter* and *dark energy*. We'll talk about data from the COBE satellite mission and more recent astrophysical observations that teach us about the early universe and the connections between cosmic structure and the microscopic world.

> The statement that the Universe arose from inflation, if it is true, is not the end of the study of cosmic origins—it is, in fact, closer to the beginning. The details of inflation depend upon the details of the underlying particle physics, so cosmology and particle physics become intimately linked together.
>
> —Alan Guth

Outline

I. This lecture looks at cosmology, the study of the structure of the universe.

   A. Cosmology may seem unrelated to particle physics; one is the study of the largest possible system and the other, the study of the tiniest. Surprisingly, these two branches of physics are intimately connected.

   B. The Big Bang that may have begun the universe was a high-energy, high-temperature event. Because high energies correspond in quantum mechanics to small distances, the universe was in some sense tiny at the moment of the Big Bang. We need to understand the rules and constituents at the particle level to learn where the universe began and where it is going.

II. Let's begin with the Big Bang, a theory that has been around for a long time. Why do we believe in this theory for the origin of the universe?

A. First of all, imagine that you're an astronomer looking at stars. If we lived in a static universe—a universe that's very large, possibly even infinite, and has been here forever—then you would expect to see many stars, all of them moving in different directions. You should be able to determine the speed of stars, which we can using the Doppler effect, and find some sort of distribution of velocities among the stars.

B. If we start looking at distant objects, such as stars in galaxies, they all appear to be moving away from us, and the greater their distance, the faster they seem to be moving away.

C. A simple explanation exists for this effect, but it requires an abstract conception of the universe.

    1. Let's think of the universe as two dimensional. Imagine that we live on the surface of a giant ball, like a big balloon, and we are ants crawling on the surface. This conception reduces our universe to two dimensions, which curve around in the shape of a ball.

    2. If our balloon is being blown up and expanding, then all the spots on it are moving apart. Every individual ant is moving away from every other individual ant on the surface of the balloon.

    3. If you could put yourself in the worldview of one of those ants, in every direction, you would see other ants moving away from you, and the farther away they are, the faster they move.

D. We know that our universe is expanding and that the stars are moving away from us. If we run the clock backward and plug in some numbers, such as the speed and distance of these stars, we can calculate how long ago it was when that far-off galaxy we're looking at was right next to us. The answer is about 15 billion years ago.

E. We can do similar calculations for other galaxies, and we arrive at the same answer. Every galaxy, every star that is distant from us now, appears to have been very close to us about 15 billion years ago.

F. We know, then, that 15 billion years ago, the universe was in a highly compressed state. Further, all the energy of the universe was there from the beginning. Energy is conserved, which means that the universe had its origins in a very tiny, very energetic situation that resulted in the Big Bang.

G. How might we demonstrate that this theory is true?

    1. If the Big Bang occurred, we should be able to see the aftereffects, such as light and radiation. Using the laws of thermodynamics, we should also be able to predict the temperature of the residue of this explosion now, 15 billion years later.

    2. The first such observations were made in the 1960s by Pensius and Wilson, who detected radiation coming toward us from all directions in deep outer space. This *cosmic microwave background* is no longer very hot. The universe has cooled down so much that the radiation is very long wavelength.

H. Another piece of evidence for why we might believe in the Big Bang theory can be found in the dust between the stars and the galaxies.

    1. The early universe was, essentially, an accelerator experiment with very high energy and very high densities. We should be able to calculate what was produced in this accelerator, including the particles and the heavier elements.

    2. The amounts of heavy elements produced should be relatively small, because the explosion was very brief. In looking at the dust in the universe, astronomers have determined that the material is precisely what we would calculate from the Big Bang scenario.

III. Let's take a moment to discuss what's going on today in cosmology as a result of data received from satellites.

A. Cosmic background radiation is turning out to be a wonderful tool to learn about the Big Bang. The Cosmic Background Explorer (COBE) was the first satellite to collect information on this phenomenon.

    1. COBE went up in the late 1980s and made measurements that fit beautifully with theoretical expectations.

    2. For example, radiation in different parts of the sky was observed to be almost identical. Remember that we're looking back in time; this identical radiation was emitted 15 billion years ago and is now traveling toward us from all directions.

B. Let's also think about variations in density in the universe.

   1. The universe is divided up into "clumps," with a galaxy here and a galaxy there and nothing in between. Where do these galaxies come from? Presumably, a galaxy is some region of dust and gas that coalesced by gravity and began to form stars.

   2. The differences in density are explained by similar differences in the early universe. Regions that were a little more dense at that time began to attract material and, thus, became even denser.

   3. The data reveal that this model of galactic formation is consistent with what must have taken place in the early universe to have a later universe that is clumped into galaxies, as ours is.

C. Other research in cosmology is looking at *dark matter*, a substance in outer space that doesn't glow.

   1. Every galaxy has some outlying stars that are almost always in some kind of orbit. By making careful measurements, we can deduce the speed and orbital path of those stars, as well as the mass in the middle that is causing the orbit.

   2. If we calculate that mass, we conclude that something else exists inside the galaxy that is not glowing but is gravitating.

   3. We don't quite know what this *dark matter* is. It may be yet-undiscovered particles, such as supersymmetric particles.

D. Another phenomenon that has been observed recently and is even more mysterious than dark matter is *inflation*. This phenomenon is related to what must have been occurring at the start of the Big Bang.

   1. The Big Bang would have required a supply of energy that inflates the universe rapidly for a short amount of time. This inflationary scenario explains a number of other observations, including the fact that the universe looks so much the same in all directions.

   2. Where would that energy come from? Possibly, from the Higgs (or a Higgs-like) field. If this field exists everywhere in the universe, it may have come from a kind of condensation after the Big Bang.

   3. We imagine that the early universe, at the start of the Big Bang, was very symmetrical. All the theories were unified. The electric and the weak forces were the same in nature, and the Higgs particle had not yet condensed out. Then, as the universe expanded and cooled, the Higgs field condensed out.

4. When a gas condenses into a liquid, it emits heat. As the Higgs field condensed, it would release some form of latent heat, which is a supply of energy. That release of heat would pour energy into all the particles, which would then expand rapidly.

5. This scenario is consistent with particle physics and would explain cosmological inflation. Further, new data suggest that we are now undergoing another period of inflation; the rate of expansion is picking up as though some cosmic field is condensing and adding energy, which is called *dark energy*.

IV. We have evidence for a Big Bang about 15 billion years ago that involved an inflationary period and high temperatures. The goal of current research is to tie all these ideas together—string theory, standard model physics, and Big Bang cosmology—to form a coherent picture of the cosmos.

Essential Reading:

Barnett, Muhry, and Quinn, *The Charm of Strange Quarks*, chapter 8.

Kane, *The Particle Garden*, chapter 12.

Recommended Reading:

Calle, *Superstrings and Other Things*, chapter 25, last two sections.

Lederman, *The God Particle*, second half of chapter 9, starting with the section on "Flatness and dark matter."

Weinberg, *The First Three Minutes*. (The whole book is good!)

Wilczek and Devine, *Longing for the Harmonies*, chapters 4–6; Ninth Theme.

Questions to Consider:

1. Can you think of a method for estimating the age of the universe? How on earth can we know that it is (roughly) 15 billion years old?

2. Cosmic background radiation is the oldest light we can see. It was emitted when the universe was a few hundred thousand years old. (Any light emitted *earlier* than that existed in a universe dense enough that the light was reabsorbed.) If we cannot *see* anything from before that

time, how can we *know* anything about the universe from before that time?

3. Does it make physical sense to ask, "What happened before the Big Bang"?

Lecture Twenty-Three—Transcript
Really Big Stuff—The Origin of the Universe

Today we're going to talk about some really big stuff: cosmology. It's the study of the structure of the universe. What are its origins? What does it look like today? Where is it headed? What's the fate of the cosmos as a whole? Now, it might seem, at first glance, like talking about cosmology is completely disconnected from particle physics. In fact, as disconnected as you could imagine—we've gone from the tiniest of the tiny to the largest of all possible things we could study, and one of the beautiful things It's very elegant, and very surprising—these two branches of physics are very intimately connected. It has been this way for a long time; and more so today than ever before, you need to understand particle physics to talk about the structure of the universe, and vice versa. What the cosmologists are learning about the structure of the universe is teaching us about limits of the standard model.

One way of thinking about it is, if the universe begins in some cataclysmic big bang, that's an extremely high-energy, high-temperature event. And, high energies correspond, in quantum mechanics, to small distances. Everything was really close together. In a sense, the universe was tiny at the moment of the Big Bang; so, particle physics becomes extremely important if you want to understand where everything begins, how the universe starts. You really need to understand the rules, and the constituents, so that you have a sense of what's going to happen next. So, I want to begin today by saying a little bit about the Big Bang, because it's a pretty wild theory. It's been around for a long time, and I want to discuss it as a scientific question. Not as an issue of a creation myth, but rather as a creation scenario that can be tested: a logical consequence, a mathematical structure, data to confirm this hypothesis.

Why do we believe in the Big Bang? There are many reasons. And, let me begin by giving you what I consider to be the three best arguments for this idea of the origin of the universe.

First of all, imagine that you're an astronomer looking at stars. Now, if we lived in a, kind of a static universe, a universe that's very, very large, possibly even infinite, and has been here forever, then you would expect to see a bunch of stars out there. Some of them running towards us, some of them running away from us. Moving every which way. It should be some sort of a distribution of velocities out there. Now, can we figure out the speed of a star? It's not, at first glance, obvious how you would do that. You use a simple physics effect called the Doppler effect, which you may have noticed if you're standing on a street corner, and a police car or an ambulance goes by, and you listen to the siren. It makes that characteristic—first a high pitch, when it's coming towards you, and then a low pitch as it runs away—Wheee-eeeen, as it goes by. And that effect, the Doppler effect, is completely understood; it's just a simple wave phenomenon. If the source of a wave is coming towards you, it shifts up in frequency. If it's running away from you, it shifts down in frequency. And light is also a wave phenomenon. Everything is both particle and wave. And in particular, light has many, many properties that are just completely understandable in terms of waves. If an object is glowing and it runs away from you, the color is just shifted a little bit. If it's running away from you, it's shifted a little bit more towards the red. If it's running towards you, it is shifted just a little bit more towards the blue in the spectrum.

So, if you look at stars carefully, and you measure their color, you can deduce their relative motion towards you, or away from you. And what was noticed a long time ago was that if you start looking at far-away objects, stars in galaxies, they all appear to be running away from us. And that's true in any portion of the sky: Northern hemisphere, Southern hemisphere, all directions; all the galaxies that we can see appear to be running away from us. And the farther away they are, the faster they seem to be running away from us. Think about that for a second. It's a weird concept. And, you can imagine the reaction of astronomers and physicists when they first started collecting this data. Edwin Hubble is famous for making the first experimental measurements of this effect; we've named the Hubble telescope after him in honor of this discovery.

You might first think, wow, we're really special. We must be at the center of the universe. Everything is rushing away from us. That's a possible explanation, but it really doesn't seem all that likely. And it doesn't really explain why things that are farther away from us are running away even faster still. There is actually a simpler explanation, and it requires an abstract conception of the universe as a whole. So let me simplify the universe for a moment. Instead of thinking of it as three-dimensional, think of it as two-dimensional.

Imagine that we live on the surface of some giant ball, like a big balloon. You paint little galaxies on the balloon, and we are little ants crawling around on the surface of this balloon. I've reduced our universe to two dimensions, and those two dimensions curve around in the shape of a ball. If that balloon is expanding, if somebody is blowing it up, then all the little spots on it are moving apart from one another. Every individual ant is moving away from every other little individual ant on the surface of that balloon. And, if you could put yourself in the worldview of one of those ants, what would you see? You would look around you, and in every direction you would see other ants, and they're all moving away from you; and the farther away they are, the faster they move. If they're twice as far away, they move away from you twice as rapidly. So every ant has a worldview that's the same as the one that we are presently observing. This makes us not special. It makes the universe quite symmetrical, and it also explains this fact that double the distance corresponds to double the speed. So this is really the origin, I think, of this idea of a Big Bang. The universe is expanding.

So run the clock backwards. If stars are running away from us now, where were they a week ago? Well, they must have been a little bit closer to us. And keep running the clock backwards, and now plug in some numbers. Look at how far away they are, and how fast they're moving, and ask how long ago was it when that faraway galaxy was right next to us, if it's been doing this motion for a long time? And the answer, nowadays, is about 15 billion years ago. About 15 billion years is the amount of time it takes for a galaxy which is presently at its observed distance, and observed velocity, to have started off right next to us.

Now, you look at another galaxy, at a different distance, and you do that calculation. You get the same number. Okay? That's again a very interesting and compelling observation. Every galaxy, every star that's far away, appears to have been very close to us about 15 billion years ago. So, what does that mean? It means 15 billion years ago, the whole universe was in some highly compressed, very small state. This is the idea of the Big Bang. We begin with everything really close together. All of the energy of the universe was there from the beginning; energy is conserved. So, if you think about that, it means that in the beginning you have this very tiny, very energetic situation, which is the Big Bang. Now, we don't yet have any of the details of the Big Bang, but this is the origin of the idea.

So, let's play the science game of what-if. Okay? What if there was a Big Bang? How could you demonstrate that this is really true? Well, the idea would be, draw some conclusions, and then check to see if those conclusions are satisfied in the world. For example, if the universe begins very hot, hot things glow, always. This is a law of physics; they emit radiation. So this early Big Bang, it's a big explosion, there should be light—among other things, there should be light—which was emitted from that explosion. And, if this Big Bang scenario says the universe is symmetrical, then that radiation should be emitted in all directions. So, any direction that you look, you should be able to see the afterglow. All right? You've got a big explosion, and then it begins to cool down; and if it's just an expanding object, the laws of thermodynamics tell you how rapidly it will cool down. So, you should be able to predict how hot is the residual of this explosion, now, 15 billion years later. You come to a conclusion and it's a quantitative conclusion. If we begin with an explosion, 15 billion years ago, then the temperature today should be—and you can express it as an energy, or what's better, as a temperature—the universe should be at a certain temperature. And so, we've gone out and looked.

In fact, the first observation of this was not intentional it was accidental. Penzias and Wilson, in the early '60s, observed this afterglow, just some radiation coming towards us from all directions from deep outer space. In fact, this radiation is coming, essentially, from the farthest you can possibly look; and it appears as though we are inside of a giant box, if you wish, that's got a certain uniform temperature. This is very compelling evidence that this scenario fits. And that background radiation is called the cosmic background radiation—sometimes called the cosmic microwave background—it's really not very hot anymore. The universe has cooled down so much that the radiation is very long wavelength, very low-energy radiation. It's not visible to the eye. You don't see a glow anymore. It's like an old fire; it starts off white-hot, and then it's red-hot, and as it cools, you can't see it with your eyes anymore, but you can still put your hands over it and feel that there is radiation coming out. And now we've waited for 15 billion years; and, of course, it's just a steady process of getting colder and colder. But, it's never going to be at zero. It's just approaching zero.

The old data of Penzias and Wilson was very crude, just a couple of data points indicating that there seemed to be this microwave radiation. We're living in a very giant, fairly cold, microwave oven, in a certain sense. Nowadays, there's much improved data. I will tell you about some of that shortly. We have sent satellites up into orbit—to get away from the noise and heat of planet Earth—and taken very, very careful measurements. And so far, it works like a champ. This idea—that you're looking at a black, dark, but warm object—is consistent with the data, at very high levels of accuracy.

The third piece of evidence that I want to mention for why we might believe that the universe begins in this cataclysm, this Big Bang, is the following.

Let's now apply the laws of particle physics, the standard model, to that early universe. What we have is a big particle-physics accelerator experiment, very high energy, very high density. We should be able to calculate, if our model is correct, what was produced. You produce electrons and positrons, and protons and anti-protons. And, some of those protons will smash together, because it's very hot, just like they do today in the core of the sun, and form heavier elements. And so, you should be able to calculate all of the nucleosynthesis, the creation of elements, in those early minutes of the universe. So you sit down and work out the details. How much hydrogen do you expect, compared to how much helium, compared to how much lithium, and so on. And, the predictions are, that you shouldn't have very many heavy elements, because the cataclysm was very rapid. It's not like the sun that has been running for billions of years, cooking if you like. Here you just have a very brief explosion and then everything spreads apart. So, you only have a short amount of time to create some heavier elements. And again, we have gone out, astronomers have looked and—typically what you do is, you look not towards stars, but out into the dark spaces between the stars, and between the galaxies; and there's evidence for dust. The light that comes through that dust gets absorbed, and indicates what is the dust made of. And it appears to be made in precisely—that material is precisely what we expect from this Big Bang scenario.

These are quantitative predictions that are verified by the data. So, I think today the Big Bang scenario is accepted. It is not quite as rigorously accepted as the standard model of particle physics itself, because the amount of data that's consistent with particle physics is enormous. The amount of data associated with the Big Bang is smaller in number and more difficult; the astronomical observations are more challenging. So, I think it's a good theory, very well believed, but more subject to modification and new ideas than the standard model of particle physics itself.

Let me tell you a little bit more about what's going on today, because we have been sending up satellites, and we continue to do so. In particular, this cosmic background radiation is turning out to be just a wonderful tool to learn about this early event, this Big Bang. The first great satellite to do this was called COBE—the Cosmic Background Explorer is the acronym. COBE went up in the late '80s and made measurements for a couple of years. And, if you look at the data, it's just so beautiful. There's a theoretical expectation—it's called the blackbody curve—and the data just lies on that curve so well that you can't even see the uncertainties. They're so small on a graph that they just fit right into the line that you draw through the data points. What they observed was some interesting facts.

For example, the radiation that you see from one part of the sky is pretty much exactly identical to the radiation you see from another part of the sky. Now realize, you're looking back in time—15 billion years. This is radiation that was emitted 15 billion years ago in the Big Bang and has traveled towards us ever since. And something that comes from one side of the universe, and that comes from the other side of the universe, and reaches us, is identical. Now that's very interesting, thought provoking—it implies that the universe began all together in one spot, because that's the only way you could imagine that stuff over there, and stuff elsewhere, would know what the temperature is supposed to be. They agree, spectacularly.

And how about variations? When you look in light, you see galaxies here, and then no galaxy, and then galaxy, and then no galaxy. The universe is clumpy, and we would like to understand the clumpiness. Right? This is part of cosmology, understanding the structure of the universe that we live in. Where do galaxies come from? Well, presumably there was some region of dust and gas that coalesced by gravity, and began to form stars. And now you ask, well, why are some parts of the galaxies more dense and other parts less dense? And so, you kind of work your way backward. You say, well, in the early universe there must have been some fluctuations. Some parts were a little bit more dense and some parts were a little bit less dense, and the parts that were more dense attracted, and so they got even more and more; and so you have a kind of a feedback which forms galaxies. So this is a model of galactic formation, and you have to ask yourself, is this consistent with the Big Bang scenario? If the Big Bang is really completely uniform, then you'd have this explosion of dust and gas that's completely uniform, and then you wouldn't have any mechanism for galaxy formation. So people have been continuing to look at this data, and have noticed that there are, in fact, tiny variations. Although it's almost completely uniform, there is a little bit more density here and there, and a little bit less density in other places. It is quite consistent with what you would need, in the early universe, to have a later universe that clumped into galaxies the way ours has; so even more quantitative data indicating that this Big Bang scenario is correct.

This is a very hot field of research. There are balloons that have been sent up with observers; there are satellites that are being launched, and will be launched in the near future. They all have acronyms: BOOMERANG is one of the balloon experiments, and PLANCK and MAP are some of the satellite experiments. You'll probably be hearing about these things,

406

because they are taking very precise data. And now, we can start to compare with the standard model, or perhaps something deeper than the standard model, because, where do those fluctuations come from?

In the standard model you have some—you can't have perfect uniformity. The Heisenberg Uncertainty Principle says there will always be fluctuations. Little particle-anti-particle pair will appear here, but not there; and there are calculations you can do, statistical calculations. What is the probability of getting a high-density region here, and a smaller-density region there? One can actually connect this physics of the cosmos to the standard model. Make numerical, quantitative predictions of what is expected for those little tiny quantum fluctuations that happened during the Big Bang and then get magnified by gravity over the years—first, turning into the observable fluctuations in the cosmic background radiation, and ultimately into the fluctuations that we see. Here's a galaxy, and there's empty space.

As we collect this data, there are many puzzles remaining. Our picture of cosmology is still, I would say, at a more primitive state than our picture of the standard model of particle physics itself. There's a lot of mysterious stuff out there that we would love to understand and we don't yet. For example, the dark matter. Let me tell you about what dark matter is, because it's cosmological in origin, but may be particle physics in action. Another connection between the very large, and the very small.

There are many evidences, scientific data points, that point to the existence of something out there, in outer space, that doesn't glow. We can't see it directly. We only infer its existence. Let me give you a simple example, although there are many of these. If you look at a galaxy, there are always some outlying stars. And, those outlying stars are all, almost always, in some kind of orbit. By making careful measurements, we can deduce the speed and the orbital path of those stars. Now, we have a theory of gravity. In fact, you can use Newton's law of gravity, and usually that's good enough; Albert Einstein's theory of gravity pretty much isn't even necessary. You could use it, but it's going to give you essentially the same results as good old Isaac Newton. If you watch an object in orbit, you can conclude, by making measurements of its orbit, what is the mass in the middle that's making it go in that orbit? We can figure out the mass of the sun, without actually going and putting it on a scale, by looking at planet Earth, or any of the planets, and seeing what our orbital motion is. It's just the laws of physics that tell you. So you look at these stars, and you calculate how much mass is there inside of their orbit, that's attracting them, preventing them from running off in a straight line. And you get a big number. And now you look with your telescopes, and you count all the stars in the galaxy, and you add up the mass of all those stars, and it doesn't even come close to what you expected. It's like one percent of what you expected. It's really radically far away. It's not, you know, as if astronomers have a difficult time; you can't really count all the stars individually. You could imagine factors-of-two mistakes, but not a factor of 100. So, apparently there's something else inside the galaxy, which is massive, which is gravitating, and yet it's not glowing.

Now, you might say, well, there's dust. And there is dust. But we know that there's dust. We can measure how much dust there is because dust absorbs light. So if you emit light, or if you absorb light, we will be able to detect it, and we wouldn't call it dark matter. So this is something else. It's something very exotic. What is it? Well, we don't know. Right? This is one of the big puzzles of cosmology, and everybody's busy, trying to figure out what is the dark matter. There are lots of ideas. It might be just a bunch of, I don't know, protons and neutrons, maybe even forming little clumps of matter, some little charcoal blocks out there, which are somehow not absorbing the sunlight. Dust is visible because it absorbs light, but allows some to pass through. If it was really black, we might not really be able to notice it, if it was also really small.

Well, that's a possibility, but now you start doing some calculations, and this Big Bang scenario doesn't predict that. I told you that we think we understand how much matter was formed in the Big Bang, and the matter that was formed, the ordinary quarks and electrons, leptons and all the ordinary stuff, adds up to about one percent of what seems to be out there. So what else could it be? Could it be anything else? Well, we already have some candidates. What about neutrinos? The universe is filled with neutrinos. Every star makes them. The Big Bang made them. There's a big, large density of neutrinos everywhere. Imagine for a moment that those neutrinos are massive. This is a possibility in particle physics; it's something that we're investigating actively today in the laboratory. If the neutrino is massive, you add up all these tiny numbers and you get a big number. And, that would be something that gravitates. There might be more neutrinos in the region of galaxies, because there's lots of gravity there. And, if they're massive, they might actually be attracted. It's an idea. And again, you have to ask yourself, well, can we verify this idea?

It turns out that the neutrinos cannot explain all the dark matter that we believe is out there. They might explain a tiny fraction of it; however, we've already put limits: the neutrino can't weigh a pound, it's a tiny object. In fact, we have very, very stringent limits. It's an extraordinary light object. So, there is just not enough mass sitting in the neutrinos to explain all the dark matter that we've hypothesized is there in order to explain other phenomena which we really can see; like the rotation of stars, the expansion of the galaxies, galaxies as a whole. All sorts of data point to dark matter. And, it points to the same amount of dark matter.

So what else could it be? Well, maybe there are other undiscovered particles, like supersymmetric particles. We've talked about those briefly. It's a very speculative idea. It's a new form of matter. The squarks, and the sleptons, and these crazy particles that are predicted by string theory. And people have been looking for them very actively, at particle-physics accelerators. If it's real, we hope to find evidence of it in future, upcoming generations of particle-physics accelerators. And if there really are supersymmetric particles, they too would exist in the universe. They would have been created in the Big Bang, which was a giant accelerator experiment 15 billion years ago; and some of them might still be sitting around, gravitating. So all of this kind of speculation is very tentative, and it hooks together astronomy, cosmology and particle physics. Everybody's working together to try to understand what this stuff is. I think that there's going to be lots of progress. I can't imagine that, over the next 10 or 15 years, there's not going to be some great discoveries that will teach us more about what this dark matter is. And so, that's something to keep your eyes open for.

The mysteries don't end here with the dark matter. There is another phenomenon going on, which is quite recently observed, and it's even more mysterious than the dark matter. I think we can imagine some new undiscovered particles out there, and we can go look for them. There's another phenomenon that happened in the early universe, and it's happening again now, and it's got the cute name "inflation". So think of the Big Bang. The naïve image of a big bang is just an explosion. And it turns out, if you work out the details, consistent with all the data that I've been talking about—in particular, this cosmic-background-radiation data—it looks now like what you really need in the beginnings of the Big Bang is a very rapid explosion. You need some supply of energy that inflates the universe very rapidly, for a very short amount of time. This inflationary scenario explains a lot of things, including the fact that the universe looks so much the same in all directions.

And where would that energy come from? Well, I can propose a mechanism consistent with the standard model of particle physics. I'm not suggesting that this is the right explanation; there are some detailed problems with it. But it might be the right conceptual explanation. Remember the Higgs particle, this bizarre part of the standard model? It is conventional physics right now. Although we don't have direct evidence for the existence of the Higgs, there's a lot of indirect evidence, and we're looking for one. The main thing, the weirdest thing, about the Higgs is that it's everywhere. It's a—we call it a condensed-out field. There is actually a non-zero value of the Higgs field everywhere. We're swimming through this stuff. And, that's why particles have mass, is because they're swimming through this Higgs field. It's not a bunch of Higgs particles; particles would be excitations, or waves, in that field. It's just the field itself.

Now, if this field exists everywhere in the universe, where did it come from? The idea in the standard model is that it condensed out. It's like water vapor that you cool. And as you cool it, it undergoes a phase change, and all of a sudden you've got liquid water. So you ask, where did the water come from? Well, it originated in this beautiful, symmetrical gas; and then, as it cooled, it condensed out. It's the same with the Higgs. We imagine that, in the early universe, at the start of the Big Bang, it was very symmetrical. All the theories unified. The electric and the weak forces are the same in nature, and the Higgs particle was not yet condensed out. And then, as the universe expands and cools, the Higgs field condenses out. And what happens when a gas condenses into a liquid? It emits some heat, some thermal energy. Farmers know about this—if they've got a crop of orange trees in Florida, and they're worried about a freak cold snap, they'll put buckets of water underneath the trees, so that if the water freezes, the emitted heat will prevent the oranges from all freezing and dying. It's a standard and understood physical phenomenon. You are releasing some latent heat. And, as the Higgs field condenses, it too would release some form of latent heat—a supply of energy. It's not creating energy out of nothing; it was there as potential and it's become manifest, but that release of heat would pour energy into all the particles, which would then expand very rapidly. This is a scenario that is consistent with particle physics. That would explain this cosmological inflation that the cosmologists independently postulated, for other reasons, and believe they have evidence really happened.

In fact, the new data, from some of these balloon observatories, is that it's happening again. The universe—you might expect—we had a Big Bang, everything is expanding, but the expansion should be slowing down, because everything's gravitating; and so you imagine that all the galaxies are kind of being pulled back in. And the question of whether the pulling back in is enough to halt the expansion, turn it around, and lead us to an ultimate crunch is an open question. But at the very least, you would expect that the expansion should be slowing down. It might be slowing down, but won't ever stop. We might continue to expand forever. But it appears, now, that we're undergoing a period of inflation again. That the rate of expansion is actually picking up, as though some cosmic field is condensing and adding energy. Well, we don't really know what's going on. It's not the Higgs; the mathematics of the Higgs doesn't explain this new expansion, this new inflation; the buzzword on the street, in particle physics and cosmology, is dark energy. It acts as though there's a new supply of energy in the universe, and since we really don't know what that dark energy is, dark energy is even more mysterious than dark matter. You could imagine some sort of particle-physics explanation, but we don't have one today. Again, an active field of research.

So let me, let me summarize with the state of the world today in cosmology. We have compelling evidence for a Big Bang, about 15 billion years ago, which involved an inflationary period, very high temperatures, particle physics in action; and, then, everything just proceeds according to the known laws of physics. The universe cools, the Higgs field is present in the universe today, the galaxies form; we see the remnants of all this stuff. It's a nice picture; it's not complete, and there are probably some holes and gaps and maybe even some fundamental ideas still missing.

If you go back early enough, the temperature gets high enough that you expect really complete unification. In fact, you expect, at some ultra-early stage of this Big Bang, that we get beyond what the standard model can tell us. Once you get up to energies which I called the plank energy, extraordinarily high temperatures, the standard-model physics begins to break down, and we really need something more fundamental—like string theory—to really tell us how this story begins. And so I think the goal, the hope, is that all of this stuff is going to tie together in a neat bundle: string theory, standard-model physics, and Big Bang cosmology, to form a coherent picture of the cosmos as a whole.

Lecture Twenty-Four
Looking Back and Looking Forward

Scope: What have we learned after more than 100 years of intense study of fundamental particles, and what puzzles still remain? What influence does particle physics have on our lives, on science, and on our philosophical worldview? We'll conclude this course with thoughts on what you might take away from this experience—a sense of physical order and an understanding of the constituents of the world. Remember that progress in science does not generally overthrow the old, but modifies and extends it, which means that you can use your understanding of the standard model as a framework in which to fit new knowledge and understanding.

> The quality of a society is indicated by the questions it asks. One of the questions is, what is man made of? The answer is matter, and it is the nature of matter that is the domain of High Energy Physics. The society that doesn't ask this question is a suffering society.
> —Sid Drell

> Scientific understanding has inherent cultural value. It has great beauty. It adds to the satisfaction of our lives.
> —Robert Wilson (to the Joint Committee on Atomic Energy quoted in *The World Treasury of Physics, Astronomy, and Mathematics*, p. 697)

Outline

I. In this final lecture, we will look back at everything we've been talking about and try to put it into a framework.

   A. First, let's ask, "What is the standard model of particle physics?" The idea of the standard model is that we can understand all the complexity and diversity of physical phenomena in terms of the point-like elementary particles from which all objects are made.

   B. These particles interact with one another through some fairly simple and well-understood forces—the electroweak force, the strong nuclear force, and gravity.

C. Speaking generically, there are only two kinds of particles, quarks and leptons. The quarks are the strongly interacting particles that form protons and neutrons and a few more exotic particles that are found only in accelerator experiments or cosmological or astrophysical phenomenon. The leptons are the lighter ones, such as the electrons or the neutrinos that come out of the sun.

D. This theory is as quantitatively successful as any scientific theory in any branch of science. Despite its success, however, the standard model is not the end of the story. Many puzzles and questions remain, even in the framework of the standard model itself.

E. Sometimes, physicists talk about discovering a *theory of everything* (TOE), which would describe the true fundamental constituents and forces. From that basis, this theory would explain the quarks and the leptons and their interactions, and so on. In other words, this theory would explain the standard model.

    1. The standard model explains how atoms are formed. An atom is quarks forming into protons and neutrons that bind together to form a nucleus, along with electrons, which are electrically attracted. The atom can be explained by the theories of QCD, to describe the nucleus, and QED, or the electroweak theory, to describe the attraction of electrons to the nucleus.

    2. This theory explains how two hydrogen atoms can bind together and form a stable hydrogen molecule and how hydrogens and oxygens can combine to form stable water molecules. In principle, you start from this fundamental theory and work your way up to a deep understanding of chemistry, biology, and other sciences.

F. Ultimately, what particle physicists are studying is the five conceptual ideas that tie together the physical world: force and energy, matter, and space and time. Those five ingredients are the ultimate components of particle physics.

II. We began this course by asking why people care about particle physics.

A. As we said, we can speak of spinoffs of particle physics. This branch of physics may be the one that is least related to our lives, but it has resulted in some of the most sophisticated, complex machinery ever designed, including equipment for medical diagnosis and treatment. Further, in a sense, the Web is a spinoff of particle physics.

B. Other spinoffs from particle physics are not technological, but intellectual, including the field theory itself. Quantum field theory is a general mathematical framework that forms the basis for doing calculations. It can be extended to describe interactions in a number of complicated systems, including those in biology and even economics.

C. Finally, we might note the reductionistic pleasure we find in studying particle physics and the idea that this study satisfies basic childlike curiosity.

III. What should you take away from this course?

A. The guiding idea here is that the world is orderly, made, at its core, of simple "ingredients" governed by simple laws.

B. To the best of all the data and observations we've made for the past 100 years, we've learned that the world is made of protons and neutrons, which form nuclei, with electrons around them. That picture will be deepened and broadened in the future, but it will not go away.

C. Your job in the future is to question the sensational scientific discoveries you hear about and tie them into the standard model.

    1. If a supersymmetric particle is discovered, you'll ask, "How are these particles related to ordinary particles? What have they done to the standard model? How have they deepened it? What do we understand now?"

    2. You should also maintain your skepticism if you hear about discoveries that fly in the face of 100 years of data.

D. A new issue in physics research today is outreach. Physicists are being asked to make a greater effort to communicate their knowledge to the public. As you hear about these exotic ideas— string theory, dark matter, dark energy, extra dimensions—you are now equipped to make some sense of them for yourself.

Essential Reading:

Barnett, Muhry, and Quinn, *The Charm of Strange Quarks*, pp. 204–207.

Kane, *The Particle Garden*, chapter 7.

Schwarz, *A Tour of the Subatomic Zoo*, chapter 9.

Weinberg, *Dreams of a Final Theory*, chapter X.

Recommended Reading:

The bibliography contains some great selections; I encourage you to look there for interesting reading. (If you haven't read anything by Feynman yet, I'd certainly recommend you pick up one of his.)

Questions to Consider:

1. What aspect of the standard model do you find most appealing, and what (if any) are you most certain will be replaced or improved upon?

2. As a final assignment, look for an article in a current magazine or on the Web that covers a development in particle physics that is *newer* than this course! How does it connect to what you have learned here? Does it change your perceptions of what you have learned?

Lecture Twenty-Four—Transcript
Looking Back and Looking Forward

In this final lecture, I want to look back at everything we've been talking about, and try to put it into a whole framework. What have we learned? Why are we interested in this stuff? Where are we headed? So, no new topics today. No new particles or new concepts of physics.

I just want to think, first of all, about what *is* the standard model of particle physics? Okay, physicists want to understand how the world works, and that means everything, from complicated things to simple things. The idea of particle physics, this branch of all of physics, is that we can understand all the complexity and diversity of physical phenomena—anything that's measurable. Everything that can be observed is understandable in terms of this framework, the standard model of particle physics, that says the world is made of little particles. Everything. Everything physical is made of little point-like elementary objects, and they interact with one another. They interact through some fairly simple and well-understood forces. In fact, there's not very many of them: the electroweak force, the strong nuclear force, and gravity. That's it. And what kind of particles are there? Well, there are really only two kinds of particles, generically: quarks and leptons. The quarks are the strongly interacting ones that ultimately form protons and neutrons, and a few more exotic particles that are found only in accelerator experiments or in cosmological or astrophysical phenomena. The leptons are the lighter ones, like the electrons that buzz around, or the neutrinos that come out of the sun. And that's the story that forms the basis.

One can imagine creating a little periodic table, kind of an analogy to the old periodic table, the one that appears in all the chemistry books, that describes the fundamental atoms of the world, like hydrogen, and helium, and carbon, and so on. Our new periodic table, of the fundamental particles, has a few entries. There's an up and a down quark, a strange and a charm, a top and a bottom. Similarly, there are six leptons. That periodic table fits on a T-shirt, so that's nice, an indication that we do have a fundamental theory of nature. On the other hand, you can't fit, on that same T-shirt, the particles or the rules. That requires a couple of pages of writing down the actual mathematical formulas which define the standard model of particle physics. If you go to the particle data tables, which is a collection of the information that we have today about particle physics, you can find precisely such a short summary of this conceptual framework for understanding the world.

It is a very, very successful theory. In fact, I would argue that there has never been, to date, any scientific theory, in any branch of science, which is so quantitatively successful. This is not hundreds, but thousands—tens of thousands of experiments over 100 years— and they are all consistent. You can do calculations, you can make pre-dictions and post-dictions; that is to say you can calculate what you expect, and compare with what has been seen, or what you will see in a future generation of accelerator—and it works. Sometimes the calculations are tough, and in those cases we can sort of calculate to maybe ten percent, and it agrees with the data. In other cases, the calculation is easier. Sometimes we can calculate with uncertainties of one percent, or a tenth of a percent, or even, in some remarkable cases, parts per billion. Twelve digits of accuracy. And again, to date, every experiment that I'm familiar with, in any branch of particle physics—that includes accelerators, it also includes radiation experiments, small-scale tabletop experiments, experiments in balloons, looking at cosmic rays, deep underground in mines—they all agree with the standard model.

It's fairly mind-boggling, the success of this theory, and makes us feel very confident that we do have a deep understanding of how the world works. The world really is made of little particles, the quarks and the leptons, and they fit together, they interact with these forces. It's nice. It's like having a worldview that we live in a solar system with the sun in the middle, and planets orbiting around it. And that, too, is a very successful picture that we have that describes nature quite quantitatively. We can use that model to send rocket ships to Mars, and beyond; and so far, except when our technology blows up, it's been successful.

Despite its wonderful successes, the standard model is not the end of the story. Physicists aren't done. There are many puzzles and many questions. Even in the framework of the standard model itself, there are some puzzles and questions. We've been talking about many of those, and, of course, there remain many more that we haven't had time to talk about. Do the neutrinos have mass, and if so, what are those masses, and how does that fit into the story? Is there a Higgs particle? Is it real, or is it just a mathematical construct that makes this stuff all tie together? Again, something that we really would like to know the answer to, and we're trying to find out, how about this violation of left-right symmetry? The weak force of nature seems to distinguish between your left hand and your right hand in some fundamental way. They are fundamentally different, as far as the weak force is concerned? Why is that? We can describe it very, very accurately. We can make quantitative predictions; but I can't tell you yet why that's true.

And, we would like to know. Why is matter and antimatter not quite the same? Why is there CP violation in the universe? Why are we here? Why is the universe mostly matter, and not mostly antimatter, or actually a mix, a fifty-fifty mix?

You look at that little periodic table, and it's got three generations. There are the light particles, up and down quarks, the electron, and the electron neutrino, and they make up the world, the ordinary, everyday world. Everything. All objects, all materials, all phenomena that we normally deal with in our lives, come from just a subset of all the available, fundamental and elementary particles of nature. Why is that? Why is it that there is a duplicate generation, and then a triplicate generation, that appear to be identical, except they're more heavy? And, why do they have the masses that they do? We just described them. We go out in the laboratory and we measure those masses; however, we don't know why the muon weighs what it does. So, I think that most physicists believe the standard model is elegant and descriptive, but incomplete. And in particular, I can prove that it's incomplete if you allow me to worry about gravity. Quantum mechanics is incompatible with the present theory of gravity that we have, and so far it hasn't mattered because quantum-mechanical experiments deal with little things where gravity doesn't matter. And the experiments—like looking at solar systems, where gravity does matter—quantum mechanics is irrelevant. And so, we haven't had to worry about that; but conceptually, we really would like some theory that ties everything together.

Physicists sometimes talk about the theory of everything. A TOE, the Theory Of Everything. I don't really like that word, but you see it all the time. I don't like the word because, boy, it makes physicists just seem really arrogant. A theory of everything! If you read that physicists have discovered a theory of everything, you think wow. They can explain the weather in three weeks, the stock market next week, they can solve the Middle East crisis, and they can tell me how to make the perfect omelet, because they have a theory of everything. Well, of course that's not what physicists mean when they talk about this theory of everything. What they mean is a theory of everything at the simplest level. And, we are making some progress towards such a theory.

The standard model is kind of a—well, it's not really a theory of everything, because of all the problems that I've described. What we're looking for is something that ties it all together, at the microscopic level. What are the true fundamental constituents, and what are the true fundamental forces? And then, from that, the hope is that one can then explain the quarks, and the leptons, and their interactions. In other words, one could explain the standard model— wouldn't that be nice—because the standard model explains how atoms are formed. An atom is quarks forming into protons and neutrons, which bind together to form a nucleus, and then electrons, which are electrically attracted. So the theory of QCD, to describe the nucleus, and QED—or actually the electroweak theory—to describe the attraction of electrons to that nucleus, is enough to explain everything you would want to know about an atom. Everything: its size, its mass, its chemical properties and its electrical interactions with other atoms. So, in principle, this underlying theory explains how two hydrogen atoms can bind together and form a stable hydrogen molecule. And, how hydrogens and oxygens can combine to form stable water molecules.

Now in principle, you start from this fundamental theory and work your way up, and now you understand chemistry, deeply. Now the chemists are interested in this, and they occasionally do the quantum-mechanical calculations to make predictions and to understand their systems; and sometimes they've got their own rules and ways of thinking about things that are effective. They work and they don't really always worry about the connections down to the fundamentals. The more complicated of a system that you're looking at, the less likely you are to want to really—in real life—follow this chain all the way down to the bottom. If you're a biologist, studying the structure and properties of DNA molecules and how they replicate, and how errors appear, and how they correct those errors—the kind of stuff that a biologist might really want to know about—it's nice for them to know that there is a standard model of particle physics which is explaining the chemistry, which is explaining the biology. It's nice, from a conceptual framework, but from a practical point of view, it's pretty rare that the biologist will really go down all the way and start to do QCD calculations. In fact, I don't think they ever have to bother doing that.

Ultimately, what particle physicists are studying is the five, sort of conceptual ideas that tie together the physical world: force and energy, matter—what things are really made of—space, and time. Those five ingredients are the ultimate components of particle physics. And, we've been talking about all of those throughout this entire course. Force and energy. I think the core of particle physics is the matter itself, what are things made of; and that's certainly been the focus in this course. But when you're actually working out the mathematical details, all five of those ingredients take pretty much equal billing. Space and time is the framework, in which the particles live, and play, and interact, and we need to understand those. Right now our standard model involves all of those ingredients, but doesn't really unify or tie them together in the most elegant way that you could imagine. And so, science progresses. We continue to learn about particle physics.

We began this course asking, "Why do people care about this stuff?" And I had some answers then, and I'll give you some answers again now; and I hope that at this point you have your own answers. Everybody cares about this for different reasons. It speaks to us in our own personal way. You can think about spin-offs; particle physics is a very, very high-tech field of scientific knowledge. I mean, we are building giant accelerators, some of the most sophisticated, complex machinery ever designed, along with the space shuttle, and the telecommunication grid in the United States. Particle-physics accelerators are really high-tech, and there are spin-offs. There have always been medical spin-offs from this field, right from day one. When Roentgen discovered X-rays, the first of a new kind of radiation, within a year doctors were taking X-rays of people's bones. And this still happens today. Particle accelerators are used by neighboring medical facilities, sometimes for diagnosis, sometimes for treatment. People use proton beams, and pion beams, and anti-electron beams in order to image and to treat various medical ailments.

The magnets that are designed to make these super-high-energy particle beams go around in circles, at a particle accelerator like Fermilab, are themselves very high-tech devices. They're superconducting magnets, which required a lot of development of new materials, and new ideas. Now, I don't want to say that the particle physicists invented or discovered superconductivity; that's a separate branch of physics. But the fact that they're using them and developing them has led to spin-offs, has led to the development of this technology, and you know, this list just goes on and on. The World Wide Web is, in a very real sense, a spin-off of particle physics. It was the desire of particle physicists to have rapid, easy communication of scientific ideas across international laboratories that led to the development of this mechanism for communicating via computer. In fact, a lot of computer technology, both hardware and software, has been very, very strongly developed by particle physicists, who have to deal with enormous quantities of data, perhaps the largest data-storage problem of any branch of anything that I can think of. If you're collecting all these particles that are spewing out into your detectors from one of these modern accelerators, it's an enormous task. You need fast computers, you need big computers; and so there's been a lot of development there.

So those are techno spin-offs, and you can kind of go out and hunt for many, many more. Many of the accelerator public relations folks are interested in letting everybody know what this branch of science is giving us in our everyday lives. But, it is pretty esoteric physics. Particle physics is the most esoteric, the least connected in a direct way to our lives, of any other branch of physics or science that I can think of. Now, we've been talking about quarks, and neutrinos, and everybody realizes that these things are abstract. You can't see them, you can't hold a neutrino in your hand. It is fundamentally mathematical. It is a framework, an abstract framework for understanding the world. So it's no surprise that particle physicists don't have the best toys, or applications, that come from the work that they're doing.

There are many other spin-offs that aren't technological, they're scientific or intellectual. For instance, the biggest one that I can think of right off the bat is the field theory itself. Okay? The field theory, quantum field theory, is a mathematical framework that forms the basis for doing all of these calculations that I keep talking about. And that framework is very general. It just describes interactions when you have a complicated system. The basic idea of field theory is, you've got particle A and particle B communicating through a medium. The medium is the electric field, which can itself create matter-antimatter pairs. But the abstract idea here is, how do you have a system with lots of pieces, where piece A needs to communicate with piece B? And when you think about it that way, you begin to realize that this quantum field theory can be extended to many other situations, and has been over the years. In fact, parts of field theory were borrowed from other branches of science and mathematics, and many have been given back.

So biologists can talk about interacting systems through field theories. Physicists who are dealing with oh, say, liquid crystals, like you might see in a modern computer display, really have a system that has a field of liquid-crystal particles. Where you care about one particle interacting with another. And, they can actually use quantum field theory to help them to understand and describe that system. Even some economists have been trained as particle physicists, and then left the field and went into the stock-market business, and they write computer models in which they try to model stock-market phenomena using principles of field theory, to some limited success. They're dealing with a very complicated system that doesn not have some simple, simple objects right underneath it. The objects underneath the stock market are people, not quarks. And so, you're dealing with an intrinsically more complicated system. The mathematics are useful, but not quite as accurate in its predictive power.

This is one of the lovely things about particle physics, and yet another kind of motivation. It's not a spin-off, but just this kind of puzzle aspect. Particle physicists are solving deep puzzles about what the world is, and what it's made of. These are puzzles that have appealed to humans for a long time. People were asking these questions thousands of years ago and writing about them, and presumably it's a natural question. To look around you, and see all the complications in the world, and ask, can't we understand all this stuff in a simple way? Doesn't it all tie together? And it's wonderful that we've been so successful. Particle physicists are lucky. All right? They deal with the absolutely simplest ingredients; and trying to understand even simple things is hard enough. Then when you try to put that together to understand more complicated things, life just gets even tougher still. So this idea of science as a puzzle—particle physicists are lucky because their puzzle pieces, the building blocks, are really easy to work with.

I can't think of any reason why it had to be this way. The universe could have been a complicated place. You look around, you see all these different phenomena, and each phenomenon might have had a different explanation. I could imagine a universe that was chaotic, and random, and unpredictable. And, I think for a long time, people believed that they lived in such a universe. It's taken a lot of experimentation, and a lot of deep thinking, to realize that everything—everything—can be understood in terms of the interactions of quarks and leptons, via these limited number of forces that's available to them.

And finally, to me, there is this reductionistic pleasure. Okay? It appeals to me, just as a human being, to think that complicated things can be understood in terms of simpler things. So, it's not that particle physics can tell me how to make the perfect omelet, but I get pleasure out of thinking that an omelet is made of a certain set of ingredients. There are eggs, and there's butter, and there's cheese; and in specific ratios, and with a certain method of production, that's how I understand an omelet, as a reductionist. I think of its ingredients, and how they fit together to form this complicated and delicious whole. And then, I can keep asking that, well, so what is butter, after all? You can break butter down into smaller and smaller pieces until you get to a molecule of butter, and then you ask what is that, and why does it feel soft, and look yellow, and taste nice? And, we can understand all of the properties of butter, in principle, by looking at the carbon, and oxygen, and so on. Molecules that form together to make it. Which in turn, are made of protons and electrons, which in turn are made of quarks. I realize it's a kind of an absurd chain when you talk about an omelet and quarks in the same sentence. But nonetheless, I personally get some sense of satisfaction at the idea, just the conceptual idea, that we can understand stuff in this reductionistic way.

I said that was the last, but maybe the really, really last motivation for me is just plain old childlike curiosity. I think we all have this. Why is the world the way it is? I just want to know. I want to go out and do experiments. And to me, it's easier for me to tackle the simple questions. There are many physicists, and scientists, who would rather look at something complicated. They would rather look at an omelet, or a human body, and understand it. Not in terms of its ingredients, but just in all the myriad ways that you can understand a complicated object. That's, you know, absolutely legitimate science, which appeals to me as well. But somehow, for me, I find this basic stuff the most intriguing and the most exciting.

With what should you walk away from this course? You've spent a fair amount of time and mental energy trying to wrap your head around these esoteric concepts. The big, guiding idea here is that the world is orderly. The world is, at its core, made of some simple things with simple laws. And so, when you look around, and you see, either phenomena in your own life—or equally likely, maybe more likely—you read about phenomena in the newspaper, on PBS shows, reading whatever kind of scientific journal articles you might pick up at the newsstand you're going to be hearing about phenomena in the world—now you have a framework. No matter what happens. And, I anticipate great discoveries, right? There have always been great discoveries throughout our lifetimes, and there will surely be many, many more, some of them very deep. Some of them rattling the foundations of the standard model, perhaps. But always, I think you can take what you've learned in this course, and connect what you're hearing to what you know about how the world is made.

The world is made of protons and neutrons, which form nuclei, with electrons around them. That's apparently the way the world works, to the best of all of the data and observations that we've been able to make for 100 years. And, I can almost not imagine that that picture will go away. It will be deepened. It has already been deepened; the protons and neutrons are now understood to be made of quarks. So does that falsify my statement that the world is made of protons and neutrons? Not at all. We just understand now what a proton is, and what a neutron is. So it, similarly, I don't think quarks are ever going to go away, just like Newton's laws that were postulated, written down, in the 1600s have never gone away. We understand them more deeply now. Albert Einstein has extended their realm of applicability from ordinary life, cars and rocket ships and ordinary stuff, to black holes and the early universe. So I would anticipate all sorts of developments which deepen and broaden our understanding of the universe. But, I don't imagine that we're ever going to toss away all this knowledge that we've gained.

It's a wild idea, and it's one that takes people a long time to fully appreciate, that when you hear about exotic stuff, it does fit into this framework. It may extend it. It may deepen it. And that's your job from now on. To ask yourself, when you hear about something on the TV—it will usually be formulated in some sensational way, because that is human nature. Any physicist who makes a new discovery of a supersymmetric particle is going to say, "We've broken the standard model!" and it's going to make headlines. And now, your job is to say, "Ah, supersymmetry, I don't remember exactly what supersymmetry was," but of course the article will give you lots of explanation, which will hook it back into the standard-model framework. How are supersymmetric particles related to ordinary particles? What have they done to the standard model? How have they deepened it? What do we understand now? What of those big puzzles that people have been working on have they now finally understood?

Sometimes you will hear about results that are so sensational that you should be skeptical. One of the values of a course like this is for you to understand what is it that people are worrying about, and what is it that people aren't worrying about? So for example, if you hear a story on the news about somebody who has discovered a perpetual-motion machine, I want all your skepticism radar to come on full alert. That's not just an extension or a broadening of the standard model; that flies in the face of 100 years of data. What you're saying is that everybody was wrong all along; all of the data is wrong. And it's like saying oh, the earth is really flat; or the earth is the center of the universe. That is in direct contradiction with hundreds of years of data, and so—I suppose it could be true; maybe the Earth really is flat, and we've been fooling ourselves all along, but I'm pretty sure that's not the case. So, it's good to be skeptical. In fact, you should be skeptical as all-heck when people talk about the discovery of supersymmetry, or some new data on the top quark. All of this stuff is speculative, and all of it needs to fit together. Science forms a web, and there are always new strands, and there are always deeper layers. But, we can't just throw away what we already know.

So, I think that you are now equipped for this. It's also beneficial that there's a change in climate that I perceive. For example, in the funding agencies now, when I apply for a research grant, they insist that part of what I do with the federal funding from my research is outreach. All scientists and physicists are now being asked to think of ways to communicate what we're learning with the public. It's a wonderful idea. I think for far too long, physicists have been going off into never-never land, working their mathematics, understanding deep truths about the world, forming this lovely, coherent picture; and, by not articulating it in words that are non-mathematical, we're sort of leaving people in the dark. We understand stuff. It's wonderful. It's great to have this sense that we understand how the world works, and I don't think that you have to be a mathematician, or a physicist, to appreciate these ideas. So I think, when you start to read about supersymmetric discoveries, there will be some effort, more effort than was made in the past, to try to explain what it is that they've discovered; string theory, or dark matter, or dark energy, or extra dimensions. All of these wild, crazy ideas that people are coming up with.

And so, I think that the framework that's set up in this course is going to hook in with the framework that you will be hearing and learning about, and allow you to make sense of all of this exotic stuff. You are equipped to evaluate what's interesting, and why it's interesting. You are equipped to understand, why do we care, and why do we believe in this stuff? Good luck, and enjoy.

Timeline

Useful online source of original papers:
http://dbserv.ihep.su/hist/owa/hw.part1

Prehistory: 400 B.C. .... Democritus proposes *atomos*, the indivisible fundamental building blocks of matter.

1861–1868 .... James Clerk Maxwell unifies electricity and magnetism in a series of papers and proposes the electromagnetic nature of light.

1869 .... Mendeleev designs the periodic table of the elements.

1895 .... Discovery of x-rays (Wilhelm Rontgen, Nobel Prize, 1901).

1897 .... Discovery of the electron (J. J. Thomson, Nobel Prize, 1906).

1896–1900 .... Discovery of alpha, beta, and gamma rays (Pierre and Marie Curie, Nobel Prize, 1903; Henri Becquerel, Nobel Prize, 1903; Ernest Rutherford, Nobel Prize, 1908).

1900 .... The concept of quanta, $E = h\,\nu$ (Max Planck, Nobel Prize, 1918). J. J. Thomson proposes a "plum pudding" model of the atom.

1905 .... Albert Einstein's "miracle year": special relativity, the photon concept with application to the photoelectric effect, and Brownian motion (Nobel Prize, 1921).

1909 .... Charge of the electron measured (Robert Millikan, Nobel Prize, 1923).

1911 .... Discovery of the atomic nucleus (Ernest Rutherford, with Hans Geiger and Ernest Marsden).

1912 .... Cosmic rays first observed (V. F. Hess, Nobel Prize, 1936, with Anderson). Cloud chamber invented (Charles Wilson, Nobel Prize, 1927, with Compton).

1913 .... Bohr model of the atom (Niels Bohr, Nobel Prize, 1922).

1915 .... General relativity (Albert Einstein).

1918 .... Emmy Noether's theorem relating symmetry to conservation laws.

1919 .... Rutherford observed first nuclear transmutation (alpha on nitrogen gives oxygen + proton). Proton as a fundamental particle is named.

1923 .... Louis de Broglie argues for wave-particle duality (Nobel Prize, 1929). Arthur Compton shows the particle-like nature of x-rays (Nobel Prize, 1927).

1925 .... Pauli formulates the exclusion principle and postulates new quantum property of the electron (Nobel Prize, 1945). This property, the spin, was introduced and quantified by Goudsmit and Uhlenbeck. Quantum mechanics developed in matrix form by Werner Heisenberg (Nobel Prize, 1932), followed by Erwin Schrödinger's wave mechanics (Nobel Prize, 1933).

1926 .... Born introduces the "probability interpretation" (Nobel Prize, 1954). The term *photon* is coined (Gilbert Lewis).

1927 .... Heisenberg's uncertainty principle.
Wigner introduces *parity*.

1928 .... Dirac's relativistic wave equation
(Nobel Prize, 1933).

1930 .... The neutrino is postulated (Pauli,
named by Fermi). Cyclotron invented
(Ernest Lawrence, Nobel Prize, 1939,
and M. Stanley Livingston). Max Born
says, "Physics as we know it will be
over in six months."

1931 .... The Van de Graaff accelerator is
invented. Dirac predicts antimatter.

1932 .... Positron discovered (Carl Anderson,
Nobel Prize, 1936). Neutron discovered
(James Chadwick, Nobel Prize, 1935).

1934 .... Beta decay theory written down
(Fermi). Yukawa's theory of nuclear
force (Nobel Prize, 1949).

1937 .... Experimental evidence for mesotron
(now called muon) by Neddermeyer
and Anderson, Street and Stevenson
(I. I. Rabi later asked, "Who ordered
that?").

1938 .... Nuclear fission of uranium observed
(Otto Hahn, Fritz Strassmann, Ottto
Frisch, Lise Meitner).

1939 .... Muon decay observed (Bruno Rossi).

1945 .... First nuclear weapons detonated.

1946 .... Big Bang theory proposed (George
Gamow, named by Hoyle later to
mock it).

1947 .... Discovery of pion and pion decay in
cosmic rays (Perkins, Lattes, and
others). Discovery of V particles in

cloud chambers at Manchester (George Rochester and Clifford Butler) (now called *kaon* and *lambda*). Lamb shift measured (hydrogen fine structure) (Nobel Prize, 1955).

1948 .... QED developed by Richard Feynman, Julian Schwinger, and Sin-Itiro Tomonaga (Nobel Prize, 1965). Pions produced in accelerators (Berkeley synchrotron).

1949 .... Kaon decay observed (Brown et al.). Spark chamber invented (J. W. Keufel).

1950 .... Neutral pion and its decay discovered (Bjorklund, Panofsky, and Steinberger).

1952 .... Delta particle discovered (Anderson and Fermi). Bubble chamber invented (Donald Glaser). Brookhaven cosmotron accelerator opens.

1953 .... Associated production (later understood as strange particles) observed.

1954 .... First "gauge theory" with charged force carriers (C. N. Yang and Robert Mills).

1956 .... Anti-proton discovered (Berkeley bevatron). Anti-neutrinos from reactors observed (F. Reines and C. Cowan, Nobel Prize, 1995, shared with Martin Perl). Strangeness proposed (Murray Gell-Mann and Kazuhiko Nishijima). Parity violation predicted by T. D. Lee and C. N. Yang (Nobel Prizes, 1957).

1957 .... Parity violation observed (C. S. Wu).

1959 .... CERN (in Europe) and Brookhaven accelerator start operations.

1961 .... Kaon "regeneration" observed. Discovery of spin 1 (vector) mesons (rho, omega, eta).

1962 .... Accelerator production of neutrino beams. Muon-type neutrino is a separate flavor (Lederman, Schwartz, Steinberger, Nobel Prize, 1988).

1963 .... Cabibbo theory of weak decays.

1964 .... Quark model introduced (independently) by Murray Gell-Mann and George Zweig (Nobel Prize, 1969, to Gell-Mann). Omega-particle discovered (bevatron). CP violation in K0 decay discovered (James Cronin, Val Fitch, Nobel Prize, 1980). Higgs mechanism proposed (Peter Higgs, Robert Brout, and F. Englert). Color and gluons proposed (Greenberg, Nambu).

1965 .... Cosmic microwave background observed (Arno Penzias and Robert Wilson, Nobel Prize, 1978).

1966 .... Stanford linear accelerator (SLAC) starts operation.

1967 .... Electroweak unification (the *WSG electroweak model*) proposed independently by Steven Weinberg and Abdus Salam, based in part on contributions of Sheldon Glashow (Nobel Prize, 1979).

1968–1969 .... Deep inelastic scattering at SLAC: Bjorken scaling observed, evidence for partons argued by Feynman (Friedman, Kendall, and Taylor, Nobel Prize, 1990). Ray Davis's Homestake experiment first sees solar neutrino deficit.

1970 .... Proposal of charm (Glashow, Iliopoulos, and Maiani: the *GIM* mechanism).

1971–1972 .... Gerard 't Hooft proves renormalizability of electroweak theory (Nobel Prize, 1999).

1972 .... Fermilab begins operations.

1973 .... QCD developed (Gross, Wilczek, Politzer, Fritzsch, Gell-Mann, Leutwyler, Weinberg, and others). Neutrino experiments verify parton = quark (Gargamelle detector at CERN). Neutral weak current observed (Gargamelle detector at CERN). CKM matrix for weak decays (Cabibbo, Kobayashi, Maskawa).

1974 .... J/psi discovered, the "November Revolution" (S.C.C. Ting at Brookhaven, Burton Richter at SLAC, Nobel Prize, 1976). SU(5) grand unified theory proposed (Howard Georgi and Sheldon Glashow) (later shown to be incorrect because of incorrect predictions of proton decay).

1975 .... Charmed baryons discovered. Tau lepton discovered (Marty Perl, at SPEAR, SLAC, Nobel Prize, 1995, shared with Reines). Quark jets seen (at DESY, in Germany).

1976 .... CERN SPS (super-proton synchrotron) starts.

1977 .... Bottom/anti-bottom meson (upsilon) discovered at Fermilab (Leon

Lederman). The standard model is now complete.

1978 .... Parity violation observed in neutral current reactions (SLAC).

1979 .... Three jets seen at PETRA (Germany), direct evidence for gluons.

1983 .... Discovery of Z and W bosons at CERN SPS (Carlo Rubbia's group, Nobel Prize, 1984).

1984 .... Superstring theory developed (Michael Green and John Schwarz).

1987 .... Supernova 1987a (detected with visible light and neutrinos!).

1989 .... Z0 copiously produced at LEP and SLC (number of neutrinos = 3).

1990 .... COBE (Cosmic Background Explorer) satellite returns high-precision data on the cosmic microwave spectrum.

1993 .... Gallium experiments confirm the solar neutrino problem. Atmospheric neutrino anomaly discovered in Japan's Kamiokande experiment. Precision Z0 measurements confirm standard model.

1995 .... Top quark discovered at Fermilab.

1996 .... SuperKamiokande neutrino detector begins operations in Japan.

2000 .... Tau neutrino discovered at Fermilab. Cosmic background radiation spectrum results from Boomerang and Maxima.

2002 .... Evidence for neutrino flavor oscillations from SNO detector in Canada.

Glossary

accelerator: Any device that creates high-speed, high-energy particles.

alpha radiation: When a material releases a stream of particles (radiation) that are, specifically, positively charged helium nuclei (two protons and two neutrons bound together). Caused by the strong force.

angular momentum: A quantitative measure of how rapidly objects turn.

antimatter: Particles that are "opposites" in most respects to ordinary particles; for example, an anti-electron has opposite sign of charge and opposite sign "electron number." However, it has the same mass and the same spin. Matter and antimatter will annihilate, producing pure energy.

associated production: The observation that the strong force conserves strangeness number; that is, in a strong interaction in which two particles are produced, if one has strangeness +1, the other will have strangeness −1.

asymptotic freedom: Quarks that are very close (that is, that interact with very high energy) feel almost no strong force. They are "free."

atom: The smallest building block of *chemistry*, an individual particle of an element. Physically, a heavy nucleus with electrons orbiting.

baryon: A strongly interacting particle with half-integer spin, such as a proton or neutron.

baryon number: A conserved quantum number, +1 for any baryon, −1 for any anti-baryon. No known violation of baryon number has yet been observed but is being investigated.

beta decay: When any particle or nucleus transforms to something else via the weak force, in the process, emitting an electron and (anti) neutrino.

beta radiation: When a material releases a stream of particles (radiation) that are, specifically, electrons. Caused by the weak force.

bevatron: A large nuclear accelerator located in Berkeley, CA. The name came from the beam energy, "billion eV," or BeV. (We would now call that a GeV.) This was the facility where, for example, the antiproton was first discovered.

Big Bang: The theory of the cataclysmic start of the universe. A point of infinite temperature and density, from which the universe was born. Currently estimated to be about 12–15 billion years ago.

boson: Any particle with integer spin (0, 1, 2...). Bosons may be in the same quantum state.

bottom quark: A heavy quark, in the "third generation," similar to a down quark (charge −1/3) but heavier.

broken symmetry: A symmetry of nature that has been hidden by accidental circumstance. For example, a magnet has a direction, even though the underlying forces are truly and precisely symmetrical. The direction is arbitrary, but once chosen, the underlying symmetry is broken and no longer apparent.

Brownian motion: The random "drunken walk" motion of tiny particles (e.g. pollen grains) drifting in a fluid. Observed by Brown in the 1800s, explained quantitatively by Einstein in 1905 as evidence of tinier (invisible) atoms bumping into the grains.

bubble chamber: A device to track charged particles (passing particles ionize a liquid, forming bubbles). Invented in the 1950s.

CERN: European Center for Nuclear Research; home of the largest, highest energy European accelerator.

charm quark: A heavy quark, in the "second generation," similar to an up quark but heavier. First found in the November Revolution in the form of a c-cbar meson.

cloud chamber: A device to track charged particles (passing particles ionize a vapor, forming droplets). Invented in the early 1900s.

color: The "charge" for the strong force. There are three colors.

confinement: The fact that quarks cannot be removed from hadronic matter. They are permanently connected into "colorless" objects.

conservation law: When some quantity remains unchanged during an interaction. For example, charge conservation states that the sum of all electric charges never changes in any particle reaction.

cosmic background radiation: Residual evidence of the Big Bang; low-energy (mostly microwave) photon "bath" we live in, which permeates the universe.

cosmology: The field of physics that studies the structure and evolution of the universe.

CP violation: A symmetry of nature that is slightly broken, effectively a "matter-antimatter" symmetry that is not exactly perfect. A topic of great current interest but not yet well understood or explained.

cyclotron: An older type of particle accelerator.

dark matter: Something that is present throughout much of the universe, carries mass, but does not emit visible radiation; therefore, we don't know what it is. Current estimates are that much of the universe is dark matter!

decay: When a particle transforms, emitting some other lower mass particles as a result. For example, a neutron decays in roughly nine minutes, emitting a proton, electron, and neutrino.

DESY: German high-energy accelerator facility.

detector: Any device designed to identify and characterize the properties of particles.

DIS (Deep Inelastic Scattering): Ultra-high energy particle experiments begun in the 1970s at SLAC. Deep means "very high energy;" the incident electrons come in and go deep into the heart of the nucleus. Inelastic means they give up much of their energy, rather than bouncing, like a rock smashing through a glass window. DIS gave us the first direct evidence for the existence of quarks inside protons and neutrons.

down quark: One of the two light quarks, with spin $1/2$ and charge $-1/3$. (The proton is two ups and a down; the neutron consists of two downs and an up.)

eight-fold way: An early theoretical model to explain regularities in the particle zoo, invented by Murray Gell-Man in 1964. It was the precursor to the idea of quarks.

electroweak theory: The unified quantum field theory of electromagnetism and the weak force. Coined by Salam; developed by Weinberg, Salam, and Glashow.

emulsion: An old, fairly simple detector technology, effectively just like photographic film.

eV: A unit of energy. 1 eV = 1.6×10^{-19} joules. It is a typical "atomic" energy scale.

family: Another name for a generation of particles; for example, the up and down quarks form the lightest family of quarks.

femtometer: A metric unit of distance, 10^{-15} meters; about the size of a proton. Abbreviated fm; sometimes called a "Fermi."

Fermilab: Highest energy accelerator operating in the United States; location of the Tevatron ring; accelerates protons and anti-protons.

fermion: Any particle with half-integer spin (1/2, 3/2, and so on), such as electrons, protons, quarks, and others.

Feynman diagram: Symbolic shorthand for quantum field theory calculations that resembles a "cartoon sketch" of the particle reaction.

field: The physical manifestation of a force of nature, present throughout space; an alternative way of thinking about forces as opposed to "action at a distance."

flavor: The name for the characteristic that distinguishes certain groups of otherwise similar particles. For example, there are three flavors of neutrino: electron type, mu type, and tau type.

force carrier: A particle which serves as the intermediary for a force of nature. For example, the photon is thought of as a particle, a tiny quantum bundle of energy which allows one charged particle to "feel" another one electromagnetically. The photon is then the electromagnetic force carrier.

gamma radiation: When a material releases a stream of high-energy electromagnetic waves (radiation) of very short wavelength.

gauge boson: The force carrier particle in any quantum field theory, such as the photon in QED, the gluon in QCD, and the W and Z in the electroweak theory.

gauge symmetry: An abstract symmetry of nature present for the three fundamental forces (strong, electromagnetic, and weak). The presence of gauge symmetry in a theory requires the existence of mass-less gauge bosons.

general relativity: Einstein's theory of gravity. Not yet consistent with quantum mechanics but spectacularly successful in all experiments to date.

GeV: A billion eV's, 10^9 eV. (See eV.)

gluon: The gauge boson (force carrier) of the strong force.

grand unification: The idea that, at very high energies, all forces of nature (strong, electromagnetic, weak, and ultimately, gravity) become one. They are different low-energy manifestations of one fundamental force. Not yet an established theory.

GUT: A theory of grand unification.

hadron: Any particle that interacts strongly. (Any object made up of quarks.)

Heisenberg uncertainty principle: A fundamental principle of quantum mechanics that certain observables (e.g. position and momentum) are intimately connected: the better you measure one of these quantities, the more uncertain the other must be. It's a quantitative statement of the idea that observations must affect the system being observed.

Higgs mechanism: The mathematics that makes the Higgs field special, breaking the electroweak symmetry, giving the W and Z bosons mass.

Higgs particle: The physical manifestation of the Higgs field; the last unproven piece of the standard model. Actively being sought.

interaction: A synonym for *force*; a way in which particles transform or perturb one another. For example, the weak interaction causes beta decay.

J/psi: The particle found in the November Revolution of 1974, a charm/anti-charm meson. Simultaneously discovered at SLAC and Brookhaven. The particle that in some ways clinched the quark model in many physicists' minds.

kaon: A strange meson (see strangeness, meson); heavy relative of the pion.

Lamb shift: A tiny splitting in the light emitted by hydrogen atoms, indicative of subtle QED effects. The experimental push in 1947 to make QED a successful theory.

lepton: A class of spin 1/2 particles that do not interact strongly. The electron, muon, tau, and the three flavors of neutrinos are all the leptons.

meson: Any strongly interacting particle with integer spin. A bound state of quark and anti-quark. Lightest example is the pion.

MeV: A million eV's, 10^6 eV. (See eV).

molecule: A chemical building block that is not fundamental but built up out of a bound state of two or more atoms, such as an H_2O (water) molecule.

muon: A fundamental lepton, in the "second generation"; a heavy version of the electron. (I. I. Rabi asked, "Who ordered that?") Decays weakly into electron, mu-neutrino, and electron/anti-neutrino.

neutrino: A fundamental lepton, with (near) zero rest mass and zero charge. Interacts only weakly. It is under intense current investigation, because it is not known what the mass is or whether the three different flavors can oscillate from one to the other.

neutrino oscillation: When one flavor of neutrino spontaneously (through quantum mechanics) changes into a different flavor. Under current study at many facilities, this is not a part of the standard model.

neutron: Partner of the proton, electrically neutral baryon that is present in essentially all nuclei. Consists of d, d, and u quark.

Noether's theorem: A beautiful and important mathematical theorem relating symmetries of physical theories with conserved qualities.

November Revolution: The discovery of the J/psi particle in November 1974.

nucleon: A generic name for neutrons and protons, the "constituents of nuclei."

nucleus: A bound collection of protons and neutrons; the center of atoms.

parity: Mirror symmetry; the observation that the laws of physics are the same when viewed in a mirror.

parity violation: The breaking of parity symmetry. Only the weak interaction violates parity; for example, the weak interaction will produce "left-handed" particles but not "right-handed" ones in some reactions.

particle physics: The study of the fundamental constituents of nature and their interactions with one another.

Pauli exclusion principle: A quantum mechanical law stating that identical fermions cannot be in the same quantum state at the same time. This principle is responsible, in a very real sense, for much of chemistry.

periodic table: Mendeleev's organization of atoms into a simple "table," in increasing order of weight, which shows the underlying structure of atoms.

photoelectric effect: Explained by Einstein in 1905; when light hits a metal, it can eject electrons. The details can only be understood by postulating some "particle-like" properties to the light.

photon: The spin-1 gauge boson of electromagnetism, the quantum of light, the carrier of the electromagnetic force.

pion: Originally postulated by Yukawa as the carrier of the strong force, it is now simply the lightest meson, a spin-0 quark/anti-quark bound state. Comes in +, −, and zero charges; the charged versions decay weakly into a muon and muon neutrino.

positron: An anti-electron (see antimatter).

proton: The fundamental baryon, building block (with neutrons) of nuclei. Composed of two ups and a down. A stable particle as far as we know (still under investigation, but if it decays, the lifetime is very long).

QCD: Quantum chromodynamics, the quantum field theory describing the color force, the strong interaction between quarks.

QED: Quantum electrodynamics, the quantum field theory describing the electromagnetic force.

quanta: Chunks, of just about anything.

Quantum chromodynamics: See QCD.

Quantum electrodynamics: See QED.

quantum mechanics: The physical theory which tells how microscopic particles behave under the influence of forces.

quantum numbers: Properties of elementary particles that come in chunks; for example, the quantum numbers of a proton are: charge 1, strangeness 0, baryon number 1, lepton number 0, and so on.

quark: The fundamental spin 1/2 constituents of all strongly interacting objects. Quarks come in six flavors: u, d, s, c, t, b (up, down, strange, charm, top, bottom). Never found isolated; they always come as q-qbar (meson) or q q q (baryon). Quarks carry electric charge and color charge and experience all fundamental forces of nature (but the strong force generally dominates). The heavier quarks are not stable.

reductionism: The philosophical principle that complex systems can be understood once you know what they are made of and how the constituents interact.

relativity: Generic term usually reserved for Einstein's theory that describes the motion of particles moving at high speeds. Modifies our conventional views of space time and gravity. (It does not mean that "everything is relative.")

renormalizable: A quantum field theory that can be easily fixed up to yield finite results.

renormalization: A mathematical technique to take a quantum field theory, which generally produces infinite results, and "fix up" a *small* number of formal "bare" quantities so that, after adding the infinite quantum corrections, you match experimental data in the end. Once you do this for the small number of quantities, every other observable is then finite and well determined.

RHIC: Relativistic Heavy Ion Collider. A nuclear physics accelerator in Brookhaven, NY, just recently opened. Produces high-energy, heavy particles to make high-density, high-temperature collisions.

SLAC: Stanford Linear Accelerator Center. A two-mile-long electron accelerator, now up to a maximum of almost 100 GeV. Extremely successful and long-lived accelerator.

solar neutrino: A neutrino emitted from the nuclear reactions in the core of the sun. Generally very low in energy, but there are a lot of them.

solar wind: A constant stream of energetic particles emitted by the sun.

special relativity: Einstein's 1905 theory describing the motion of particles moving at high speeds. Modifies our conventional views of space and time. (*Special* means the theory is limited to observers moving with steady velocity and ignores gravity.)

spin: A quantum property of particles. It is somewhat analogous to "rotational angular momentum," how fast something is spinning, but applies even to point particles. Spin is quantized. Particles can have spin 1/2, 3/2, 5/2, and so on (fermions) or 0, 1, 2, and so on (bosons). (Those numbers are really expressed in units of h/2 pi, where h is Planck's constant, 6.63×10^{-34} Joule × sec.) Spin can also be thought of as describing degrees of freedom of a particle; for example, spin 1/2 particles have two degrees of freedom (up or down).

SSC: Superconducting Supercollider, the failed U.S. mega-project that would have produced protons at extreme energies to search for the Higgs and learn about electroweak symmetry breaking.

standard model: The theory of fundamental particles. It consists of a list of the particles and their properties (the quarks and leptons and the Higgs, their charges and masses) and the theories that describe their interactions, specifically, QCD and the electroweak theory.

strange quark: A second-generation quark, charge –1/3, heavy relative of the down quark.

strangeness: A quantum number of hadrons that effectively just counts how many strange quarks are in the object (times –1, technically). Strangeness number is conserved by all forces except the weak force.

string theory: A tentative fundamental theory of nature in which particles are described not as points, but as tiny strings, living in 10 (or more) dimensions. This theory appears to give finite results, unifies all four forces, leads to general relativity, and may also yield the standard model as its "low-energy limit." Under active investigation as the ultimate theory of nature; stay tuned.

strong interaction: One of the four known fundamental forces; binds quarks together. (May also refer to the residual force that binds protons and neutrons together.) Responsible for alpha decays.

superconductor: A metal that conducts electricity with zero resistance. Used to make ultra-powerful magnets; needed in high-energy accelerators to bend particles.

supersymmetry: A hypothesized additional symmetry of nature that relates fermions and bosons. Not yet experimentally verified. If supersymmetry exists, every particle should have a supersymmetric "partner," probably heavier, differing by 1/2 unit of spin. Supersymmetric partners of fermions are named by adding an *s* in front (selectron, squark, and so on), while partners of bosons are named by adding an *ino* on the end (photino, gluino, Wino, and so on).

symmetry: When a system appears unchanged after some specific change is made to it (such as rotating a cube by 90 degrees or replacing a proton with a neutron in a strong interaction).

synchrotron: A circular accelerator using electric fields to speed up charged particles and magnetic fields to make them travel in a circular path. A synchrotron is designed to handle even relativistic particles.

tevatron: The name for the latest accelerator facility at Fermilab, in Chicago. TeV stands for "Terra-eV" or "trillion electron volt," a measure of the highest energies the protons in the ring can achieve. This is the facility where the top quark and tau neutrino were discovered.

TOE: Theory of Everything. The dream of particle physicists, an ultimate theory that unifies all four fundamental forces and all particles. String theory is a candidate. It is still only a theory of subatomic particles, so the name is misleading. You will not be able to figure out tomorrow's weather any better with a TOE.

top quark: The most massive quark of all, the third-generation partner of up and charm. Found at Fermilab in 1995 but predicted and expected since 1975.

translation symmetry: The symmetry of nature which states that the laws of physics are the same at any place in the universe. (Think of "translation" as meaning "moving sideways.") This does not say that conditions are the same—clearly the inside of the sun is hotter than my house, but the laws of nature are still the same there.

unification: The goal of physicists to find a deep connection between forces. Electricity and magnetism were unified by Maxwell in the 1860s; they are both manifestations of one underlying "electromagnetic field."

up quark: One of the two light quarks, with spin 1/2 and charge +2/3. (The proton is two ups and a down; the neutron consists of two downs and an up.)

vacuum fluctuations: The vacuum refers to empty space. Even in empty space, quantum mechanics allows for the bizarre possibility that particle anti-particle pairs can spontaneously spring into a tenuous, "virtual" existence for a brief time before re-annihilating into nothing. The bubbling background, which exists everywhere, is called "vacuum fluctuations."

Van de Graaff accelerator: One of the earliest particle accelerators, this device built up very high voltages (in order of 1 million volts). Often seen today in science museums and physics lecture demonstrations.

W or W-boson: A carrier of the weak force. W's are electrically charged; for example, a neutron can decay into a proton and a W-boson. The W-boson then decays into an electron and an electron anti-neutrino. Weighs about 80 times as much as a proton.

weak boson: A generic term for the W's and Z's.

weak interaction: One of the four fundamental forces of nature, responsible for beta decay, and the only force that neutrinos feel. All particles feel the weak interaction.

x-rays: High-energy electromagnetic radiation. Often used as a synonym for gamma rays, although x-rays connote slightly lower energy radiation.

Z, or Z-boson: A carrier of the weak force. Z's have no electric force; they transmit the "weak neutral current." For example, when a neutrino scatters from another particle, it can emit a (virtual) Z particle, much like an electron scatters by emitting a (virtual) photon. Weighs about 90 times as much as a proton.

Biographical Notes

A biographical list of "key contributors" to the development of particle physics is almost impossible, because the number of contributors is huge! Although famous physicists often get sole credit for their accomplishments, the great discoveries are inevitably part of a web of scientific progress. Truly significant contributions come from both brilliant and more mediocre scientists, not to mention the support from graduate and sometimes undergraduate students, technicians, lab assistants, and others. Most of the discoveries in physics have a complex lineage; historians can (and do) quibble about the attributions and origins of almost every idea in the field. Some ideas get "rediscovered" or further developed, then attributed to the one who was better able to spread the word. What follows is an extraordinarily abbreviated list of some of the most famous names in the field. The large number left out is painful to this "biographer"—especially the more recent contributors, whose number is large and growing.

To learn more, a starting point is with Nobel Prize winners. Even among this select group, I have left many out! The Nobel site, www.nobel.se/physics/ laureates, is a good starting place. (Another good site for more biographies is www.gap-system.org/~history/BiogIndex.html.) Of course, many of the reference texts for this course also contain extended biographies of various subsets of key players.

Niels Bohr (b. 1885, d. 1962, Nobel Prize 1922). A Danish physicist who lived much of his life in Copenhagen, Niels Bohr is known as the first physicist who made the transition from classical to quantum understanding of atoms. He was infamous for his mumbling; many people recount stories of listening intently to the great Niels Bohr, only to have no idea afterward what he said. As a postdoctoral researcher, he worked briefly in Ernest Rutherford's lab starting in 1911. He expanded Rutherford's classical "solar-system" model of an atom into what is now known as a semi-classical, or hybrid quantum model. Quantum mechanics did not yet exist, but Bohr's model of the atom introduced many of the qualitative ideas of quantum physics and helped provide a stepping stone to the more fully developed theory of the 1920s and 1930s. Ten years later, Bohr helped create and direct the Institute of Theoretical Physics in Copenhagen (now known as the Niels Bohr Institute). Although Bohr himself did not make so many direct significant contributions to the quantitative development of quantum mechanics, his leadership, philosophy, and guidance were of essential and fundamental importance. Bohr helped develop ways of interpreting quantum mechanics; in fact, we now refer to the *Copenhagen interpretation* of quantum mechanics when discussing how to think about quantum physics. His famous debates with Einstein helped further establish quantum mechanics.

Marie Curie (b. 1867, d. 1934, Nobel Prize[physics] 1903, Nobel Prize [chemistry] 1911). One of the most famous pioneers in the early days of modern physics and inventor of the term *radioactivity*, Marie Sklodowska left her native Poland in 1891 (where opportunities for advanced studies were extremely limited for women) to study math and physics in France. She married Pierre Curie in 1895, a like-minded, similarly energetic and brilliant scientist. Marie became interested in Becquerel's new discovery of "uranium rays" and decided to pursue a Ph.D. in this field. She maintained her fascination with radioactivity throughout her professional career. Curie was the first person to understand, based on her research, that radioactivity was not a chemical phenomenon, but linked to the atomic nucleus. She worked closely with her husband, Pierre (until his untimely death in a street accident in 1906), often in extremely difficult circumstances, arduously isolating, then studying radioactive elements, including polonium (which Curie named after her native Poland). She isolated a tenth of a gram of pure radium from literally tons of raw material, all worked and processed by hand. She received her Ph.D. in 1903, the same year she got the Nobel Prize! She received her second Nobel (the first person ever to do so) in chemistry, for her work in isolating and studying the radioactive element radium. (Her daughter, Irene, would later win the Nobel Prize in chemistry for the discovery of artificial radioactivity.) Marie Curie died of leukemia at the age of 67, possibly the result of massive exposure to radioactive materials during her career.

P. A. M. Dirac (b. 1902, d. 1984, Nobel Prize 1933). Paul Dirac was an extraordinarily brilliant man, raised in England with a Swiss father and British mother. He was a man of few words. As a child, his father insisted that only French should be spoken at the dinner table. As a result, Dirac never said much! Acquaintances joked that his vocabulary was "yes," "no," and "I don't know." Still, his papers were beautifully written; he was inspired by the beauty of mathematics. He had 11 papers in print by the time he got his Ph.D. at age 24. Along with developing the new theory of quantum mechanics, Dirac struggled with combining relativity with quantum mechanics; he wanted his quantum equation to satisfy Einstein's special theory of relativity. The Dirac equation is an elegant mathematical marriage of those two theories. This equation predicted the existence of antimatter, which was later experimentally discovered.

Dirac worked briefly with Bohr as a postdoc in 1926 and received his Nobel Prize in 1933 (along with Schrödinger). His mother went with him to Stockholm; he was 31 at the time. Dirac made a large number of other significant contributions during his career, including the introduction of an extraordinarily powerful mathematical notation, now called *bra's* and *ket's* for the little brackets he used, which is indispensable in practical quantum mechanical calculations. Dirac held the Lucasian chair of mathematics at Cambridge University (held earlier by Isaac Newton and, currently, by Stephen Hawking) for 37 years.

Albert Einstein (b. 1879, d. 1955, Nobel Prize 1921). Born in Germany, Einstein is surely the most brilliant and famous physicist of the 20th century. (There are countless biographies available; I would urge you to read more about this fascinating man. His own writings are often surprisingly accessible, as well.) His contributions to physics are extraordinarily broad and far-ranging and, in several cases, profoundly changed the way we conceptualize the physical world. In 1905 (his "miracle year"), Einstein was a young man working in a patent office, doing physics in his spare time; he had not been able to secure a university position. He published three papers that year, any one of which would have put him on history's top 10 list of physicists. One of those papers was on the special theory of relativity, which changed our understanding of the nature of space and time. Another was on Brownian motion, a paper that effectively put to rest the longstanding debate, both physical and philosophical, regarding the existence of atoms. The third was on the photoelectric effect, which was in a real sense the birth of quantum mechanics. That paper extended and generalized Max Planck's ideas. Einstein was the first to appreciate the true depth of the idea that photons are quanta, or "chunks," of energy. (Technological applications of this concept abound today, from electric eyes to solar panels.) Roughly 10 years later, Einstein published his work on general relativity, in which he explained and described a theory of gravity that once again revolutionized how we conceive of space and time.

In his later life, Einstein struggled to find a grander "unified theory," with no success. He also argued frequently against the quantum theory that he had helped give birth to, saying famously, "God does not play dice with the universe." This was not a religious commentary, but a statement that he found the random nature of quantum mechanics philosophically unacceptable. Although he was, in the end, demonstrably wrong in this attitude, his constant questioning of the then-young theory of quantum mechanics played an important role in its early development.

Enrico Fermi (b. 1901, d. 1954, Nobel Prize 1938). Enrico Fermi was an Italian physicist, famous as perhaps the last physicist who was both a brilliant experimentalist and world-class theorist. Fermi paved the path to nuclear power (and weapons) and, along the way, played a central role in the development of both quantum mechanics and nuclear physics. Fermi's self-confident but casual and open style was in sharp contrast with the somewhat elitist approach of German physicists of that era and heavily influenced the following generation of physicists, many of whom worked or studied with him. (Fermi was also famous for his "Fermi problems," realistic puzzles involving logic, numeracy, and basic physics principles; his ability to make quantitative estimates of the solution to such vague but real-life problems was truly impressive.)

The number of physics terms with *Fermi* as an adjective is mind-boggling: Fermi constant, Fermi statistics, Fermions, the Fermi (a unit of length, roughly the size of a proton, 10^{-15} m), Fermi gas, Fermi sea, Fermi energy, Fermi momentum, Fermilab, Fermium (element number 100 in the periodic table), and the list goes on. Fermi named the neutrino and developed the first mathematical theory of the weak interaction (beta decays). He studied the physics of neutrons early in his career and pions, later. He escaped fascist Italy (his wife was Jewish), using his Nobel Prize in 1938 as a tool to emigrate to the United States. Once there, he worked on experiments that led to the first successful controlled nuclear chain reaction (under the sports stadium at the University of Chicago!). After the war, he fought the development of the H-bomb on ethical grounds. Fermi died very young, of cancer, still in his prime as a physicist.

Richard Feynman (b. 1918, d. 1988, Nobel Prize 1965). Born and raised in New York and one of the most engaging and charismatic figures in recent physics history, Richard Feynman can be characterized, in my opinion, as the greatest physicist of the second half of the 20th century. Irreverent, egotistical, brilliant, a true scholar and teacher, Feynman made contributions to a wide variety of branches of physics. His Nobel Prize work was for the development of QED, the quantum theory of light. One of the great aspects of that work was his invention of the *Feynman diagram*, a simple cartoon-like sketching device that represents a quantitative mathematical formula. These diagrams are ubiquitous now, not just in QED but in almost any branch of modern physics, allowing one to think qualitatively about complex physical processes while retaining the mathematics "underneath" the pictures. Feynman played a central role in the ongoing development of particle physics after QED, including his insightful interpretation of high-energy data from SLAC (the Stanford Linear Accelerator) in terms of *partons*, later to become quarks.

A number of biographies of Feynman are available (see the Bibliography). His own writings on physics are inspiring and delightful. In the 1960s, Feynman decided to "rewrite the book" on freshman physics. The result, his famous *Feynman Lectures*, is on the shelf of almost any professional physicist. However, by his own admission, the lectures were not a complete success with the CalTech freshmen for whom he had intended them. (His autobiographies may annoy you because of his ego and occasional lack of respect for others, but if you can get past that, they are fascinating and entertaining.)

Murray Gell-Mann (b. 1929, Nobel Prize 1969). Murray Gell-Mann a professor first at Chicago, then at CalTech, was a prime driver in our understanding of the strongly interacting particles, the quarks. A colleague of Feynman, Gell-Mann's style was quite different; Gell-Mann was literate and a bit sophisticated in his tastes, wearing fine clothes, but still with a great sense of humor and whimsy. His broad interests led one colleague to say, rather facetiously, "Murray has no particular talent for physics, but he's so smart he's a great physicist anyway." He invented the concept of *strangeness*, later understood to signal the presence of a new, heavier form of matter (the *strange particle*). He developed a mathematical framework based on symmetry, which he called the *eight-fold way*, to explain the properties of the rapidly expanding "zoo" of particles being discovered at accelerators in the 1960s. (The name was borrowed from Eastern mystical teachings, a reference that Gell-Mann later came to regret when his rigorous mathematical theories were sometimes misinterpreted as esoteric or philosophically vague.) Gell-Mann's development of the consequences of nuclear symmetries led to his articulation and naming of the *quark model*, which forms the underpinnings of our modern theory of QCD (quantum chromodynamics, the theory of the strong force.) His choice of particle name came from James Joyce's *Finnegan's Wake*, with a passage beginning "three quarks for Muster Mark!"—an appropriate name for what seemed, at the time, like a perfectly absurd particle of nature.

Werner Heisenberg (b. 1901, d. 1976, Nobel Prize 1932). A German physicist who was the first to create a mathematical theory of quantum mechanics at the young age of 23, publishing the work in 1925. Heisenberg argued that a theory should refer only to quantities that can be *observed*, an idea that played a guiding role in the development of quantum physics. Heisenberg was very much a mathematician; he nearly failed to get his Ph.D. because of his weakness with laboratory work. He joined Niels Bohr's Institute in 1926 for a productive year. In 1927, Heisenberg developed his famous *uncertainty principle*, which states that one cannot, in principle, have precise simultaneous knowledge of the momentum and position of a particle. (This principle is, in fact, a rigorous, mathematically derivable statement that contains useful quantitative, predictive power. It is far more than just a curious philosophical idea.) During World War II, Heisenberg spent five years directing the unsuccessful German atomic bomb project, a period in his life which remains somewhat controversial. (There has even been a recent Broadway play about this.) He was imprisoned by the allies briefly after the war but was returned to Germany in 1946, where he established the Max Planck Institute and served as its director until his retirement.

James Clerk Maxwell (b. 1831, d. 1879). A Scotsman, born without privilege or high social rank, Maxwell worked in the field of mathematical physics, and electricity and magnetism, during the 1800s, when the scientific community was tackling this "exotic" field with great vigor. Maxwell was especially intrigued by the discoveries of Michael Faraday (himself a man with humble beginnings) who had introduced the concept of *force field* as a physically relevant entity. Maxwell succeeded in mathematically describing *all* phenomena of electric and magnetic origin in a set of four relatively simple equations, now called *Maxwell's equations*. Most of those had been developed over the previous decades by others, but Maxwell organized and formalized them, adding a key component based not on experiment, but on his own aesthetic mathematical sense of symmetry, intimately and permanently unifying electricity and magnetism. Maxwell discovered that these equations led to the phenomenon of *traveling electromagnetic (EM) radiation*—moving at the speed of light— and with this, realized the deep connection to optics, as well.

Today, Maxwell's equations and the corresponding unification of forces are regarded as one of the grand highlights of human intellectual achievement. They form the basis of electrical engineering and modern optics and have survived the discoveries of modern physics in the 20th century essentially unscathed. They paved the way for the discovery of relativity (being fully relativistic equations, even though Maxwell didn't appreciate that!) and form the classical underpinnings of QED, the quantum theory of light. The study of Maxwell's ideas generally forms the second half of any standard college-level introductory physics course.

Isaac Newton (b. 1643, d. 1727). The father of physics, indeed of all modern science. In many ways, I can think of few individuals of the last 1000 years who had more direct and profound influence on the human condition. Isaac Newton's masterwork, the *Principia*, articulated not only a number of physical laws, but also the scientific method itself. Newton's laws describe and explain motion and gravity. When faced with the need to solve the equations he developed, Newton *invented* the calculus required to solve them. Newton's central laws are *universal*, applicable to any system in any circumstance. Even today, their accuracy and power is extraordinary. Although Newton's laws must be extended under extreme conditions (such as for objects traveling near the speed of light), they still form the basis for much of modern technology. Newton did both theory and experiment; his research touched and formed the roots of many branches of modern physics, including optics, thermodynamics (heat), fluids, and more. Students in freshman physics learn about Newton's work in their first semester (then repeatedly, with further depth, as they progress). The metric unit of force, the *Newton*, is named in his honor. Newton was not a pleasant or easy man. He had a big ego, never married, and had many disputes over intellectual priorities during his life. However, in an uncharacteristic but famous quote, he said, "If I have seen further, it is by standing on the shoulders of giants."

Emmy Noether (b. 1883, d. 1935). Although not a physicist, this mathematician played a significant role in the early conceptual understanding of modern theoretical physics. Born and raised in Germany, Noether persevered in a system heavily biased against women in academia. She studied mathematics at university but was only allowed to sit in on classes unofficially. Even then, professors had to give permission individually. She studied under some of the great mathematicians whose names are also linked with developments of modern physics: Hilbert (whose work is now applied in quantum mechanics), Klein (general relativity and field theory), and Minkowski (special relativity). Noether later did her Ph.D. work with a mathematician whose work would also become closely tied to future physics developments, Paul Gordon (quantum mechanics and gauge field theories). Normally, such a brilliant and promising student in Germany would get a still higher degree ("habilitation"), but this was forbidden for women at the time. Noether helped her father, also a mathematician, and continued working, doing research, and supervising doctoral students, all without pay. (Hilbert, in attempting to get her a faculty position, said, "I do not see that the sex of the candidate is an argument against her admission as Privatdozent. After all, we are a university and not a bathing establishment.") Noether's publication list and reputation grew rapidly, but it took 12 years until she successfully petitioned to obtain the habilitation.

In 1915, she published the work that is most directly relevant to particle physicists, now known as *Noether's theorem*. Albert Einstein praised this work, calling it "penetrating mathematical thinking." The bulk of Noether's further research was in a field known as abstract algebra, which has also played a role in quantum mechanical theory. By 1922, she started receiving a small salary but was still without tenure. (Noether was also a Jew, a Social Democrat, and an outspoken pacifist, none of which was much help in that regard.) She left Germany in 1933 for the United States, as a result of Nazi dismissal of Jewish faculty. She spent several years teaching and working as a visiting professor at Bryn Mawr in Pennsylvania. Emmy Noether died quite young, of complications after a surgery. Albert Einstein, in a tribute to Noether in the *New York Times* on May 5, 1935, wrote:

> In the realm of algebra, in which the most gifted mathematicians have been busy for centuries, she discovered methods which have proved of enormous importance ... Pure mathematics is, in its way, the poetry of logical ideas ... In this effort toward logical beauty, spiritual formulas are discovered necessary for deeper penetration into the laws of nature.

Wolfgang Pauli (b. 1900, d. 1958, Nobel Prize 1945). Born in Vienna, Wolfgang Pauli was in many respects the theoretical leader of the development of quantum mechanics in the 1920s and 1930s. Pauli even impressed Albert Einstein when, in his sophomore year in college, he published a brilliant review article explaining special relativity for a mathematical encyclopedia. Pauli spent a year with Bohr in Copenhagen (1922–1923) and, soon after, developed his famous "Pauli exclusion principle," which states that no two quantum particles (of spin 1/2) may exist in the same quantum state. This law of nature has tremendous applications. It explains, among many other things, the chemistry and properties of the periodic table of the elements and the structure of atomic nuclei. In 1931, Pauli predicted the existence of a new particle of nature, later called the neutrino, although it was several years before he would actually publish a paper about it. Pauli was always a sharp thinker and, ever skeptical, able to poke holes in anyone's theory. He was quite intolerant of sloppy thinking, striking fear in the hearts of visiting speakers when he sat in the front row. He is famous for his quips, such as stating that a paper was "not even wrong." To a postdoc, he once said, "I don't mind your thinking slowly. I mind your publishing faster than you think." Still, for his brilliant, critical, and honest thinking, Pauli has been called by some "the conscience of physics."

Max Planck (b. 1858, d. 1947, Nobel Prize 1918). A German physicist who started his career studying thermodynamics (heat), a branch of physics that was well established by the late 1800s. When Planck entered university at age 16, he was told that physics was essentially a complete science with little prospect of further developments. Fortunately, he decided to study physics despite the bleak future for research that was presented to him! He got his Ph.D. by age 21. Planck's fame arises from his work on the question of electromagnetic radiation emitted by hot objects. It had been established that classical physics (Maxwell's equations, combined with thermodynamics) not only couldn't explain the data but also gave nonsensical, infinite results. After what Planck called "the most strenuous work of my life," he was able to describe the data by postulating that electromagnetic radiation was *quantized* in its interaction with matter, following a characteristic formula he derived. ($E = hf$, which states that the energy of a bundle of light is a constant, h, times the frequency of the light). This was the birth of quantum mechanics. The idea that energy comes in chunks, or bundles, *quanta*, was a stroke of genius, a thoroughly nonclassical and unexpected concept. Interestingly, Planck's contributions to quantum theory didn't continue after this and, in fact, he was not convinced by the developments in quantum theory that arose from his work. The constant h that appears in the equation he developed, and in essentially all quantum mechanics formulas since, is now named *Planck's constant*, in his honor.

Ernest Rutherford (b. 1871, d. 1937, Nobel Prize 1908). Born in New Zealand but living his life in England, Rutherford was a brilliant experimentalist, the first of the modern breed of particle physicists. He was a bear of a man, with a loud voice and a strong personality. Working at the Cavendish Laboratories, Rutherford was the first person to identify the types of radiation emitted by naturally radioactive substances. He recognized that radioactivity represented the transformation of elements from one type to another. Rutherford's experiments with Hans Geiger and Ernest Marsden (an undergraduate) helped him formulate the classical but thoroughly modern picture of an atom as a composite object with a tiny, heavy nucleus at the center and electrons in orbit. His famous quote about this experiment is: "It was quite the most incredible event that ever happened to me in my life. It was as if you fired a 15-inch artillery shell at a piece of tissue paper and it came back and hit you." Rutherford later showed that he could send alpha particles onto nitrogen and convert atoms into oxygen, the first deliberate "transmutation of elements" accomplished by humans. In his later life, Rutherford remained active, administrating the Cavendish Labs and guiding many future physicists in their early careers. Shortly before his death, Rutherford said, "Anyone who expects a source of power from the transformation of these atoms is talking moonshine." (The first nuclear chain reaction was constructed shortly after his death.)

Abdus Salam (b. 1926, d. 1996, Nobel Prize 1979). Born and raised in Pakistan, Abdus Salam was a mathematically precocious child, getting his Ph.D. by age 25. His major work and contributions were in the mathematical development of what is now called the standard model, in particular, the unification of the weak and electromagnetic forces of nature. He coined the phrase *electroweak theory* and made significant contributions to issues in modern quantum mechanics. Salam predicted, on the basis of his theories, the existence of the weak neutral current (a new form of fundamental interaction) and of W and Z particles (carriers of the weak force), before they were experimentally identified. In many aspects, his work paralleled, but was independent of, the theoretical efforts of Steven Weinberg. Salam left Pakistan in 1957; spent many years as a professor in Imperial College, London; and later went on to direct the International Center for Theoretical Physics in Trieste, Italy, where he made great strides in improving the connections and opportunities for third-world physicists to participate in world-class research. A pious and devout Muslim, Salam also worked for many years with various UN committees to encourage the advancement of science and technology in developing nations.

Erwin Schrödinger (b. 1887, d. 1961, Nobel Prize 1933). An Austrian with broad intellectual interests and a distinctly unconventional personal style, Schrödinger is most famous for the equation that bears his name, the fundamental wave equation of quantum mechanics. Schrödinger served in the German army in World War I and submitted two theory papers from the front. As a young physicist, he was asked to give a colloquium on the thesis of de Broglie, and a physicist in the audience commented that he thought this way of talking was rather childish: "To deal with waves, one has to have a wave equation." Schrödinger found that equation in 1926. He is said to have developed the ideas when on holiday with a Viennese girlfriend. (He had many lovers, not always secret from his wife.) Although his work on quantum mechanics followed Heisenberg's, it was in a form that was mathematically more accessible to other physicists. His equation is still the way that quantum mechanics is generally first taught to students today. (He later proved that it is mathematically equivalent to Heisenberg's methods.) Schrödinger's equation was spectacularly successful and opened the way to quantitative (versus qualitative) calculations with the new theory of quantum mechanics. He left Germany in 1933 because of his discomfort with the growing Nazi presence, later returned to Vienna, and ended his career in Dublin. Interestingly, Schrödinger, too, had a hard time with the philosophical implications of the theory he helped develop. He said, "Had I known that we were not going to get rid of this damned quantum jumping, I never would have involved myself in this business."

Steven Weinberg (b. 1933, Nobel Prize 1979). Steven Weinberg was one of the central figures in the development of the electroweak theory and the standard model. He made the first quantitative prediction of the mass of the W and Z bosons, well before their experimental discovery. His work has spanned the broadest spectrum of issues in particle physics, including strong and weak interactions, neutrino physics, astrophysics, cosmology, and grand unification. He has written several books aimed at the general public and has been a passionate, articulate, and prominent spokesman for support and dissemination of high-energy physics. He is currently a professor at the University of Texas in Austin.

Bibliography

Essential Reading

Barnett, R. M., Muhry, H., Quinn, H. R., *The Charm of Strange Quarks*, New York: Springer-Verlag, 2000. Very readable and up-to-date; a great collection of pictures, anecdotes, and well-organized summaries and explanations. A few symbols and equations, on a level roughly similar to Cindy Schwarz's book (college level but not overly technical; perhaps a slightly higher assumption about the reader's tolerance for "light" math).

Kane, Gordon, *The Particle Garden: Our Universe as Understood by Particle Physicists*, Cambridge: Perseus Publishing, 1996. An excellent text that summarizes many of the central ideas of contemporary particle physics without getting into technicalities. Well written, aimed at a general audience.

Schwarz, Cindy, *A Tour of the Subatomic Zoo*, New York: AIP, 1996. Used as a textbook in an undergraduate course for non-science majors with no physics background, this book is perhaps not quite so entertainingly "readable" as some others, but it is accessible and contains some *slightly* more technical details for the reader who wants to go a little beyond the purely descriptive, qualitative, story level.

't Hooft, Gerard, *In Search of the Ultimate Building Blocks*, Cambridge: Cambridge University Press, 1996. A firsthand account from one of the key theorists who helped create the standard model ('t Hooft won the 1999 Nobel Prize for these contributions). He has tried hard to write a book without math or technical detail and succeeds for the most part; the level is accessible (perhaps on a par with Krauss), although I can't call it easy (especially the last few chapters). But he has a great talent for making analogies with real-world (and, thus, comprehensible) phenomena to explain esoteric ideas.

Weinberg, Steven, *Dreams of a Final Theory*, New York: Vintage Books, 1993. A firsthand account from a key theorist who helped create the standard model (Weinberg won the 1979 Nobel Prize for his contributions). This book is certainly not a textbook (no equations or math whatsoever), nor a typical "summary of particle physics," but it waxes a bit more philosophical about the goals and ideas of modern physics. You may find Weinberg philosophically provocative, but this is a highly literate and pleasurable book.

Recommended Reading

Bodanis, David, $E=mc^2$, New York: Walker and Co., 2000. Bodanis is not a physicist, and this book is not about particle physics, but it is a more personal tale of Einstein's famous equation, which of course plays a central role in our understanding of particles and their dynamics. The book is *far* from technical; it's not really even centered on the physics, but rather on the stories, characters, and history leading up to (and following) the development of the equation. The guide for further reading at the end is a good one, with nice descriptions.

Bromley, D. Allan, *A Century of Physics*, Heidelberg: Springer, 2002. This lovely book consists of very short (paragraph-long) discussions of key events and people in the development of physics over the last 100 years, with lots of great pictures. It's a "tour book," beautiful to look at and fun to peruse.

Calle, Carlos, *Superstrings and Other Things: A Guide to Physics*, Bristol: IOP, 2001. Very readable text at a level accessible to a non-science undergraduate. It covers a lot of physics, not just particles. Calle very occasionally uses numbers and equations, at a similar (or slightly higher) level than the Schwartz or Barnett texts, but still nowhere approaching a "physics text" in terms of math requirements. Nice summary boxes and interludes, good selection of illustrations and explanations.

Close, Frank, *The Cosmic Onion*, College Park: American Institute of Physics, 1983. A slightly older book and one of the earlier attempts at "popularizing" particle physics, this book is a bit more terse and technical than the others I've described. It is explicitly intended for "prospective science undergraduates" or "[any] non-scientist prepared to think." It's somewhat tougher going than, for example, Cindy Schwarz's book or Gordon Kane's, but if you want to go a little further into the details, this is an excellent choice.

Close, Frank, Marten, Michael, and Sutton, Christine, *The Particle Explosion*, Oxford: Oxford University Press, 1987. A colorful book, great pictures, nontechnical, complete, lots of history, very descriptive and detailed.

Davies, Paul, *The Last 3 Minutes*, New York: Basic Books, 1996. The title is a takeoff from Steven Weinberg's *The First 3 Minutes*. This book covers a readable selection of topics in modern cosmology (as modern as a decade-old book can be now).

Ferris, Timothy, ed., *The World Treasury of Physics, Astronomy, and Mathematics*, New York: Little, Brown and Company, 1991. A collection of articles by and about scientists. Not all particle physics (even ranges to math and astronomy, of course!) but very readable and interesting, a nice selection of essays. Not technical at all, but because it is a collection, each article has a different style and level. Some of the short biographies are charming.

Feynman, Richard, see sample list of titles below. If you become interested in Feynman (it's nearly impossible not to!), I recommend just about anything written by Feynman himself. For entertainment, you can't beat his "autobiographies." Start with *Surely You're Joking, Mr. Feynman: Adventures of a Curious Character*, 1985 (Norton and Co.), followed by *What Do You Care What Other People Think? Further Adventures of a Curious Character*, 1988. For some real physics from the master, I recommend *6 Easy Pieces*, Helix Books, 1994, a compilation of six lectures about topics in modern physics Feynman gave to his freshman students at CalTech. These are truly brilliant. Another similar text, this time about more classical questions in physics, is *The Character of Physical Law*, MIT Press (1985). Feynman also wrote *QED: The Strange Theory of Light and Matter*, Princeton University Press (1985), in which he attempts to explain QED to non-experts. An interesting book, but I believe it's still a little hard. I'm not sure if Feynman completely succeeded with his goal on this one, but it's definitely worth a look if you really want to dive deeper.

Franklin, Allan, *Are There Really Neutrinos: An Evidential History*, New York: Perseus Books, 2003. This book is different from most in this list, being written by a historian and philosopher of physics (also a particle physicist). It focuses on the history of neutrinos, addressing the fundamental question: Why do we believe in the reality of particles like the neutrino? Could the neutrino be a social construct rather than something real? An interesting perspective. Mild warning: The book is technical in many places, definitely aimed above the "novice general public" in my opinion.

Fraser, Gordon, *The Quark Machines: How Europe Fought the Particle Physics Wars*, Bristol: Institute of Physics Publishing, 1997 (paperback). Nontechnical summary of more recent developments, with an unusually international perspective, a good deal of discussion of the people involved, good list of references for further reading.

Gamow, George, *Mr. Tompkins in Paperback*, Cambridge: Cambridge University Press, 1993. A whimsical book, very amusing, although a bit dated. Mr. Tompkins is an "interested layperson" who attends lectures on modern physics for the public, but alas, he inevitably falls asleep partway into the lecture. This book describes his ensuing dreams, in which the laws of physics being discussed are "visualized" in some dramatic way. For example, when Mr. Tompkins falls asleep while learning about relativity, in his dream, the speed of light is only about 25 miles/hour; thus, relativity effects become "visible" as he walks around town or gets in his car. In the lecture about quantum mechanics, Planck's constant (which tells the strength of quantum effects) is huge, and billiard balls on Tompkins's pool table suddenly "fuzz out," then tunnel off the table. Very creative way of presenting physics, although the topics covered are not exactly the most "modern" particle physics.

Glashow, Sheldon, *Interactions: A Journey through the Mind of a Particle Physicist and the Matter of This World*, New York: Warner, 1988, and *The Charm of Physics*, College Park: AIP Press, 1991. Two books by a Nobel Prize winner, both light in style, entertaining, filled with stories, from a somewhat personal perspective. Both books also contain a great deal of particle physics. Recommended.

Gleick, James, *Genius: The Life and Science of Richard Feynman*, New York: Pantheon Books, 1992. Not about the physics, but a biography of Richard Feynman, well written, a good glimpse at this remarkable character.

Greene, Brian, *The Elegant Universe*, New York: Vintage Books, 2000. Greene has tried to meet a great challenge—to write a "popular" book about the most abstract and mathematically complex physical theory of them all, string theory. He does a great job, but it's not always the easiest going. I suspect that the casual reader will often be left a little dizzy. Still, Greene manages to "return to reality" enough to allow you to continue reading and continue getting a lot out of the book right up to the end. This is hard stuff, and Greene does an amazing job of presenting it without getting technical.

Hawking, Stephen, *A Brief History of Time* and *The Universe in a Nutshell*, New York: Bantam 2001. Stephen Hawking is one of the most celebrated and well-known contemporary physicists. His books (sometimes considered "coffee-table books," in the sense that some people buy them but don't actually read them) are comprehensive, up-to-date, and full of wonderful ideas about modern physics. However, I often find that when I get to material about which I'm not an expert, Hawking can lose me; I imagine this could be a problem throughout for the non-physicist reader. I also find that Professor Hawking has some strong ideas that sometimes reach well beyond particle physics, to the realm of philosophy and interpretation, which the reader should be aware of. Interesting books nonetheless.

Kevles, Daniel J., *The Physicists*, Cambridge: Harvard University Press 1995. History of particle physics, more focused on the earlier years. A little dry and dense, but lots of interesting details and anecdotes. I found the introduction/preface on the rise and fall of the SSC especially interesting.

Krauss, Lawrence M., *Fear of Physics*, New York: Basic Books, 1993. Delightful book. Not about particle physics per se (although that's clearly his focus). More a book about how physicists think, how they approach problems. As the book progresses and Krauss gets more into the details and explanations of modern particle physics ideas, the reader may become a little lost because of the completely nontechnical approach (all examples are analogies or simple pictures), but it's a creative and insightful book. Definitely recommended. I just loved the opening chapters.

Landsberg, Peter T., *Seeking Ultimates: An Intuitive Guide to Physics*, Bristol: IOP Publishing, 2000. A popular-style account of many of the fundamental ideas of physics, not limited to particle physics. Some personal philosophy is also mixed in with the book; the author believes deeply that physics is, and always will be, incomplete, but that it is the journey that is most important. Quite readable and a little broader in scope than many of the other books I have included.

Lederman, Leon, *The God Particle*, Boston: Houghton Mifflin, 1993. Whimsical, with a light style and lots of bad puns, a Nobel Prize-winning experimentalist's take on the significance, content, and meaning of particle physics. This is not a book filled with technical explanations or details; it's more a collection of stories and insights, with lots of history and fascinating introduction. Very entertaining reading and still up-to-date. (I find his curious renaming of the Higgs to be completely inappropriate and somewhat annoying, but then, he's the one with the Nobel Prize.)

Leighton, Ralph, *Tuva or Bust*, New York: W.W. Norton and Co., 1991. A less scholarly but more personal work about Richard Feynman. A breezy and delightful story of Feynman as a friend and human being.

Pais, Abraham, *Inward Bound*, Oxford: Oxford University Press, 1988. A slightly more technical book, aimed at scientists (though not necessarily physicists). It's a little dense, very comprehensive, but quite readable if you are comfortable with a few equations and a little mathematical reasoning.

Park, Robert, *Voodoo Science*, Oxford: Oxford University Press, 2000. This book has nothing directly to do with particle physics, but it's a fascinating story of "the road from foolishness to fraud." It's about "bad science," with thoughts on how to distinguish real science from the marginal. The author, Robert Park, is strongly opinionated; if you disagree with him, you may find his strength of conviction challenging, but I think it's a good representation of where most working physicists stand. Great reading, very interesting, entertaining, and thought provoking.

Riordin, Michael, *The Hunting of the Quark*, New York: Simon and Schuster, 1987. A historical narrative focusing on the 1960s–1970s, with many anecdotes and stories about the people and events, as well as the physical ideas. No math but occasional overuse of jargon. The chapters on the November Revolution are among the best of the lot, though.

Taubes, Gary, *Nobel Dreams: Power, Deceit, and the Ultimate Experiment*, New York: Random House, 1987. The "human side" of particle physics, with all the hubris, ego, deceit, confusion, and passion exposed. Much more entertaining reading than most of the other references I've provided, although by its nature, less physics and more soap opera!

Wearth, S., and Phillips, M., eds., *History of Physics: Readings from Physics Today*, Heidelberg: Springer, 1987. A variety of articles, although nothing really state of the art. I especially enjoy "The Birth of Elementary Particle Physics" (Laurie Brown and Lillian Hoddeson, from *Physics Today*, April 1982.)

Weinberg, Steven, *The First 3 Minutes*, New York: Basic Books, 1993. An excellent introduction to the Big Bang and the origin of our universe. Although there have been some important theoretical developments since this book was written, it is still relevant. The level of the book is above the "purely popular," a bit more on the level of an undergraduate text (but suitable for a non-physics major).

Wilczek, Frank, and Devine, Betsy, *Longing for the Harmonies*, New York: Norton, 1987. Although the book is 15 years old, it is still quite relevant, because the authors do not try to teach the details of modern physics, but rather, to give a taste of the type of aesthetic and intellectual issues involved. They focus on the "musical metaphors" of science. The authors believe that modern physics is not commonsense, but neither is it mystical or incomprehensible. This book is chock full of gems, lots of wonderful insights about the world that help clarify what modern physics is about. I recommend it highly.

Web Sites

http://www.nobel.se/physics/laureates/index.html. The Nobel Prize page for physics, with detailed links if you're curious to follow up on any of the people or ideas.

http://press.web.cern.ch/pdg/particleadventure/other/suggestedreading.html. An excellent collection of additional readings from CERN, the European center for particle physics. Very comprehensive.

http://www.pparc.ac.uk/Ed/ppres_books.asp?Pv=1. PPARC is a particle physics site from the United Kingdom. This site offers a collection of links with many book suggestions (with summaries and "level descriptions"). A useful source for finding further readings.

http://www.fnal.gov. Fermilab's central Web site; this is the premier particle physics experimental facility in the United States. The facility has a solid public outreach program, and its Web site is well organized and interesting, with lots of material aimed at the non-physicist. If you are interested in getting some K–12 students involved in learning about particle physics, the link http://quarknet.fnal.gov is a great resource.

http://www.slac.stanford.edu. SLAC (Stanford Linear Accelerator Center) is the "other" premier U.S. particle physics experimental facility. This site has a great deal of information for the public, and the www.slac.stanford.edu/history/ link is a nice one.

http://particleadventure.org/. Comprehensive and educational site, highly recommended. Created by the Particle Data Group. The site is aimed at, perhaps, the high school level, but I find it informative and educational for any interested readers. Another related link is http://www.cpepweb.org/particles.html.

http://public.web.cern.ch/Public/. This is the Web site of CERN (the premier European particle physics facility) aimed at the public. Lots of easily readable information about particle physics here.

http://www.sno.phy.queensu.ca/. The SNO collaboration's Web site, a good place to find its latest news and summaries of other experimental efforts.

http://pdg.lbl.gov. The central repository summarizing all standard model information, compiled by the Particle Data Group of Berkeley.

http://cpepweb.org. The "wall chart" of particle physics. This is from the Particle Data Group. The links to the group's more technical Web pages contain an incredible summary of all the detailed quantitative knowledge we have accumulated regarding all fundamental particles and the laws of physics.

http://dbserv.ihep.su/compas/contents.html. A chronology of particle physics ideas, with links to original papers. Very comprehensive for those interested in historical aspects.

http://www.aps.org/resources/particle.html. American Physical Society page with many links (and descriptions of those links), heading to a variety of resources appropriate for everyone from students in grades K–12 to interested members of the public to working physicists.

http://www.hep.ucl.ac.uk/~djm/higgsa.html. In 1993, the then U.K. Science Minister, William Waldegrave, issued a challenge to physicists to answer the questions "What is the Higgs boson, and why do we want to find it?" on one side of a single sheet of paper. Here is David J. Miller's prize-winning "qualitative layperson's explanation." Other replies can be found at http://hepwww.ph.qmw.ac.uk/epp/higgs.html.

http://www.sns.ias.edu/~jnb/. John Bahcall's Web site, with a focus on solar neutrinos (check the link on the left for "popular accounts").

http://scienceworld.wolfram.com/physics/. Eric Weisstein's "World of Physics," with short biographies of just about every physicist I've ever heard of. Nice historical/biographical reference, although rather brief in most cases.

http://dept.physics.upenn.edu/~erler/electroweak/index.html. An up-to-date summary of quite technical information about the status of the standard model, maintained by Jens Erler and Paul Langacker at the University of Pennsylvania Some of it may be readable, but for the most part, it is designed for particle physicists.

http://www.aip.org/history. Excellent site, very readable, with a small (but growing) collection of historical hypertexts of some key physicists and physics topics.

http://www.powersof10.com/. Not exactly particle physics, but a wonderful site to help think about the world at different distance/size scales.

Notes

Notes

Notes

Notes